Rae Royle

66 75

Developments in Precambrian Geology 5

PRECAMBRIAN BANDED IRON-FORMATIONS

Physicochemical Conditions of Formation

DEVELOPMENTS IN PRECAMBRIAN GEOLOGY
Advisory Editor B.F. Windley

Further titles in this series

1. B.F. WINDLEY and S.M. NAQVI (Editors)
   Archaean Geochemistry
2. D.R. HUNTER (Editor)
   Precambrian of the Southern Hemisphere
3. K.C. CONDIE
   Archaen Greenstone Belts
4. A. KRÖNER (Editor)
   Precambrian Plate Tectonics

DEVELOPMENTS IN PRECAMBRIAN GEOLOGY 5

PRECAMBRIAN BANDED IRON-FORMATIONS

PHYSICOCHEMICAL CONDITIONS OF FORMATION

Y.P. MEL'NIK
Institute of Geochemistry and Mineral Physics, Kiev, U.S.S.R.

Translation from the Russian by *Dorothy B. Vitaliano*

ELSEVIER SCIENTIFIC PUBLISHING COMPANY
Amsterdam – Oxford – New York 1982

ELSEVIER SCIENTIFIC PUBLISHING COMPANY
Molenwerf 1, 1014 AG Amsterdam
P.O. Box 211, Amsterdam, The Netherlands

Distributors for the United States and Canada:

ELSEVIER NORTH-HOLLAND INC.
52, Vanderbilt Avenue
New York, N.Y' 10017

Library of Congress Cataloging in Publication Data

Mel'nyk, I͡U. P.
Precambrian banded iron-formations.

(Developments in Precambrian geology ; 5)
Translation of: Fiziko-khimicheskie uslovii͡a obrazovanii͡a dokembriĭskikh zhelezistykh kvart͡sitov.
On t.p.: Institute of Geochemistry and Mineral Physics, Kiev, U.S.S.R.
Includes bibliographical references and index.
1. Quartzite. 2. Iron ores. 3. Geology, Stratigraphic --Pre-Cambrian. I. Title. II. Series.
QE475.Q3M4413 553'.3 81-15134
ISBN 0-444-41934-9 AACR2

ISBN 0-444-41934-9 (Vol. 5)
ISBN 0-444-41719-2 (Series)

Printed in The Netherlands

Preface to the English edition

Work on the book "Physicochemical conditions of formation of the Precambrian ferruginous quartzites", published in Russian in Kiev ("Naukova Dumka" Publishing House, 1973), was finished in 1971–1972. In the ensuing years numerous scientific works have been published, both in the U.S.S.R. and abroad, containing new data on the genetic position of the banded iron-formations (BIF), the mineralogical composition and chemistry of iron cherts, the geochemistry of the stable isotopes, and the evolution of the hydrosphere and biosphere. Of special interest were the results of detailed investigations of slightly metamorphosed BIF in Canada, Australia, South Africa, and some other areas, and also new finds of traces of ancient organisms and organic compounds in Archean iron-formations. Naturally, the new edition of my book in English would be incomplete without taking account of these data.

In addition, in 1971–1977 I carried out new mineralogical and geochemical investigations on the BIF of the Ukrainian shield, and also experimental and theoretical study of the mechanism of precipitation and crystallization of iron-hydroxide and silica sediments under different conditions, and of the stability of the main minerals of the BIF—goethite, hematite, magnetite, greenalite, minnesotaite, grunerite, fayalite, etc.—at high temperatures and high pressures. In connection with the further development, refinement, and supplementation of a system of consistent thermodynamic values of the minerals, precipitates, and ions of iron and associated elements, numerous diagrams of mineral equilibria characterizing the physicochemical conditions of deposition, diagenesis, and metamorphism of the BIF have been essentially brought up to date. In addition, clearer new diagrams have been constructed, in which values of activities (concentrations) or molar fractions (percentages) are used directly to characterize the variations in composition, instead of the logarithms of those values. These diagrams are supplemented by model schemes in which we have attempted to demonstrate the development of the iron-ore process in time and space.

Substantial changes have also been made in the layout of the book; in particular, new sections have been added describing the textural and structural particulars of the BIF, the chemical composition of the rocks and minerals and their stability in the face of granitization. We considered it advisable to give an appendix summarizing the thermodynamic values used in the calculations, with briefly stated reasons for choosing them (for the iron minerals). In accordance with the wishes of "Elsevier Scientific Publishing Company", a subject index is given and the system of citing the published sources has been changed. At the same time it turned out that the examination of controversial problems of the theory of metamorphism could be substantially shortened without harm (the more so, as several of the author's papers have been published in English translation by this time), and also the sections dealing with the metamorphism of (calcium-magnesium) carbonate and some ultrabasic rocks could be dropped entirely.

I hope that all these changes and additions, on which much time and effort unfortunately had to be expended, will prove advisable and make the monograph more complete and more useful for all geologists interested in the physicochemical aspects of mineral and ore formation.

I especially thank Prof. C. Klein (Indiana University, USA) for his constant attention to and support of my investigations, for his valuable comments on this revised English edition and for his assistance rendered the translator, and also Mrs. Dorothy B. Vitaliano, who in 1975 translated the monograph into English for the U.S. Geological Survey, which helped make a wider circle of geologists acquainted with this work. I also am grateful to Prof. R. Garrels (Northwestern University, USA), Dr. A. Trendall (Geological Survey, Australia), and Prof. E. Alexandrov (Queens College, USA) for their interest in the work and discussion of the problems under consideration.

Preface to the Russian edition

The problem of the formation of the Precambrian banded iron-formations (BIF), repeatedly the subject of discussion, continues to be pressing. The present interest in it is explained not only by the ever-increasing importance of iron cherts as economic ores, but also by the special position they occupy in the geologic history of the Earth.

Banded iron-formations are known on all continents and invariably are a component part of the metamorphic basement complexes of all ancient shields; the total iron reserves in them are enormous. First appearing in the Archean or beginning of the Proterozoic, these unique banded rocks, consisting of systematically interbedded ferruginous and cherty bands, are maximally developed in the middle of the proterozoic and then practically disappear. Judging from available data, the intensive development of the first forms of life, the appearance of free oxygen in the atmosphere, and an irreversible change in the hydrosphere are related precisely to this turning point. The combination of factors which caused the massive deposition of purely chemogenic iron-chert sediments, uncontaminated by clastic products, was never again repeated.

Despite the substantial number of publications devoted to the study and substantiation of various points of view, many problems remain controversial: the sources and paths of transport of the iron and silica, forms of migration, causes and conditions of deposition, composition of the original sediments, character of diagenetic changes, and particulars of metamorphism have not been adequately clarified. Practically every possible version of genetic hypotheses has been proposed and discussed at different times, thus the problem is not to propose a new hypothesis, but to determine more specifically the conditions of ore formation. For this a new approach to the solution of complex geologic problems is needed, based on quantitative evaluation of the natural phenomena using modern physicochemical methods, including experimental investigation of equilibria and modeling, thermodynamic analysis of mineral equilibria in a wide range of temperatures and pressures, and study of the distribution of free carbon in iron cherts and of the stable oxygen isotopes in the minerals.

Physicochemical analysis of the conditions of mineral formation has been limited to a substantial extent by the lack of reliable thermodynamic values, particularly those of the rock-forming silicates and of compressed gases at high temperatures. The development of a system of consistent values prior to this work made it possible to broaden substantially the possibilities of applying thermodynamic methods to the analysis of various natural processes. In particular, the forms of transport of the elements in low-temperature waters have been examined, the conditions under which precipitates of different composition are deposited when equilibrium parameters change have been determined, and the role of possible geochemical barriers has been evaluated. A new type of high-T and -P diagram has been constructed, on the basis of which information can be obtained on mineral equilibria in the iron cherts being studied and on the evolution of the composition of the liquid at different stages of metamorphism. The thermodynamic data are tied in with the results of geological and geochemical investigations of Precambrian iron cherts and of recent processes of migration, deposition, and diagenesis. In those cases where mathematical methods were not effective (for instance, in studying colloidal forms of transport, the properties of amorphous and extremely dispersed sediments, etc.), special experimental investigations were made.

The variations in distribution of oxygen isotopes in the oxides and carbonates of iron-formations have been examined in detail and the complex origin of the main ore mineral—magnetite—has been demonstrated. Isotopic investigations were supplemented by a special study of the chemistry of the rocks of the cherty-iron formations as a whole and of the individual layers, bands, and minerals constituting them.

As a result, a physicochemical model for the formation of the BIF is proposed which is consistent with modern ideas on the evolution of sedimentation and volcanism and of the atmosphere, hydrosphere, and biosphere in the Precambrian. This model, which proposes a mainly volcanic source for the iron and silica and a biochemical and chemical mechanism of deposition, is the most likely but not the only possible one. Other versions, or different interpretations, are not ruled out, but it is perfectly obvious that in any genetic postulates, the specific physicochemical data must be taken into account. It is also quite understandable that in a work which is a first attempt at physicochemical analysis of the entire geological cycle—source of the material → transport → deposition → diagenesis → metamorphism—not all the problems have been worked out in sufficient detail and not all the evidence is conclusive; far from it. Further investigations in this direction are needed, including not only determination of the role of the individual parameters in ore formation, but also direct experimental modeling of the process.

The author is deeply indebted to Academician Ya. N. Belevtsev of the Academy of Sciences of the Ukrainian SSR, for many years director of joint investigations of the Precambrian cherty iron-formations, who in every way supported the introduction of modern experimental, thermodynamic, and isotopic methods into the science of the origin of ore deposits. Of very great help in the work on the monograph were M.A. Yaroshchuk, A.A. Drozdovskaya, I.P. Lugovaya, and K.A. Vorob'yeva, scientific colleagues at the Institute of Geological Sciences and later at the Institute of Geochemistry and Physics of Minerals of the Academy of Sciences of the Ukrainian SSR, and also E.R. Solov'yeva and A.A. Sarapina, who edited the manuscript.

Contents

Chapter 1

Precambrian banded iron-formations, their distribution and their age, spatial, and genetic relationships

Introduction

The Precambrian iron-formations, of which banded ferruginous or iron cherts and jaspilites are the most important component, are known on all the continents. The rocks constituting these formations are very diverse in lithologic composition and origin: from typical volcanic (both basic and acid) to purely sedimentary (clastic, chemogenic). The iron-formations are usually related to complexly deformed geosynclinal complexes—eugeosynclinal or miogeosynclinal—although they are also encountered in slightly altered sediments on old platforms.

The largest areas where Precambrian iron-formations are developed are the East European platform (the Krivoy Rog, Kremenchug, Belozerka, and other districts of the Ukrainian shield, the Voronezh, Baltic, and Central Kazakhstan shields), the Lake Superior area on the North American platform, Minas Gerais in Brazil, Hamersley in Australia, Singhbum in India, etc.

BIF have been found in the Precambrian beginning with the oldest strata, 3500 m.y. old. The rocks of the Konka-Belozerka zone of the Ukrainian shield and of the Pilbara Block of the Australian platform have an age of 2700–3500 m.y.

An age of 2000–2700 m.y. has been established for the iron cherts of the Bazavluk zone of the Ukrainian shield, the Olenogorsk deposit on the Baltic shield, the Singhbum deposit in India, and the Algoma-type BIF of the Lake Superior area.

The rocks of the Krivoy Rog–Kremenchug zone, of the Kursk group in the Kursk Magnetic Anomaly (KMA), of the Animikie group of Lake Superior, and of the Hamersley formation in Australia are dated to 1700–2000 m.y.

Younger BIF are also known, including post-Precambrian ones, related chiefly to volcanogenic rocks (Kalugin, 1969; Novokhatsy, 1973).

Up to now there has been no generally accepted classification of these rocks based on lithologic association. Semenenko et al. (1967) distinguish four types of geosynclinal cherty iron-formations. Gross (1961, 1965, 1973) distinguishes a sedimentary cherty iron-formation of Lake Superior type and a volcanogenic cherty iron-formation of Algoma type which includes the keratophyre, metabasic, and ultrabasic formations*. In many cases it is difficult to assess the importance of volcanogenic material in the make-up of the rhythmically interbedded strata, as for example in the upper group of the Belozerka area of the Ukrainian shield. Often the interrelationships of the different formations in time and space also remain in dispute.

Geologic position of banded iron-formations

Let us briefly describe some of the main areas of development of Precambrian BIF of different formational types. Main attention will be paid to the general regularities of distribution of the iron-formations and to their age, spatial, and genetic relationships, rather than to the details of the geology of the individual areas. In so doing we will retain the terminology and nomenclature of the rocks used by the various authors.

East European platform

On the East European platform, BIF are concentrated mainly in the western part, in the Ukrainian, Baltic, and Voronezh shields, and also in Kazakhstan in the eastern part of the platform.

Iron-formations of the Ukrainian shield

In the Ukrainian shield, BIF are developed in a number of synclinorium zones in different structural levels of the Precambrian. They are spatially unconnected, which makes it difficult to correlate them and establish their common genetic regularities. Usually several zones of development of BIF are distinguished (Semenenko, 1973), differing in location (Fig. 1) and time of formation (Fig. 2).

The Konka–Belozerka synclinorium, which was stabilized in the first Precambrian megacycle 2700 m.y. ago, is considered to be the *first and oldest zone.* Rocks with an absolute age of 3500 m.y. have been established in this zone. The cherty iron-metabasite* and cherty iron-ultrabasite* formations of the Lower Konka group were developed essentially at that time. In the final stages of development of the zone, basic volcanism was superseded by acid volcanism and the rocks of the cherty iron-keratophyre* formation of

* See footnote, p. 5.

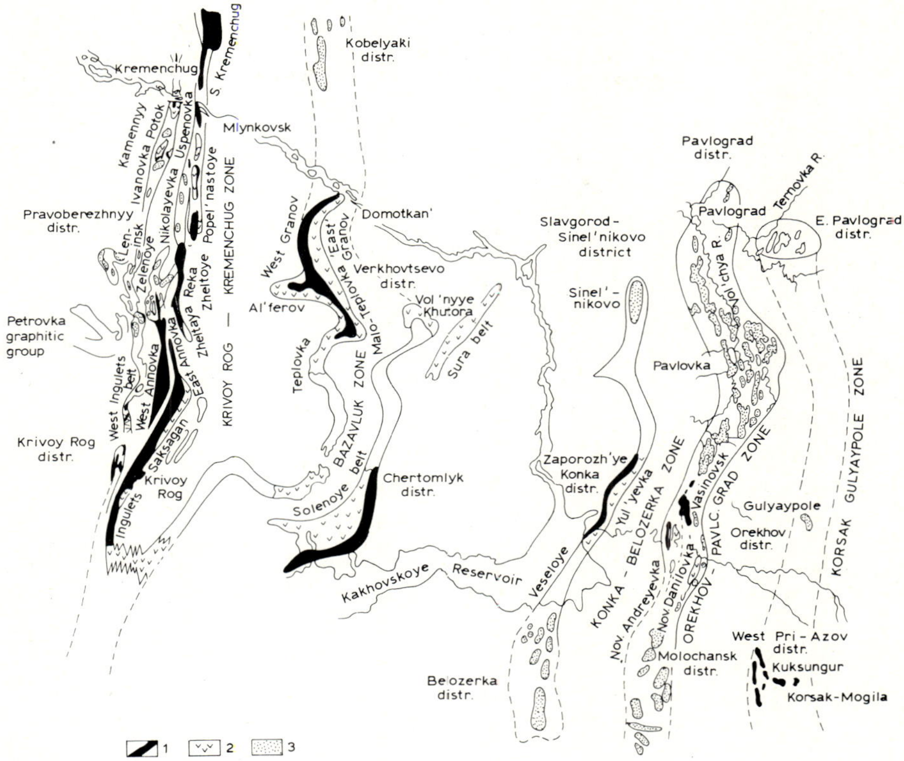

Fig. 1. Sketch map of the cherty iron-formations of the Ukrainian S.S.R. (after Semenenko et al , 1956). Groups: *1* = Saksagan; *2* = Metabasite; *3* = undifferentiated Saksagan and Metabasite.

the Upper Konka group were deposited. According to Semenenko et al. (1956) the rocks of this zone were metamorphosed to schists and hornstones and reach the rank of gneiss only in the marginal parts of the zone at the contact with granitoids.

The *second zone* takes in the Bazavluk synclinorium, which began to develop in the period of 2700–2800 m.y. and stabilized 2300 m.y. ago. In this zone the cherty iron deposits consist of a cherty iron-metabasite* formation in the Lower Bazavluk group and a cherty iron-keratophyre* formation in the Upper Bazavluk group. The rocks of this zone are very diverse in rank of metamorphism: from phyllite to pyroxene-gneiss.

The *third zone* is confined to the Orekhov–Pavlograd synclinorium, developed 2300–1800 m.y. ago. Semenenko (1973) correlates the Korsak and

* See footnote, p. 5.

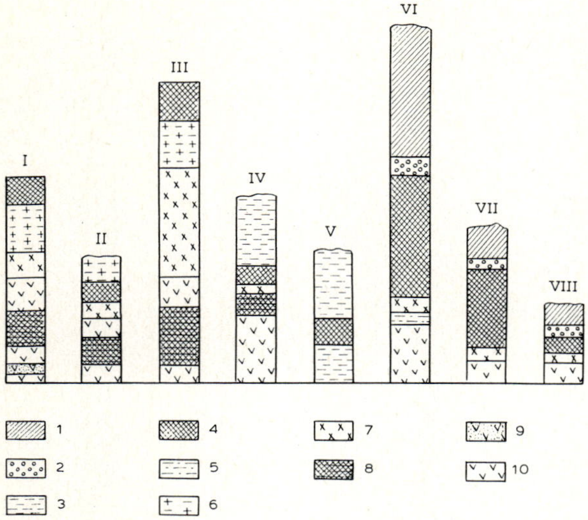

Fig. 2. Comparative schematic sections of the cherty ironformations of the Ukrainian shield. Symbols: *1* = schists and gneisses, carbonate rocks of the upper suite; *2* = sandy suites (sandstones, conglomerates); *3* = schist suites; *4* = cherty iron-schist suites; *5* = schist-sandy suites (arkoses, phyllites); *6* = schist-keratophyre suites (keratophyres, tuffs, schists); *7* = schist-ultramafic suites (talc, talc-magnesite, and green schists); *8* = cherty iron-metabasite suites; *9* = quartzite-sandstones in metabasites; *10* = metabasite suites.
Districts: *I* = Konka; *II* = Belozerka; *III* = Verkhovtsevo; *IV* = Orekhov; *V* = Gulyaypole; *VI* = Saksagan; *VII* = Kremenchug; *VIII* = Ingulets. Compiled from data of Semenenko et al. (1959, 1967).

Mangush zones of the Pri-Azov area with this zone. Both cherty iron-metabasite and cherty iron-gneiss formations are present in the Orekhov–Pavlograd synclinorium, the former being limited stratigraphically to the lower part of the section and the latter overlying it. A feature of the iron cherts of this zone is the more intensive metamorphism under amphibolite and granulite facies conditions (Kalyayev et al., 1968; Glevasskiy et al., 1971). Less metamorphosed rocks occur only in individual synclines, for instance the Gulyaypole.

The fourth and main zone is the Krivoy Rog–Kremenchug, developed 1800–2000 m.y. ago. The accumulation of rocks began here, as in the other zones, with the eruption of metabasic and ultrabasic rocks, which then probably were locally succeeded by rocks of the keratophyre series. Most of the iron cherts, however, are not related to the initial but rather to the middle stage of development of the zone, when the sedimentary cherty iron-formation proper of the Middle suite of the Krivoy Rog group was formed. Maximum development of iron cherts is characteristic of the Saksagan and Kremenchug basins (Fig. 3).

No meta-igneous rocks have been definitely established in the sections of

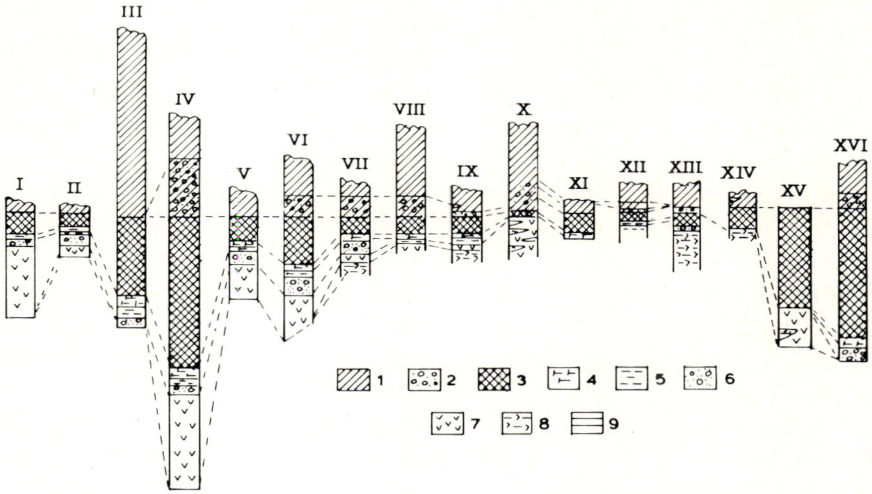

Fig. 3. Correlation of sections of the cherty iron-formations in the Krivoy Rog–Kremenchug structural-facies zone (after Semenenko et al.).
Saksagan group. Suites. Upper: 1 = upper subsuite (schists and gneisses); *2* = lower sandstone subsuite above the ore (sandstones, conglomerates, schists). *Middle: 3* = interbedded iron cherts, schists, gneisses. *Lower: 4* = talc, talc-actinolite, anthophyllite schists, ultramafics; *5* = schists, phyllites, and gneisses of lower suite; *6* = quartzite-sandstones of lower suite.
Metabasite group: 7 = amphibolites, spilites, greenschists, amphibole gneisses (undifferentiated); *8* = gneisses; *9* = iron cherts.
Districts (numerals on illustration): *I* = Ingulets South; *II* = Ingulets North; *III* = Saksagan South; *IV* = Saksagan North; *V* = Annovka; *VI* = Zheltaya Reka; *VII* = Zheltoye; *VIII* = Popel'nastoye; *IX* = Zelenoye; *X* = Ivanovka; *XI* = Mlynkov; *XII* = Nikolayevka (east limb); *XIII* = Nikolayevka (west limb); *XIV* = Uspenovka; *XV* = Kremenchug South; *XVI* = Kremenchug North.

the Middle suite, a fact which obliges one to attribute the cherty iron strata to the remote type of formation (Semenenko et al., 1959) or even to believe that the original material of the sediments was the product of subaerial weathering (Belevtsev, 1947; Gershoyg, 1947; Bordunov, 1964). Depending on the typical paragenetic associations of the rocks, Semenenko (1973) distinguishes these types of cherty iron-formations on the Ukrainian shield: cherty iron-schist (sedimentary), cherty iron-keratophyre (or jaspilite-leptite), cherty iron-metabasite and cherty iron-ultrabasite*.

The *cherty iron (sedimentary) formation* (CIS) is the most uniform in facies. The cherty iron beds are traced for 10–20 km; their thickness reaches 100–300 m and their iron content reaches 30–35%. According to Semenenko

* For brevity, these frequently recurring terms will henceforth usually be designated by acronyms: CIS for cherty iron-sedimentary formation, CIK for cherty iron-keratophyre formation, CIM for cherty iron-metabasite formation, and CIU for cherty iron-ultrabasite formation (Translator's note).

et al. (1959) these rocks were formed under conditions of prolonged hydrothermal activity of submarine volcanoes, with limited introduction of ashy material and widespread development of chemogenic deposits (remote type of formation). On the Ukrainian shield the Krivoy Rog–Kremenchug cherty iron strata of the Middle suite and also the rocks of the Upper suite of the Upper Konka group of the Belozerka syncline are of this type. Sometimes the CIS pinches out along the strike and is replaced by the CIK (Konka syncline).

The *cherty iron-keratophyre formation* (CIK) is characterized by the paragenetic association of acid dacite-rhyolite lavas, tuff-keratophyres, and cherty iron deposits. The formation is more uniform in facies. It is believed that the cherty iron sediments were formed near volcanic centers at times when volcanism was dying out. Semenenko assigns the deposits of the Upper suite of the Upper Konka group in the Konka syncline, and also the rocks of the Upper Bazavluk group in the Teplovka belt in the Verkhovtsevo syncline, to the CIK.

The *cherty iron-metabasite formation* (CIM) was formed in regions of submarine volcanism of basic composition. The iron cherts in the formation are not uniform in facies and amount to 10–30% of the total thickness of the sequence. The formation is extensively developed in the lower metabasic series of the Konka, Belozerka, Verkhovtsevo, and Sura synclines.

The *cherty iron-ultrabasite formation* (CIU) is not very widespread; only in the Konka syncline and in the Kudashev sector of the Verkhovtsevo syncline are low-grade banded cherty iron intercalations of no great thickness encountered, interbedded with apo-ultrabasic schists.

Other ideas are known concerning the genetic classification of the cherty

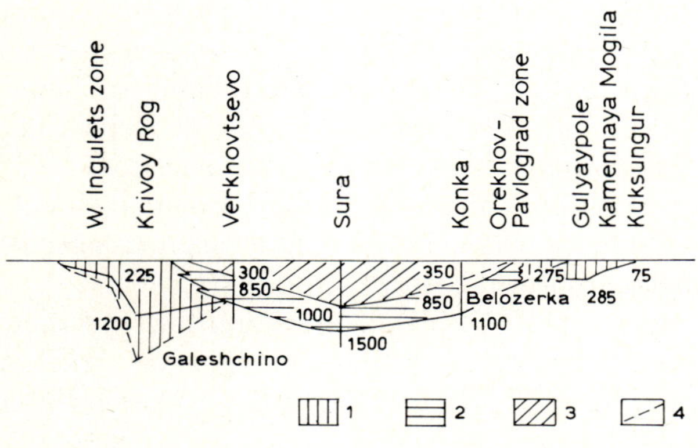

Fig. 4. Compensation diagram of the jaspilite formation (after Kalyayev): *1* = jaspilite formation of sedimentary type in border zones; *2* = jaspilite formation of volcanogenic-sedimentary type in interior zones; *3* = overlying volcanogenic-clastic subformation in interior zones; *4* = thickness of jaspilite formation in Galeshchino and Belozerka synclinoria. Numerals on illustration— total thicknesses of formation.

iron-formations of the Ukrainian shield and their spatial and temporal relationships. Kalyayev (1965) distinguishes three types of cherty iron-formation: sedimentary, volcanogenic-sedimentary, and the jaspilite subformation within the spilite-diabase formation.

The sedimentary jaspilite formation, the stratotype of which is the Krivoy Rog iron-ore sequence, is characterized by an association with pelitic and psammitic sedimentary rocks whose sources were products of subaerial weathering. The sediments of this formation were deposited in the Krivoy Rog–Kremenchug marginal depression and also in individual parageosynclinal zones: the Orekhov–Pavlograd, Western Pri-Azov, Mangush, etc.

The volcanogenic-sedimentary jaspilite formation was formed in intrageosynclinal synclinorium zones; its stratotype is the cherty iron-formation of the Belozerka syncline.

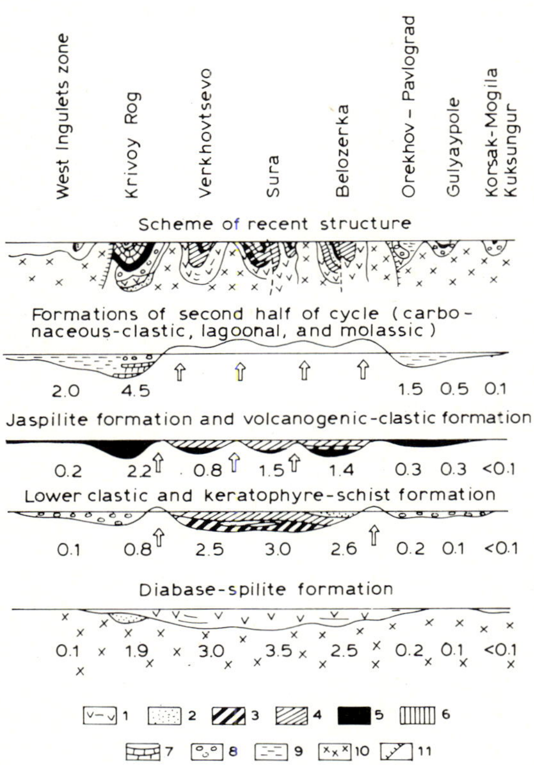

Fig. 5. Scheme of development and structure of the Greater Krivoy Rog geosyncline (after Kalyayev): *1* = metabasites and paraschists with minor horizons of ferruginous quartzites; *2* = quartzites-sandstones; *3* = ultramafics; *4* = micaschists, metabasites, tuffogenic rocks and apo-keratophyres; *5* = iron cherts, jaspilites, schists and volcanogenic rocks of the jaspilite formation; *6* = carbonaceous-graphitite schists; *7* = carbonate rocks; *8* = conglomerates; *9* = undifferentiated gneisses, including quartzites and marble; *10* = migmatites, granites, gneisses; *11* = abyssal faults. Numerals on illustration—approximate thicknesses, in km.

The jaspilite subformation developed in the spilite-diabase formation is known both in intrageosynclinal synclinorium zones and in the inner zones of the Krivoy Rog–Kremenchug marginal depression.

The interrelationships of these types of jaspilite formations are represented on the composite diagram (Fig. 4) and in the scheme of development of the Greater Krivoy Rog geosyncline (Fig. 5). From these plots it follows that the sedimentary-volcanogenic and sedimentary jaspilite formations are temporal analogs but were formed in different structural zones of the Greater Krivoy geosyncline: the former in the inner zones (Verkhovtsevo, Sura, Konka) and the latter in the outer zones of the marginal depressions (West Ingulets, Krivoy Rog–Kremenchug, Orekhov–Pavlograd, and Pri-Azov). The sources of the material in these types of formations also are different: in the former the products of volcanism predominate, in the latter, the products of subaerial weathering, forming a Precambrian geosynclinal flysch. Thus all the cherty iron-formations belong to the same geosynclinal cycle, to its pre-orogenic half, and the sedimentary and sedimentary-volcanogenic jaspilite formations were formed after deposition of the jaspilite sub-formation of the diabase-spilite formation.

The theoretical difference between the structural-stratigraphic schemes considered is the distinction of one or several geosynclinal cycles. There remain many problems that are unclear and controversial, but in most works (Belevtsev, 1962; Dobrokhotov, 1969; Semenenko, 1973) it is accepted that by the time the metabasic rocks of the Krivoy Rog–Kremenchug, Orekhov-Pavlograd, and Brusilov-Odessa zones were deposited, the other structural-facies zones were already stabilized. A new geosynclinal cycle began with eruption of the metabasic rocks of the Krivoy Rog–Kremenchug zone.

Iron-formations of the Voronezh shield

BIF occur in stratigraphic complexes in the Precambrian Voronezh shield. Layers of highly metamorphosed iron-rich rocks, of no great thickness, have been established in the old gneisses and migmatites of the Oboyan' group, which most investigators assign to the Archean (Zaytsev, 1966; Plaksenko, 1969).

In the *Mikhaylovka* group, which is correlated with the Verkhovtsevo series of the Ukrainian shield (Zaytsev, 1966), iron cherts are interbedded with keratophyres and their tuffs and orthoschists, and are underlain by orthoamphibolites.

BIF are very extensively developed in the deposits of the Lower Proterozoic *Kursk* group, mainly in the middle suite. The iron formations are associated with metasediments—phyllites. The iron cherts of the well studied Mikhaylovka area of the KMA constitute several discontinuous layers forming two belts: a northeastern one 450–500 km long and a northwestern one

600 km long. These belts skirt the arch uplift of the crystalline basement, as it were. The total thickness of the iron cherts of the Kursk group is of the order of 1100–1200 m (Illarionov, 1965). Iron cherts consisting of heavily banded and concretionary martite-hematite quartzites are also known among the rocks of the upper suite (Plaksenko, 1966).

Plaksenko (1969) distinguishes these cherty iron-formations: cherty iron-gneiss (Oboyan' group), cherty iron-metabasite (Mikhaylovka group and lower suite of the Kursk group), cherty iron-schist (middle suite of the Kursk group and cherty iron-clastogenic (upper suite of the Kursk group). A volcanic source is assumed for the material of the cherty iron-metabasite formation; the formation of the cherty iron-schist formation is not connected with volcanism, but is explained by the arrival of material from the weathered layer on Archean iron-bearing rocks. Plaksenko considers these two main formations to be separate, one succeeding the other in time in a process of irreversible development of iron-ore deposition in the Precambrian, rather than temporal analogs.

Iron-formations of the Baltic shield

On the Baltic shield, BIF are known in three major structural-facies zones: the Karelian zone of the Karelides the Kola–Norwegian zone of the Karelides, and the vast zone of the Svecofennides in the southwestern part of the shield. The first two zones are separated by the Belomorides, which most investigators consider to be older than the Karelides. (Polkanov, 1939; Chernov et al., 1970). The question of the age relationships of the Svecofennides of Sweden and Finland to the Karelides is in dispute.

In the *Karelian zone of the Karelides,* BIF are known in northern Finland in the spilite-diabase formation and are associated with basic volcanic rocks —pyroclastics, basaltic lavas and agglomerates—in northern Norway; graphite schists and limestones are occasionally encountered in the sections.

In the *Kola–Norwegian zone of the Karelides,* situated in the central part of the Kola peninsula and extending into northern Norway, iron cherts are associated with gneisses, schists of basic composition, amphibolites, and paracharnockites of the Kola group. In this sequence, which can be assigned to the volcanogenic-sedimentary formation of the spilite-diabase series, both metasedimentary and meta-igneous rocks occur (Bondarenko and Dagelayskiy, 1968). In the Lake Imandra area the cherty iron-formations of the Kola group are associated with volcanic rocks of more acid composition, similar to the rocks of the porphyry-leptite formation of central Sweden (Goryainov, 1969). In the West Karelian zone iron cherts are restricted mainly to the Gimoly group of Lower Proterozoic age: 2000–2600 m.y. (Chernov et al., 1970). The Gimoly group lies on the Archean basement of the Karelides and is overlain by the Parandovo and Bol'shoye Ozero group, where iron cherts also are known. The largest deposits of iron-formation—the Kostomuksha and Mezhozero—are restricted to large synclinal structures preserved in the Proterozoic granites and migmatites. The iron cherts of the upper sedimentary cycle are associated with the products of acid volcanism: lavas, tuffs, Hälleflintas.

In the Svecofennides, BIF are known in central and northern Sweden (Kiruna area) and in southern Finland, also associated with volcanogenic rocks. Formational group of the same type, similar to the Proterozoic formations of Karelia, are found in all these areas. Acid volcanic rocks of the porphyry-leptite formation, and sometimes carbonate rocks, are noted in paragenesis with iron cherts; there are no typical primary clastic formations. In the Kiruna area the largest deposits of iron-formation and apatite-bearing iron ores occur in paragenesis with volcanic rocks of syenitic and rhyolitic composition and belong to the cherty iron-leptite porphyry formation (Formozova, 1965). Chernov et al. (1970) correlate this formation at Kiruna with the sequence of basic volcanics and iron cherts of the Bol'shoye Ozero group, which lies with angular unconformity on the leptite-porphyry cherty iron-formation of the Gimoly group and constitutes the upper part of the Lower Proterozoic section of Karelia.

Despite the diversity of the BIF and the complexity of their interrelationships with the enclosing rocks, some general features have been established. In a detailed formational analysis, Chernov showed that the cherty iron-formations constitute the lower parts of the sections both in the Karelides and in the Svecofennides, and that their formation reflects the initial stages of geosynclinal development of the Baltic shield. In many areas the BIF are underlain by thick piles of conglomerates which rest on an old basement more than 2600 m.y. old. The BIF are overlain by volcanic rocks of the spilite-diabase series or by clastic flyschoid formations formed in the final stages of geosynclinal development.

Iron-formations of Central Kazakhstan

Precambrian iron cherts occur in several large structures of the area—the Karsakpay and Betpak-Dala belts, the west limb of the Maytyube anticlinorium, etc. (Rozanov and Filatova, 1967).

In the Karsakpay synclinorium the BIF are restricted to the Karsakpay group, Early Proterozoic in age. The rocks of the Karsakpay group constitute a narrow north–south belt extending almost uninterruptedly for 200 km. Four macrorhythms, united into individual suites, are distinguished in this series. These macrorhythms reflect cyclicity in the manifestation of basaltic volcanism: in each rhythm the lower part of the suite consists of porphyritoids after tuffs and lavas of basaltic, less often of andesitic, composition, and the upper part consists of quartz-sericite schists, marbles, and iron-rich rocks. A feature they have in common is that they are restricted to packets of originally clastic rocks.

In the *Maytyube anticlinorium* iron cherts are known among the deposits of the Satan suite, which consists mainly of clastic formations: mica-quartz schists, phyllites, conglomerates. The Satan suite lies stratigraphically above the Karsakpay series and belongs to the Upper Riphean or Lower Vendian (Rozanov and Filatova, 1967).

Among the iron cherts of the Karsakpay synclinorium, Markov (1959a) distinguished three types of sequences and correspondingly three types of formations.

The association of BIF with basic volcanic rocks is typical of the first type; it can be correlated with the jaspilite formation of the Algoma series in the Lake Superior district, which was formed in the greenstone formation. The rocks of the Verkhovtsevo group of the Ukrainian shield are also considered analogs.

The second type is distinguished by the association of BIF with metasedimentary rocks—the analog of the jaspilite formation of the Saksagan group of the Krivoy Rog and the Animikie group of Canada; this formation follows the greenstone and jaspilite formations of Keewatin type in time.

The third type is characterized by the association of thin and rapidly-pinching-out iron cherts with clastic sediments, carbonate rocks and carbonaceous shales. These deposits belong to the remote jaspilite formation, developed either at the same time as the jaspilite formation of Krivoy Rog type or after it (analog of the rocks of the upper suite of the Saksagan district).

Other ideas are known concerning the relationships of the cherty iron-formations. Rozanov and Filatova (1967) believe that in formational respects the Karsakpay group is unique and is a greenschist spilitic iron-ore formation consisting of rocks of three groups: volcanogenic, volcanogenic-sedimentary, and sedimentary. This formation accumulated in geosynclinal conditions in an aquatic environment in a period of steady subsidence of the basin floor.

North American platform

The main regions where Precambrian cherty iron-formations occur are on the Canadian shield, chiefly in the southwestern part of it.

The oldest iron cherts (2300–2700 m.y., Kenoran orogeny) are spatially and genetically related to the volcanogenic greenstones of the Algoma group. On this basis Gross (1973) distinguished the volcanogenic cherty iron-formation of Algoma type. The rocks of this formation are extensively developed in the Michipicoten region (Churchill, Helen, and other districts).

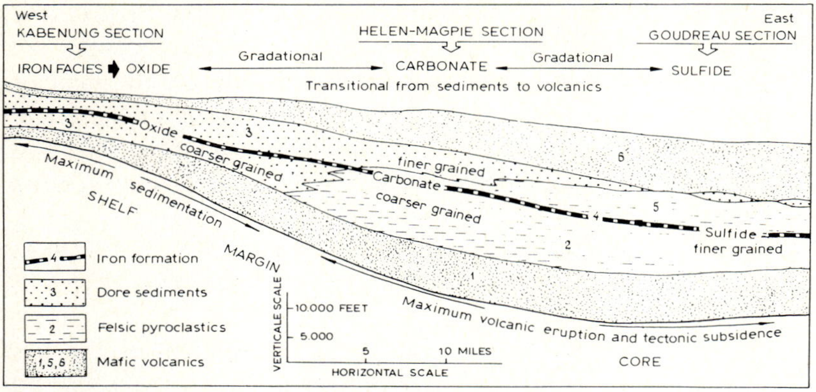

Fig. 6. Reconstructed east–west section in Michipicoten basin (Goodwin, 1973).

In the Churchill district iron cherts are associated with rhyolites, metadiorites, andesites, dacite breccias and tuffs, carbonate rocks, and graywackes (Goodwin, 1973). Also typical of the Helen district is the relationship of the cherty iron sediments to volcanism, and a relationship has been established between the composition of the sediments, their location in the sedimentary basin, and the amount of volcanic products (Fig. 6). The facies details shown in the sketch reflect the regularities of deep deposition of volcanic derivatives on the slopes of the original sedimentary basin (Goodwin, 1961; Goodwin and Ridler, 1970). The BIF which occur in the volcanic rocks are a combination of three facies—banded chert, a sulfide member, and a

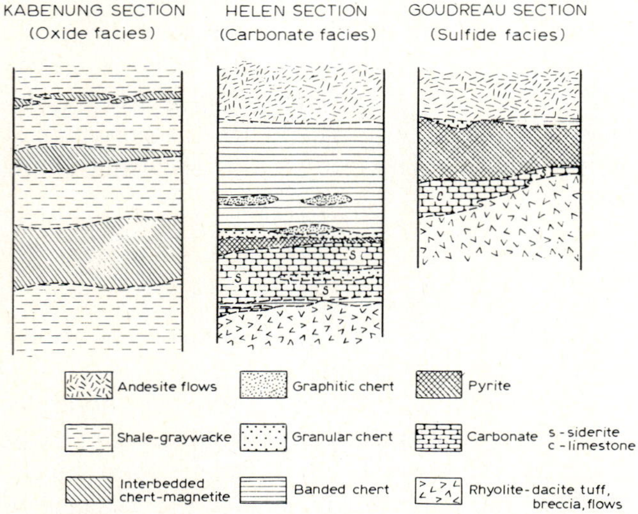

Fig. 7. Typical sections of oxide, carbonate and sulphide facies (Goodwin, 1973).

carbonate member (Fig. 7, Helen section and Goudreau section). The banded chert member consists of alternating bands of cherty material and iron minerals, chiefly siderite with magnetite, pyrite, and pyrrhotite. It gradually grades into lenticular sulfide members which in turn also are gradually replaced by the underlying carbonate member. In the western part of the district (Kabenung section) a sedimentary association is developed, consist-

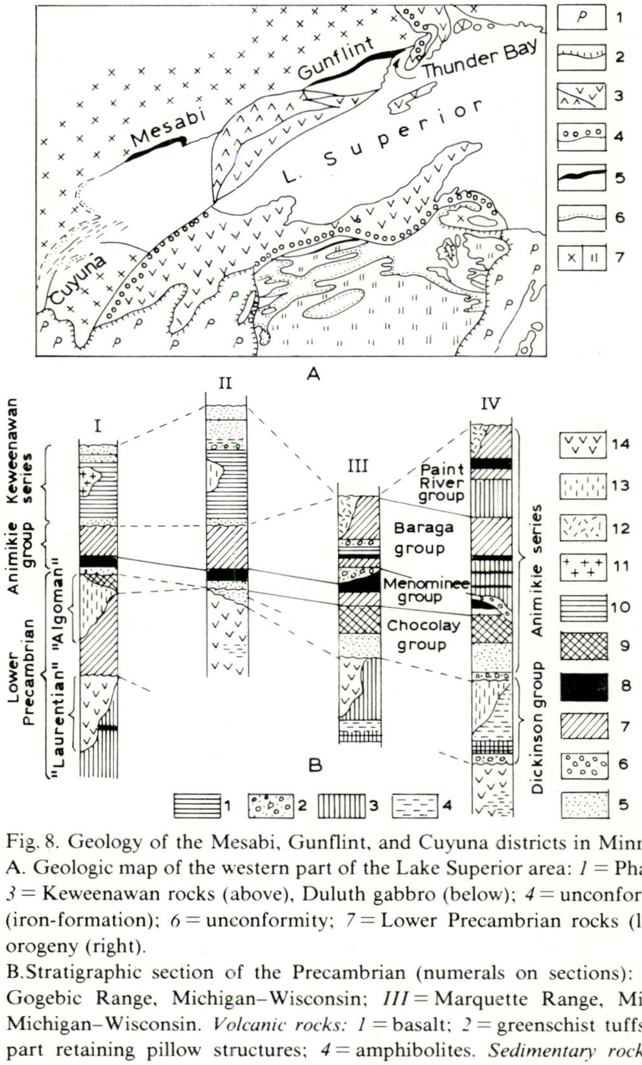

Fig. 8. Geology of the Mesabi, Gunflint, and Cuyuna districts in Minnesota (after Morey).
A. Geologic map of the western part of the Lake Superior area: *1* = Phanerozoic cover; *2* = unconformity; *3* = Keweenawan rocks (above), Duluth gabbro (below); *4* = unconformity; *5* = Animikie series or group (iron-formation); *6* = unconformity; *7* = Lower Precambrian rocks (left); rocks subjected to Penokean orogeny (right).
B. Stratigraphic section of the Precambrian (numerals on sections): *I* = Minnesota and Ontario; *II* = Gogebic Range, Michigan–Wisconsin; *III* = Marquette Range, Michigan; *IV* = Menominee system, Michigan–Wisconsin. *Volcanic rocks: 1* = basalt; *2* = greenschist tuffs and breccias; *3* = greenstones, in part retaining pillow structures; *4* = amphibolites. *Sedimentary rocks: 5* = sandstones and quartzites; *6* = conglomerates and arkoses; *7* = graywacke shales; *8* = iron-formation; *9* = dolomites; *10* = shales, probably including a certain amount of volcanic rocks. *Crystalline rocks: 11* = gabbros and granites of post-Middle Keweenawan age; *12* = granitic rocks of post-Animikie age; *13* = gneisso-granites of pre-Animikie age; *14* = granite-gneisses of established or presumed early pre-Animikie age.

ing of alternating bands of chert, magnetite, and jasper, which occur in graywackes and argillites. Thus the thick oxide facies of the western part of the district, rich in cherty material, gradually changes eastward into a thick carbonate facies containing cherty material in the center, and then into a thin sulfide facies poor in cherty material.

An interesting BIF of Algoma type is found in northwestern Canada on the Yukon–Mackenzie divide (Gross, 1972). Traces of mud flows, tuffs, ash, conglomerates, and syngenetic breccias—i.e., products testifying to deposition in a tectonically active basin with explosive volcanic activity—have been established in these cherty iron-formations.

The iron cherts of the Animikie group, extensively developed in the Lake Superior district (Fig. 8) in the states of Minnesota (Cuyuna, Mesabi, and Gunflint belts), Michigan and Wisconsin (Marquette, Gogebic, Menominee, and other belts), are younger (Huronian orogeny, consolidation 1700–1900 m.y. ago) (Gundersen and Schwartz, 1962; Gair, 1972). The rocks of the Animikie group lie on rocks of the Algoma group and are overlain by quartzites and greenstones of the Keweenawan group (see Fig. 8). Within the Animikie group itself, iron-rich and barren quartzites, paraschists, tuffs, breccias, and dolomites are interbedded and succeed one another in facies. Besides the Lake Superior district, analogous rock associations occur in the Labrador trough, in the eastern part of the Canadian shield. Gross assigned these iron cherts to the Superior type, characterized by the development of unstratified cherty and iron-rich deposits with granular or oolitic texture. Sediments of Superior type were formed in the late Precambrian in miogeosynclinal conditions of a continental shelf, succeeding a volcanogenic cherty iron-formation of Algoma type, deposited in a eugeosynclinal setting.

South America

Deposits of Precambrian BIF, the rich iron ores of Cerro Bolívar in Venezuela (Ruckmick, 1963) and in the states of Pará and Minas Gerais in Brazil (Dorr, 1969; Tolbert et al., 1973), are known. In the Sierra das Carajás area in Pará, iron cherts were found not long ago in the rather inaccessible forests of the Amazon basin, and are not well known.

In Minas Gerais there is a large deposit; the iron cherts form a 500-km belt on the edge of the Brazilian shield. The Rio das Velhas group, the age of which is older than 2800–3000 m.y., and the Minas Gerais group, 1800–1300 m.y. old and lying unconformably on the gneisses of the Rio das Velhas group, are distinguished.

In the Rio das Velhas group, iron cherts form thin lenticular bodies in Archean gneisses. The association with metavolcanic rocks and graywackes is typical of them.

The Minas Gerais group is divided into the lower and upper clastic strata, and also the middle chemogenic Itabira strata (Dorr, 1969). The BIF are traced for a distance of about 200 km and fill a deep symmetrical syncline which plunges to the northeast. The iron-formations consist of unbanded ferruginous varieties and thin-banded hematite-carbonate itabirites.

Indian platform

On the Indian platform, iron cherts constitute part of the Dharwar system, 2000–2700 m.y. old, which forms extensive synclinal zones and belts on the periphery of the Indian shield.

The iron cherts are of two main types (Krishnan, 1973)—thin-banded hematitic jaspilites and banded magnetitic cherts. Jaspilites constitute the thick Singhbhum strata (Bihar and Orissa), the Drug, Rajpur, Bastat (Madhya Pradesh) and Goa (Mysore) deposits. In the Singhbhum deposit the banded hematite rocks are underlain by shales, calcareous sandstones, conglomerates, and phyllites, beneath which lie tuffs and basic lavas. The total thickness of the formation is 2000–2500 m, up to 1000 m of which is jaspilite. A very low degree of metamorphism is typical—of the ore minerals, hematite predominates, magnetite is rare, and silica is sometimes represented by opal. The age and genetic relationships of the hematite and magnetite rocks found in Goa, Madhya Pradesh, and other areas are unknown. Krishnan (1973) considers the magnetite rocks to be the more metamorphosed analogs of the hematite rocks and unites them both into the Dharwar formation. However, ideas exist that the magnetitic rocks are older. Volcanism presumably is the source of the iron of the Precambrian BIF of India.

African platform

On the African platform, iron cherts are known in several structural positions which, however, are not unequivocally distinguished. The oldest iron-formations (2900–3400 m.y.) are restricted to greenstone belts and consist of banded chert, banded ferruginous chert, and BIF proper (Beukes, 1973). These rocks are intimately associated with cycles of volcanic activity and are component parts of them. In ultrabasic rocks of the greenstone belt, the BIF may interfinger along the strike with various types of rocks in one and the same cycle. Under these conditions the cherts and BIF consist of lenticular bodies, and several lenses may be restricted to one stratigraphic zone. Analogous structure is observed in basic and acid rocks, but the cycles are more completely developed.

The younger Kuruman and Penge iron-formations are component parts of a more prolonged (between 1950 and 2300 m.y.) and more important period of chemical deposition in the basin of the Transvaal system. The BIF are underlain by dolomites; in the transition zone there are interbedded carbonaceous limestones, banded ferruginous cherts, and carbonaceous-argillaceous shales with pyrite, that is, typical carbonate shelf deposits.

Australian platform

Iron cherts are known in three structural positions on the Australian platform. The oldest are in Western Australia in the Pilbara and Yilgarn Blocks, which were stabilized 2700–3000 m.y. ago. The iron cherts are not spatially persistent and not very thick.

The second group (1700–2000 m.y.) constitutes the Hamersley BIF in Western Australia, which occur in a syncline between old blocks. The thick strata of the formation ($\sim$2700 m) consist of iron-rich rocks interbedded with clay shales, dolomites, lavas, and tuffs. The iron-formation is underlain by conglomerates, metabasites, pyroclastics, and pillow lavas and is overlain by quartzites, conglomerates, graywackes, and dolomites (Macleod, 1966). Slight metamorphism, the temperature of which corresponds to the normal geothermal gradient, is typical of the cherty iron-formations (Trendall, 1966).

The third group consists of the BIF of the Cleve Metamorphics (South Australia), with an age of about 1780 m.y.

A direct relationship to volcanic processes has not been established for the cherty iron-formations of Australia, but Trendall (1973a) suggests that the source of the ore material was rhythmic volcanism.

Relationship of volcanic and sedimentary processes in the formation of the banded iron-formations

From a comparison of the general regularities of the geologic position of the BIF of the world it follows that among these formations there are those which quite obviously are related to volcanogenic rock complexes and those for which no such relationship is established spatially. Study of specific objects of the first type led to the development of a number of hypotheses concerning the relationship of the sources of the material to different kinds of volcanism in the formation of the iron ores. Van Hise and Leith (1911), Polovinkina (1952), Shatskiy (1954), Markov (1959a, b), Goodwin (1962), Tochilin (1963), Gross (1965, 1973), Semenenko et al. (1967), Goodwin and Ridler (1970), Krishnan (1973) and Trendall (1973b) assumed such a relationship of the sources of the iron to volcanism. Advances in understanding of the role of volcanism in sedimentation and in ore formation in general contributed to the propagation of this hypothesis (Zelenov, 1960; Naboko, 1963; Butuzova, 1969; Dzotsenidze, 1970).

Study of specific regions in which iron cherts and volcanic rocks are spatially unconnected naturally led to the development of hypotheses of an exogene source of the iron, unrelated to volcanism. Thus there arose the hypothesis that the iron cherts were formed from material supplied to the sedimentary basin in the course of intensive subaerial weathering. Svital'skiy (1924), Piatnitskiy (1924), Strakhov (1947), James (1954), White (1954), Belevtsev (1957) and Plaksenko (1966) shared this hypothesis.

These two types of hypotheses characterize the extreme points of view. However, detailed study has shown that there often are transitions between cherty iron-formation related to volcanogenic complexes and cherty iron-formation which at first glance has no such relationship. The existence of transitions has made it possible to postulate a remote indirect relationship to volcanism and to presume a volcanogenic source of the material even for those iron-chert complexes in which there are no volcanics at all. This point of view borders on the first group of hypotheses and requires an answer to a number of controversial partial questions of the problem: how is the formation of the iron cherts related to volcanism in space; does this relationship change in time, does the character of volcanism itself change, and how does this affect the accumulation of cherty iron sediments?

The difference in approach to the answer to these questions has been reflected in the classification of cherty iron-formations.

Shatskiy's (1954) ideas concerning the classification of volcanogenic and volcanogenic-sedimentary formations greatly influenced the development of ideas concerning Precambrian iron-ore formations. Shatskiy distinguished two formational types: volcanogenic-sedimentary and sedimentary. The nature of the interrelationship between these formational types of jaspilite is depicted in the scheme:

Cherty jaspilitic ↑ ——————————————— ↑ Remote jaspilitic

Volcanogenic spilitic-keratophyric or greenstone → Cherty jaspilitic → Remote jaspilitic

The significance of Shatskiy's scheme is the fact that it presumes heterogeneity of the cherty jaspilitic formation, which can arise both from products of submarine volcanism and as a result of transport of materials from the weathering of volcanic rocks on land. It is assumed that the remote jaspilitic formation could have been deposited syngenetically with the volcanogenic formations, but in parts of the geosyncline far from volcanic centers, and also could have come after the deposition of the products of volcanism in the course of development of the geosyncline. Practically all known specific cherty iron-formations fit into this scheme.

Semenenko et al. (1959, 1967) assigned a large role in the formation of the iron cherts of the Ukrainian shield to processes of volcanism. According to his formational scheme, a direct relationship to submarine volcanism is established for the CIM, CIU, and CIK, from the paragenetic association of iron cherts with volcanic rocks.

For the CIS, to which the rocks of the Krivoy Rog–Kremenchug zone belong, Semenenko also postulates the existence of a relationship to volcanism, but he believes that this relationship is a remote one. The cherty iron-formations of the Krivoy Rog–Kremenchug zone were deposited in the peripheral part of the geosyncline, far from centers of intensive volcanic activity and in periods when emission of lava and ejection of pyroclastic products was superseded by introduction of iron and silica via fumaroles. The closer relationship of the iron cherts of this formation to volcanism of acid composition is confirmed by the geochemical investigations of Grigov'yev (1971), who postulates the existence of volcanic centers even within the Krivoy Rog–Kremenchug zone itself. New investigations made by Yaroshchuk and Onopriyenko (1973) have shown that effusive and explosive volcanogenic material of basic composition participated in the deposition of the upper iron-rich horizon of the Middle suite of the Krivoy Rog group; redeposited ashy sediments have been established right in the iron-formations.

Kalyayev (1965) attaches great importance to processes of volcanism in the formation of the iron cherts of the volcanogenic-sedimentary jaspilite formation and also of the rocks of the jaspilite subformation of the spilite-diabase formation of intrageosynclinal synclinorium zones. However, he postulates that the sedimentary jaspilite formation of Krivoy Rog type was formed without the participation of volcanic processes. He regards the Krivoy Rog formation as a typical Precambrian geosynclinal flysch, deposited as the product of subaerial weathering in the marginal parts of a large geosynclinal zone.

Plaksenko (1966) believes that in the evolution of the Precambrian geosynclines the accumulation of the CIS was separated in time from the accumulation of the volcanogenic rocks and the volcanogenic cherty iron-formations related to them.

In separating the formation of the CIS from volcanic processes and not even accepting their remote connection, Plaksenko was proceeding from a postulate of Strakhov (1960, 1965). According to this postulate, the sedimentary products related to volcanism could not be characterized by the features that have been established in the well-known iron cherts of the KMA: great thickness and good persistence in the sections, systematic sedimentogenic profile, and certain geochemical features that are intrinsic to pellitic and psammitic sediments.

However, Rozanov (1971) relates the accumulation of iron cherts to volcanic processes and believes that a "lateral" series can be made of the known cherty iron formations, in which the order of formation reflects the distance of the regions of iron accumulation from regions of intensive manifestation of volcanism. This relationship will be expressed in a gradual reduction in the role of solid products of volcanism in the composition of the formations. In that arbitrary lateral series the Kursk and Krivoy Rog formations occupy an end position and will correspond to Shatskiy's remote chert formations.

Chernov et al. (1970), who studied the cherty iron-formations of Karelia, concluded that they are related to volcanism not only of basic, but mainly also of acid composition. In turn, on the basis of the composition of the parent lavas, they distinguished spilite-diabase and leptite-porphyry formations among those of eugeosynclinal type, formed simultaneously but in different paleotectonic conditions. A large part of the formations of the spilite-diabase series of the Baltic shield is confined to the junctions between geosynclinal depressions and central massifs. The leptite-porphyry series of geosynclinal formations is characterized by a close association of acid and basic volcanics with iron cherts and less often with limestones and clastic sediments (Fig. 9).

The formations of the leptite-porphyry series were formed on a comparatively rigid intrageosynclinal massif. Deposition of the leptite-porphyry cherty iron-formations took place in relatively shallow-water conditions and

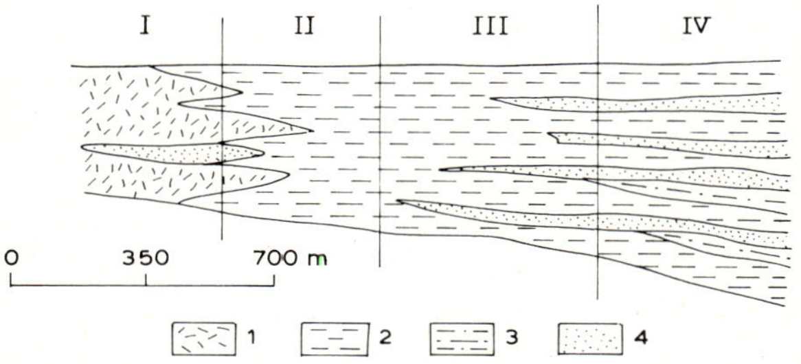

Fig. 9. Sketch of facies relationships of the paragenetic associations of rocks of the leptite-porphyry cherty iron-formation (after Chernov et al.): *1* = plagioporphyries, hälleflintas, volcanic breccias; *2* = quartz-biotite tuffogenic schists with intercalations of amphibole-garnet and biotite-garnet schists; *3* = graphitic quartz-biotite schists rich in sulfides; *4* = iron cherts. Numerals on map — rock associations: *I* = porphyry -iron chert; *II* = tuff-porphyry paragenetic; *III* = tuff aluminous-iron chert paragenetic; *IV* = tuff schist-iron chert paragenetic.

was accompanied by intensive processes of acid volcanism, as a result of which thick sequences of pyroclastic and cherty iron sediments were produced (see Fig. 9). As a result, two major epochs of iron accumulation are noted in the evolution of the Proterozoic, each of which corresponded to a certain stage of tectonic development. In the Lower Proterozoic geosynclinal period (2600–2000 m.y. ago) the large cherty iron-formations of the volcanogenic-sedimentary series, which later formed the very large deposits of iron-rich rocks of the Baltic and Ukrainian shields and of the KMA, were formed over the vast territory of the East European geosynclinal region, characterized by intensive submarine volcanism.

The Middle Proterozoic orogenic and subplatform stages of development of the Precambrian of the East European geosynclinal region (2000–1750 m.y. ago) were characterized by accumulation of sedimentary-chemogenic and clastogenic types of cherty iron-formation. In that period ore formation went on most intensively in the Krivoy Rog and Kursk depressions; on the Baltic shield only the iron cherts of the Jatulia series in southern Karelia were deposited at that time.

Investigators in foreign regions where Precambrian iron cherts are developed have likewise established their close relationship to volcanics of different composition, but they also have distinguished regions in which cherty iron-formations are developed where this relationship is less obvious, and therefore may be debatable.

As has already been mentioned, Gross (1965, 1973) distinguished two types of iron cherts on the Canadian shield, differing in textural-structural features, paragenetic associations, and also in sources of the material.

The Algoma type is characteristic of the Early Precambrian. The Michipicoten district can serve as an example. The iron cherts are not easily traced in it; the facies are very variable over short distances; up-section and along the strike the iron cherts are superseded by andesites, tuffs, pyroclastics, slates, black carbonate shales with sulfides. The iron cherts of this type are banded; no granular structures have been established in them. Gross believes that iron cherts of that type were deposited in eugeosynclinal conditions and were closely related to volcanism. Certain facies of the iron cherts were formed depending on the distance to the volcanic center: in the immediate vicinity of the volcanic centers were the sulfide, silicate, and carbonate facies; the acid facies were far from them and might occur among clastic sediments, but the source of the material in them was volcanic all the same. Goodwin (1973a, b) also relates the formation of iron cherts of Algoma type to greenstone sequences, and within them, to the formation of felsic pyroclastics and related sedimentary rocks.

In particular, the sedimentary rocks of the Michipicoten series are chiefly of volcanogenic origin, which is confirmed by the discontinuous bedding, poor sorting of the material, local rhythmicity, and presence of erosional channels. The presence of pillow structures in the lava flows also indicates submarine deposition. The genetic relationship of the iron formation is believed to be direct and clearly expressed, the main concentrations of iron occurring above the thickest members of acid volcanic rocks (see Figs. 6 and 7). Eruptive vents yielding a coarsely fragmental volcanic breccia of acid composition in the central part of the region coincide with the region of development of carbonate facies of BIF; to the west of it is the shallow-water oxide facies, and to the east the deep-water sulfide facies. Goodwin believes that the last two facies arose in the break between deposition of the underlying acid pyroclastics and of the overlying basic effusives.

The second type of iron chert distinguished by Gross (1965, 1973) is the Lake Superior. Iron cherts of this type usually have a granular texture and are associated in the section with quartzites, dolomites, black shales, and conglomerates. Syneresis cracks, stylolites, ripple marks and cross-bedding are observed in them. These iron cherts are developed mainly in the Lake Superior district and are typical of the Middle and Late Precambrian. Gross believes them to be continental-shelf sediments, developed along the margins of old platforms. The source of the material for the dolomites, argillites, and some other members of the cherty iron-formation was rocks being eroded on land, and the source of Fe and Si for the iron-formations was volcanic belts within the sedimentary basin. That idea connects the two types of iron cherts of the Canadian shield—the Algoma and Superior—and defines a causal relationship between them, i.e. it enables the Superior rocks to be regarded as the remote counterpart of the cherty iron-formation of Algoma type. That standpoint is supported by the fact that there nearly always are volcanogenic products in the section of the cherty iron-formations of the Lake Superior region. In the Mesabi and Cuyuna districts they are tuffaceous slates, in the Gunflint, basaltic lavas (Morey, 1973); volcanic products have also been established at Gogebic (Gair, 1973). Recent factual material on the Lake Superior iron cherts makes it possible to relate their formation to volcanic processes and to be critical of the earlier hypothesis of James (1954), according to which these rocks were formed only as products of subaerial weathering.

For some time past similar ideas have been expressed concerning other areas where BIF are developed.

Dorr (1973) believes that both in South America and in other parts of the world there is a close spatial and temporal relationship between the carbonate facies of an iron-formation and volcanic rocks.

Trendall (1973b) mentions that the very old iron-formations of the Yilgarn and Pilbara Blocks in Western Australia were deposited between successive sedimentary cycles, each of which began with thick volcanic (effusive and intrusive) sequences of basic and intermediate composition, containing sedimentary rocks interbedded with them, and ending with intermediate and acid lavas, breccias, agglomerates, and tuffs with minor layers of graywacke, shale, siltstone, and sandstone.

On the basis of the review of present ideas on the relationship of the processes of volcanism and sedimentation in the formation of the Precambrian iron cherts, several general conclusions can be drawn:

(1) For most of the Precambrian cherty iron-formations a relationship to volcanic processes is postulated.

(2) The nature of that relationship varies and depends on proximity of the volcanic centers to the basins of deposition of the iron cherts. Even in those cases where the relationship is believed to be very remote, manifestations of volcanism are a necessary condition for the accumulation of iron-formation.

(3) The existence of large purely clastic sedimentary BIF is doubted by most investigators.

(4) The question whether the role of volcanic processes in the formation of the iron cherts diminished from Archean to Proterozoic, and correspondingly the role of products of subaerial weathering increased, remains controversial.

Apparently there occurred a general reduction of the geosynclinal systems and processes of volcanism in post-Proterozoic time, but a definite relationship to volcanic processes is also observed for post-Proterozoic BIF.

Composition of banded iron-formations

Obligatory and definitive for BIF of all types is the presence of thin-bedded (banded) rocks characterized by relatively uniform chemical composition, similar textures and structures, and somewhat more diverse mineral associations. Variations in the latter are caused chiefly by primary authigenic-mineralogic zoning or by differences in metamorphism. Geochemically, a general feature of these rocks is similar contents of silica (SiO_2) and iron (Fe) in the face of substantial variations in the ratio of FeO to Fe_2O_3; other rock-forming components, including Al_2O_3 and alkalis, occur in small amount; the only exceptions are MgO—a usual isomorphic substitute in iron carbonates and silicates—and CaO in some types of carbonate facies. The main structural feature of the BIF is the typical thin banding, produced by an alternation of barren chert and ore (hematite, magnetite, iron carbonates and silicates) layers. Sometimes mixed semi-ore layers are distinguished, having fairly clear-cut contacts with barren and ore layers. Occasionally massive iron cherts are encountered which have all the characteristics of typical BIF except the banding.

It should be mentioned that up to now there are no generally accepted names for these rocks—in the world literature completely identical rocks are called "ferruginous cherts", "banded quartzites", "tactonites", "itabirites", "jaspilites", "microquartzites", etc. Repeated discussions of the problems of terminology and nomenclature (Petrov, 1964; Belevtsev and Goroshnikov, 1969; Illarionov, 1969; Khodyush, 1971; Brandt et al., 1972; Alexandrov, 1973; Bayley and James, 1973) have not led to any agreement. This problem was discussed at the international symposium on "Genesis of Precambrian iron and manganese deposits" (Kiev, 1970), in the proceedings of which (UNESCO, Paris, 1973) a very complete table of the Russian and English terms used was published.

At the suggestion of "Elsevier Scientific Publishing Company", the term "ferruginous quartzite" used earlier has been replaced by the broader term "banded iron-formation" (BIF). The status of these rocks has been successfully shown by Beukes (1973) in a diagram which it is advisable to give here (Fig. 10). For the iron-rich shales (schists) often associated with BIF proper, especially of Superior or Krivoy Rog type, it is advisable to use the terms "iron-rich shales" or "ferruginous shales" (or schists).

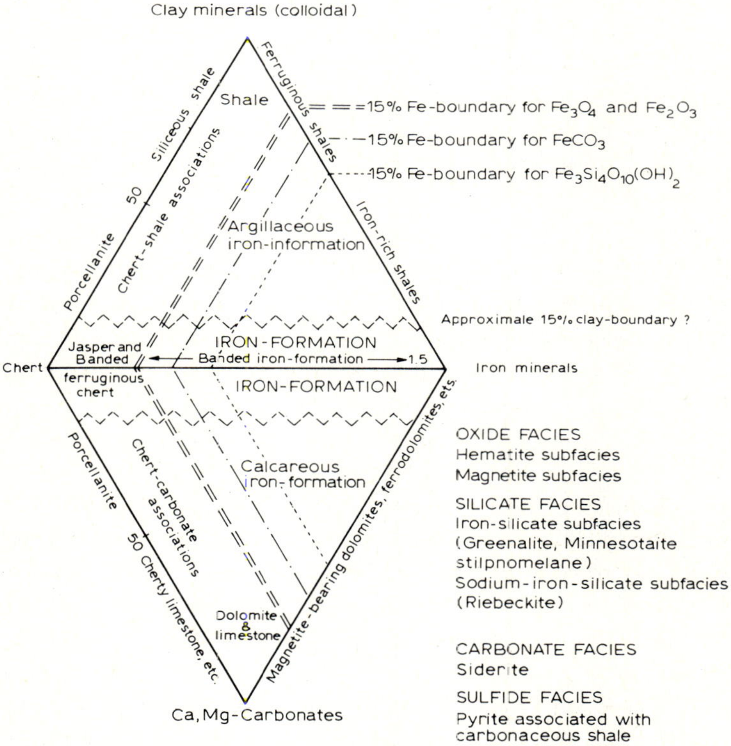

Fig. 10. Classification of iron-formation and related rocks (Beuker, 1973).

Textural and structural features

As has already been mentioned, the banding or layering is the most striking structural feature of typical BIF. Numerous investigations made both in the U.S.S.R. (Belevtsev, 1962; Gershoyg, 1965, 1968; Khodyush, 1967a), a review of which has been given by Alexandrov (1973), and abroad (Trendall, 1973; Bayley and James, 1973; Beukes, 1973) have shown that in most cases the banding proves to be complex—the macroscopically easily distinguished ore and barren bands (first order layers), from a few mm to several cm thick (maximum 10–20 cm), in turn show very fine internal banding. In the ore bands, second-order banding is produced by rhythmic alternation of so-called elementary ore bands (from fractions of a mm to 1–2 mm) and still thinner quartz bands (tenths and hundredths of a mm). The barren quartz bands of the first order nearly always contain disseminated ore along the bedding (hematite, magnetite, siderite) or thin continuous or discontinuous ore bands of higher order.

In iron-rich shales (schists) there also is a rhythmic alternation of barren (quartz or jasper-chert) and shaly (silicate) bands, banding of higher orders being typical only of the barren layers; in them a dissemination along the bedding and thin bands of siderite or magnetite are encountered. As the iron-rich shales (schists) grade into BIF the banding becomes more complicated due to the appearance of typical ore (magnetite) layers, which in some cases replace carbonate-quartz layers and in others replace shaly layers (Troshchenko, 1970); the rocks rich in carbonate and magnetite also are rhythmic.

The structural picture of the BIF is emphasized not only by differences in composition and thickness of the layers, but also by textural characteristics. In metamorphosed BIF granoblastic or hornfels textures are very common in the barren quartz and semi-ore siderite-quartz bands; equidimensional quartz grains with undulating boundaries constitute their matrix. Alternation of higher-order bands, consisting of grains of different size but of the same texture, within layers is a usual phenomenon in slightly metamorphosed rocks.

And finally, there should be mentioned the disruptions of the banding, lenticular and lenticular-undulating aspect of barren layers, cross- and undulating-cross-bedding, brecciation on bedding, and erosion or omission of quartz bands with thinly stratified ore material lying unconformably on them. All these structural features of the ferruginous quartzites, which have been described in detail by Gershoyg (1965), are in his opinion evidence of the comparatively shallow depth of the basin of deposition, where waves and currents periodically had an effect. In addition to microbanding and banding, Beukes (1973) also refers the relict oolitic textures, ripple marks, cross bedding and channel structures of redeposition in the BIF of the greenstone belt of Southern Africa to features of sedimentary deposition. Gray chert and red jasper nodules found in these rocks are believed to be colloidal accretions or lumps of silica that have been reworked by water. Post-deposition, pre-lithification slump folds, bordered on both sides by undisturbed BIF, are typical. Such structures indicate that the sedimentary material probably was in a very plastic state at the time of deformation.

However, a post-sedimentation, metamorphic origin of some of the textures cannot be ruled out, in particular the lenticular and boudinage type of disturbances of the bedding (Tochtuyev, 1967).

Granular textures are common in the slightly metamorphosed BIF of the Mesabi and Gunflint districts of Lake Superior (Bayley and James, 1973; French, 1973), the Kuruman and Penge formations of Southern Africa (Beukes, 1973), and the Midternaes area in southwestern Greenland (Appel, 1974). The granules are round or ellipsoidal, up to 0.5–1.0 mm in size; they do not occur in more metamorphosed rocks and are practically unknown in the BIF of the U.S.S.R. The granules are of complex mineralogic composition—they consist of iron silicates (greenalite, minnesotaite, and stilpnomelane) and cherty material and magnetite in highly variable proportions. Many granules contain tiny spherulites in their interior. In some cases the rocks containing the granules grade gradually into oolitic rocks in which the oolites are rimmed with hematite (Bayley and James (1973) give typical photomicrographs). They also mention that the bedding of the rocks containing the granules is usually undulating or very irregular; layers 2.5 cm thick may pinch out in a distance of a few centimeters. Mengel (1973) established that in the BIF of the Superior Basin the size of the granules which consist of iron minerals, like that of the quartz grains, corresponds to medium and coarse sand, and that they are medium rounded and moderately sorted. This permits the assumption that they were deposited from the same current. By means of correlational analysis it was established that the thickness of the iron-ore layers is a function of the grain size of the original clastic material. Similar observations were also made in due course in the Krivoy Rog (Besedin, 1957), but the hypothesis of the clastic (terrigenous) origin of the quartz has not been developed further.

The data that have been given are convincing evidence of the primary sedimentary formation of the BIF in relatively shallow-water basins. In this case, undisturbed banding and microbanding indicate very quiet deposition, and such features as granules, oolites, lenticular and undulating attitude of the layers, ripple marks and intraformational folding and brecciation reflect both the action of waves and currents and irregularity of the bottom relief.

Chemical characteristics of the rocks

The bulk chemical composition of the BIF differs sharply from the composition of any igneous, sedimentary, or metamorphic rocks. According to Nikol'skiy's data (1970), for this formation as a whole the clarke of the concentration of iron oxides is 11; of silica, 0.8; of titanium and aluminum, 0.3; of sodium, less than 0.3; and of potassium, less than 0.4. According to our data the average composition of the well-studied BIF of the Krivoy Rog group (banded iron-formations and iron-rich schists of the Middle suite interbedded with them) is close to the average composition of the BIF of the world, which is readily seen from the graph (Fig. 11) plotted according to Taylor's method (1972) in logarithmic coordinates. For all the major components of the BIF except MgO and to a lesser extent SiO_2 and MnO, a great contrast is observed in comparison with both the average composition of the Earth's crust and of sedimentary and metamorphic rocks, and with the

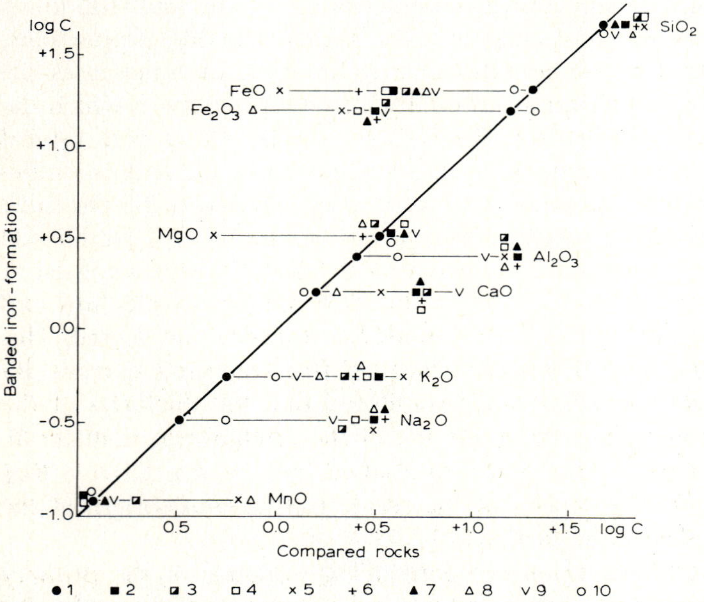

Fig. 11. Comparison of average contents of main components (log *C*, in weight %) in the BIF of the Krivoy Rog and of some volcanogenic-sedimentary formations: *1* = average composition of BIF of the Middle suite of the Krivoy Rog group (Saksagan district) (our data); *2* = average composition of the Earth's crust (Vinogradov, 1955); *3* = average composition of sedimentary rocks of geosynclines (Ronov and Yaroshevskiy, 1971); *4* = average composition of metamorphic para-rocks of shields (Ronov and Yaroshevskiy, 1971); clarkes of volcanogenic-sedimentary formations (Nikol'skiy, 1970); *5* = clastic liparitic; *6* = clastic andesitic; *7* = flyschoid-andesitic-basaltic; *8* = biotite paragneisses (metamorphosed flyschoid); *9* = spilite-keratophyre; *10* = banded iron-formation (BIF).

composition of sedimentary-volcanogenic formations. The main feature of the BIF is the substantial concentration of just the one element—iron—in the face of markedly low contents of Al_2O_3, CaO, TiO_2, Na_2O and K_2O. Predominance of MgO over CaO and of K_2O over Na_2O is also typical of the BIF of the Ukrainian shield. The only exceptions are some carbonate rocks of the BIF containing dolomite and ankerite, where $CaO \geqslant MgO$, and also alkalized silicate rocks containing riebeckite, with a sharp predominance of Na_2O over K_2O. Manganiferous varieties of BIF, found in some areas of India, Canada, and Brazil and confined chiefly to carbonate facies, constitute a special group.

The overall geochemical particulars of the formation indicate processes of deep differentiation of the material in the formation of the iron-formations.

It is characteristic that regardless of their age, the formation they belong to, and the degree of metamorphism, iron-formations themselves have similar chemical composition. This particular feature of the rocks has provided grounds for some investigators (Khodyush, 1971) to distinguish a single cherty iron-formation in the Precambrian, which spatially but not genetically is related either to volcanogenic or to clastic sandy-shaly formations.

If the volatile components (H_2O and CO_2) are not taken into account, it turns out that 99% of the iron-formations, proper, of the BIF consist of SiO_2, Fe_2O, FeO, Al_2O_3, MgO, and CaO and 97–98% of SiO_2, Fe_2O_3, FeO, Al_2O_3, and MgO, the last two components constituting not more than 10–15%. To compare such rocks it is convenient to use ratios (molar ratios) of iron content (F), alumina content (A), magnesium content (M), and calcium-alkali content (C) (Semenenko et al., 1956), determined from relationships of the type:

$$M = \frac{MgO \cdot 100}{Al_2O_3 + MgO + CaO + (FeO + Fe_2O_3)} \ (\%)$$

In a number of cases, it can be limited to three coefficients—F, A, and M—for rocks with $MgO > CaO$.

In studying the BIF of the main structural-facies zones of the Ukrainian shield, a surprising similarity of these geochemical characteristics of iron-formations of different mineralogic composition and age was established (Mel'nik, 1975). The iron-formations proper fall into a relatively narrow field bounded by values of the ratios: $F = 76$–96, $(M + C) = 5$–0, $A = 0$–5, and for the iron-rich schists: $F = 76$–80, $(M + C) = 13$–21, $A = 2$–9. A substantial number of the hematite, hematite-magnetite, siderite-magnetite and even some silicate-magnetite rocks contain no aluminum ($A < 2$) and are distinguished by high iron content ($F = 84$–96).

These particulars of the chemistry of the iron-formations which are the most important and most typical aspect of the BIF also emerge rather clearly when the distribution of the essential components in a typical section of the Middle suite of the Krivoy Rog group (BIF of Superior type) is considered; they are made up of systematically interbedded, fairly thick (10–200 m) shaly (schists, ferruginous schists, cherts, and low-grade iron-formations) and iron (iron ores, including jaspilites) layers or horizons. We note that in some cases, assigning the iron cherts to a shaly or iron horizon proves to be very arbitrary, which has made it necessary to combine several layers of similar chemical composition into one member.

The iron and shaly horizons differ mainly in Fe_2O_3 and Al_2O_3 content; silica usually is higher in the shaly bands; differences in the content of such secondary components as TiO_2, K_2O, and S are fairly clearly noticeable; the FeO, MgO, and Na_2O contents are similar. A convenient criterion for distinguishing between the iron and shaly horizons may be the $FeO:Fe_2O_3$ ratio, the value of which varies within 1–2 in the iron and 2.5–7.5 in the shaly varieties. To a lesser extent the MgO:CaO ratio is indicative; it is always greater than 1.0, and in the shaly horizons falls within 2.5–4.5. The high iron content of the shaly horizons, in which the FeO content sometimes is even higher than in the associated banded iron-formation, is very remarkable, and some correlations, in particular of MgO and FeO, are also observed.

More detailed study of the chemistry of the various iron-formations and schists showed that these two groups of rocks are clearly subdivided on the basis of relationships of the *A*, *F*, and *M* coefficients. On the diagram (Fig. 12) all the banded iron-formations fall into a very narrow field (1), within which the hematite and hematite-magnetite rocks and thin-bedded

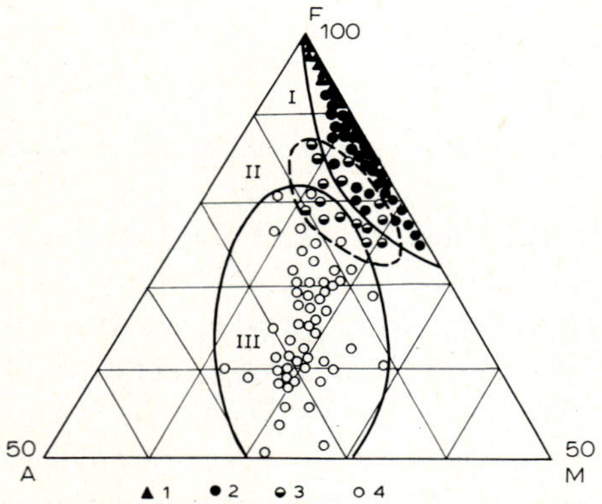

Fig. 12. Petrochemical characteristics of the BIF and associated schists of the Middle suite of the Krivoy Rog group (relationship of AFM ratios): *1* = hematite and hematite-magnetite rocks; *2* = magnetite, siderite-magnetite, and silicate-magnetite rocks; *3* = transitional strata, schist intercalations in layers (horizons) of BIF; *4* = various schists constituting schist layers (horizons) (Mel'nik, 1975).

jaspilites are characterized by very high F values ($F = 87-100$) and minimal A values ($A = 0-1.5$); the other banded iron-formations (magnetite, siderite-magnetite, and silicate-magnetite) hardly differ among themselves (2). Only when shaly bands appear does the A value increase (3–10) while a high iron content is maintained ($F = 75-87$)—a field of interbedded shaly and iron thicknesses is distinguished (3). The shaly bands from a few cm to tens of cm thick which occur in the iron horizons also fall here. The field of the iron-rich schists which constitute the shaly horizons proper is rather clearly delimited (*II*) and its overlap with the field of the alumina-free iron-formations (*I*) is explained only by the presence of analyses of mixed rocks.

To compare the contents of the essential components in the banded iron-formations, schists of different composition, and metamorphosed volcanic rocks, another graph was plotted—Fig. 13—on a typical Taylor (1972) diagram. The average chemical composition of typical siderite-magnetite and magnetite rocks was taken as the basis. Here too a similarity of the banded iron-formations and schists is noted in the contents of SiO_2, Fe, and FeO, and differences in the contents of Fe_2O_3, Al_2O_3, TiO_2, and CaO. It is typical that all the schists are similar to the amphibolites

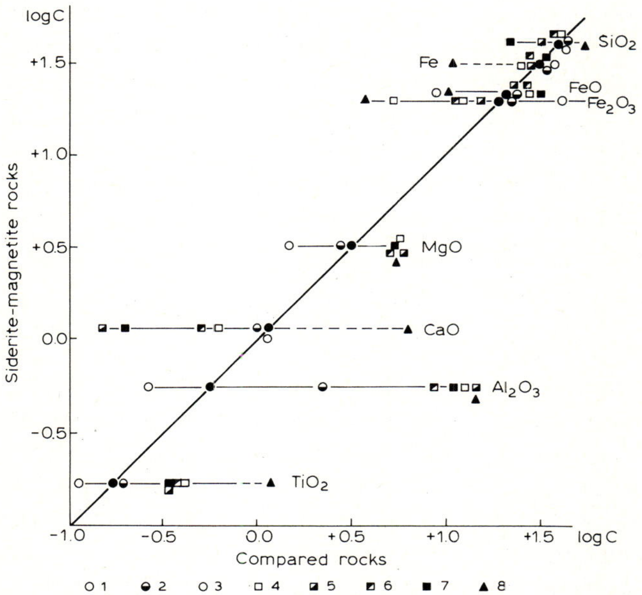

Fig. 13. Comparison of average contents of main components (log C, in weight %) in siderite-magnetite (magnetite) rocks and some other ferruginous rocks and schists of the Middle suite of the Krivoy Rog group: *1* = siderite-magnetite and magnetite rocks; *2* = siderite-silicate (chlorite, biotite, amphibole)-magnetite rocks; *3* = hematite and hematite-magnetite rocks; *4* = siderite-chlorite schists; *5* = chlorite schists; *6* = chlorite-amphibole and biotite-amphibole schists; *7* = garnet-amphibole and garnet-chlorite(biotite)-amphibole schists; *8* = amphibolites (metabasites) (Mel'nik, 1975).

(metabasites) in MgO and Al_2O_3 content, but sharply different from them in Fe, FeO, and Fe_2O_3 (the schists are enriched) and SiO_2, CaO, and TiO_2 content (the schists are poorer). The hematite and hematite-magnetite banded iron-formations and thin-bedded jaspilites are geochemically very different from the schists and, naturally, from the amphibolites and other metamorphosed volcanic rocks.

Thus the geochemical characteristics of the banded iron-formations are so specific that they are a reliable criterion for their diagnosis even in the case of fine interbedding with various clastic sediments, volcanogenic or mixed rocks. Very simple geochemical recalculations and plotting of the diagrams makes it possible to make the comparison more visual, particularly for rocks of different mineralogic composition and degree of metamorphism.

Chemical characteristics of the major minerals

The mineralogic composition of the BIF is determined by the chemical composition of the original rocks and the extent of their metamorphism. Study of the minerals of the BIF faces considerable difficulties, especially in the slightly metamorphosed rocks, which consist of a finely divided aggregate of silicates, carbonates, and iron oxides. It is practically impossible to separate pure monomineralic fractions from such aggregates and one often has to use analyses of fractions contaminated by impurities, or even of bands of rock, and determine the composition of the minerals by means of recalculations. The highly metamorphosed rocks usually consist of larger grains and aggregates, the minerals themselves are more uniform and contain fewer impurities, and the available information on the composition of olivines, pyroxenes, and in part on garnets and amphiboles is more reliable. Recently, information on the composition of the minerals of the BIF and primarily of the finely divided silicates and carbonates has been considerably extended due to use of the electron microprobe.

The monograph by Semenenko et al. (1956) remains the most complete summary of the petrography and mineralogy of the BIF of the Konka, Verkhovtsevo, and Pravoberezhnyy ("Right-Bank") areas of the Ukrainian shield. The first systematic description of the iron-rich rocks and ores of the Krivoy Rog was made by us (Mel'nik, 1962a), based on a summary of published data and the results of our own investigations. Information on the minerals in the rocks of the Krivoy Rog–Kremenchug zone is also given in the monographs by Yurk and Shnykov (1961), Bordunov (1964), and Dobrokhotov (1964).

Besides these monographs, a substantial number of papers dealing with the mineralogy of or containing information on individual minerals of the iron-formations have been published in recent years; a review of these was given by us in a special work (Mel'nik, 1975). And finally, a fundamental collective monograph, "Mineralogy of the Krivoy Rog Basin", in the compilation of which the author participated, was published in 1978 (Chief editor: Lazarenko).

Of the works devoted to the mineralogy of the BIF in other areas of the world, mainly in the U.S.A. and Canada, there should be mentioned the investigations of Kranck (1961), Gross (1961), Davis (1966), Bonnichsen (1969, 1975), French (1968, 1973), Floran and Papike (1975), and especially Klein (1966a, b, 1973, 1974), Immega and Klein (1976), Klein and Fink (1976), and Klein and Bricker (1977).

On the basis of these data, fairly complete and reliable information has been obtained which makes it possible to demonstrate certain general regularities and particulars of the chemistry of the essential minerals, and also to distinguish the main mineral associations in iron-formations of different rank of metamorphism.

In the study of the ore minerals, no essential differences have yet been found in the chemical composition of magnetite and hematite from different regions. Therefore no mineralogic-chemical description of the iron oxides is given here. Besides quartz and relatively rare Ca-Mg carbonates, among the essential minerals in the banded iron-formations magnesian-iron silicates (greenalite, minnesotaite, amphiboles, pyroxenes, and olivines) are of fundamental importance. Chlorites, garnets, and to a lesser extent micas (biotite and stilpnomelane) are common in the schists and intermediate rocks; a certain content of alumina is always necessary for the production of these minerals.

Carbonates are extremely widespread in the banded iron-formations and schists of low-rank metamorphism. Sideroplesites predominate in the com-

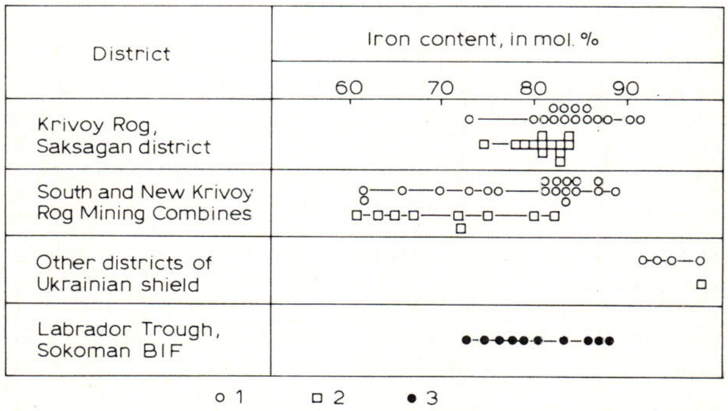

Fig. 14. Variations in iron content of carbonates of the siderite-magnesite series: *1* = in siderite and siderite-magnetite rocks of the BIF; *2* = in iron-rich schists; *3* = in slightly metamorphosed silicate-carbonate rocks of the BIF (Klein, 1974; Mel'nik, 1975; Klein and Fink, 1976).

position in the Krivoy Rog area; in other areas, for instance in the Pravoberezhnyy, nearly pure siderite occurs (Fig. 14). Substitutions of MnO and CaO are not typical—the content of the rhodochrosite molecule is within 0.2–2.6%, and of the calcite molecule 0.1–1.8%. The iron contents of the carbonates in the schists and iron-formations differ insignificantly, although no carbonates containing more than 84% $FeCO_3$ have been established in the schists of the Krivoy Rog, but the iron content of the carbonate in the siderite-magnetite rocks sometimes reaches 89–91%; however, the minimum iron content is 62–73%, and in the hematite-magnetite jaspilites it drops to 37–47%. In the latter case, the presence of dolomite in the samples analyzed is not ruled out. Analogous rocks from the Pravoberezhnyy area (Ukrainian shield) contain high-iron carbonates—up to 92% $FeCO_3$.

In other areas, siderite with a high content of substitutions is found in the rocks of the carbonate facies of the BIF. Thus, in the slightly metamorphosed Sokoman BIF of Canada, siderite associated with manganese-rich oolites contains up to 28.2% $MnCO_3$ and 5.1% $CaCO_3$ (Klein, 1974). Siderite with a low content of $MnCO_3$ (up to 6%) has an iron content of 72–89% (Fig. 14). However, more magnesian varieties occur—up to 38% $MgCO_3$ (Klein and Fink, 1976). In a review paper, Klein and Bricker (1977) give the approximate compositional range of siderite as:

from $(Mg_{0.3}Mn_{0.1}Fe_{0.6})$ to $(Mg_{0.2}Fe_{0.8})$ to $(Mg_{0.2}Mn_{0.2}Fe_{0.6})CO_3$

of dolomite-ankerite as:

from $Ca_{1.0}(Mg_{0.8}Fe_{0.1}Mn_{0.1})$ to $Ca_{1.0}(Mg_{0.5}Fe_{0.2}Mn_{0.3})$ to $Ca_{1.0}(Mg_{0.4}Fe_{0.6})(CO_3)_2$

and of calcite as:

$Ca_{0.9}(Fe,Mg,Mn)_{0.1}CO_3$

Practically all the varieties of carbonates in most of the BIF of the world fall within these limits.

Greenalite is a layered silicate with the idealized composition of $Fe_3^{2+}Si_2O_5(OH)_4$, that is, the pure iron analog of serpentine. Greenalite occurs in large amounts in the slightly metamorphosed or unmetamorphosed silicate and silicate-carbonate Sokoman BIF of Canada, and in those of Biwabik, Minnesota, southwestern Greenland, the Hamersley group of Western Australia, and several other areas. No such slightly metamorphosed BIF have been found as yet in the U.S.S.R. and the description of greenalite is given on the basis of published data.

Greenalite usually occurs in granular varieties of BIF and is rare in the banded rocks. Most investigators attribute greenalite to primary iron silicate which was formed as a result of diagenetic crystallization of an iron-bearing silica gel. Until recently the composition of greenalite was not precisely determined. For a sample from Mesabi, French (1973) gives the formula

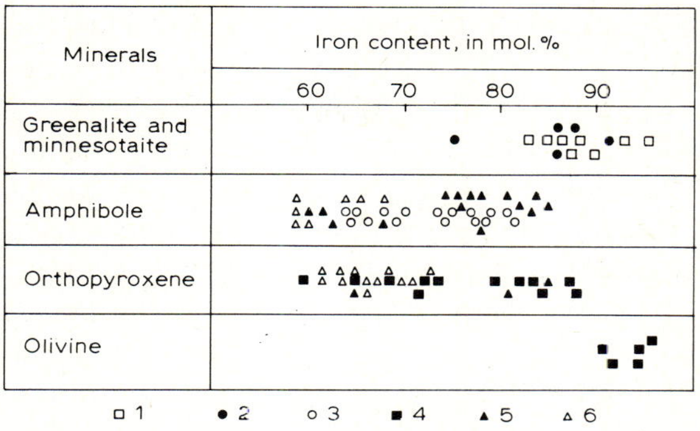

Fig. 15. Variations in iron content of greenalite, minnesotaite, amphibole, pyroxene, and olivine in silicate rocks of the BIF and iron-rich shales (schists). *Slightly metamorphosed BIF* (Biwabik, Minnesota; Gunflint, Ontario; and Sokoman, Labrador Trough; Brockman and Roper River, Australia): *1* = greenalite; *2* = minnesotaite (Klein, 1974). *Moderately and highly metamorphosed BIF: 3* = Krivoy Rog district; *4* = Mariupol' district; *5* = other districts of the Ukrainian shield (Mel'nik, 1975); *6* = Tobacco Root Mountains, Montana (Immega and Klein, 1976).

$Fe_9^{2+}\ Fe_2^{3+}\ Si_8O_{22}(OH)_{12} \cdot 2\,H_2O$, and for Australian samples, $9(Fe,Mn,Mg)O \cdot 0.2(Fe,Al)_2O_3 \cdot 8\,SiO_2 \cdot 8\,H_2O$. Greenalite from the Sokoman BIF (Klein, 1974) is characterized by high iron content—$F = 82–95$ (Fig. 15), low alumina content—$A = 0.2–2.3$, and almost total absence of calcium and alkalis. Klein and Bricker (1977) give the following approximate compositional range of greenalite:

$$(Fe_{4.0}Mg_{1.0} \text{ to } Fe_{5.3}Mg_{0.2})Al_{0-0.2}Si_4O_{10}(OH)_8$$

and it is assumed that all the iron occurs only in the form of Fe^{2+}.

Minnesotaite. Pure minnesotaite—$Fe_3Si_4O_{10}(OH)_2$—is the iron analog of talc and occurs along with greenalite or without it in unmetamorphosed and slightly metamorphosed BIF. In most BIF minnesotaite is assigned to the secondary late diagenetic or early metamorphic minerals. As is seen from the formula, minnesotaite should contain substantially more silica than greenalite. Analyses of natural minerals give the following limits for high-iron varieties (SiO_2, in weight %): greenalite 32–36%, minnesotaite 49–52%. The other chemical characteristics of these primary iron silicates are similar—the contents of alumina, calcium, manganese, and alkalis are insignificant—and the iron content falls within 75–91% (Fig. 15), although low-iron varieties ($F = 42–45$) also occur. Klein and Bricker (1977) give the approximate compositional ranges of minnesotaite as:

$$Mg_{1.7}Fe_{1.3} \text{ to } Fe_{2.8}Mg_{0.2}Si_4O_{10}(OH)_2$$

We note that high-iron minnesotaite apparently is typical of the low-alumina silicates of the rocks of the BIF.

Amphiboles in metamorphosed low-alumina silicate and carbonate-silicate iron-formations which have not been altered by metasomatic processes are represented only by magnesian-iron cummingtonite, containing 59–86% of the grunerite molecule. The other components are not characteristic and are (in weight %): Al_2O_3—0.00–2.72 (1.00); CaO—0.00–2.40 (1.04); MnO—0.00–1.65 (0.61); $Na_2O + K_2O$—0.00–1.41 (0.39); TiO_2—0.00–0.38 (0.11) (the average of 25 analyses of cummingtonites from the BIF of the Ukrainian shield is given in parentheses). It is not ruled out that some of the impurities are related to inclusions of chlorite and garnet in the material analyzed. The similarity in iron content of the amphiboles of this series from different areas (Fig. 15), where cummingtonite is associated both with siderite and with garnet, hornblende, and orthopyroxene, is striking.

Alkali amphiboles of riebeckite type constitute a special group; in iron content and content of impurities (including alumina) they are similar to cummingtonite, but differ in the high Na_2O content (up to 7%). In the Krivoy Rog basin, especially in its northern part, riebeckite occurs very widely, being an essential mineral of the metasomatites along with aegirine and albite. In normally metamorphosed BIF which were not involved in the processes of alkali metasomatism, cummingtonite is formed instead of riebeckite. Published data on other areas are contradictory. Finds of riebeckite and crocidolite in slightly metamorphosed rocks (Klein, 1974) are explained by metamorphism of Na-rich bulk compositions.

Orthopyroxenes, found only in highly metamorphosed rocks, have an iron content of 60 to 88%; the latter value is close to the maximum iron content of natural pyroxenes of the ferrosilite series (Fig. 15). In the BIF of the Ukrainian shield the content of minor components in the pyroxenes is not high: Al_2O_3—0.15–1.60 (0.78); CaO—0.49–2.11 (1.15); MnO—0.07–1.56 (0.35); $Na_2O + K_2O$—0.08–0.47 (0.26); TiO_2—0.05–0.19 (0.10) (average of 12 analyses in parentheses). Pyroxenes from other areas where BIF are developed have a low content of impurities except for the manganiferous varieties (up to 3–6% MnO) found in the Tobacco Root Mountains and in the Carter Creek area of the Ruby Mountains, Montana (Immega and Klein, 1976); the iron content of these pyroxenes is within 60–62%.

Olivine is relatively rare and close to pure fayalite in composition—91–96% Fe_2SiO_4; the content of impurities, including Al_2O_3 and CaO, is not more than 1.5%. Less iron-rich olivines do not occur, which is explained by their thermodynamic instability in iron-formations, which always contain an

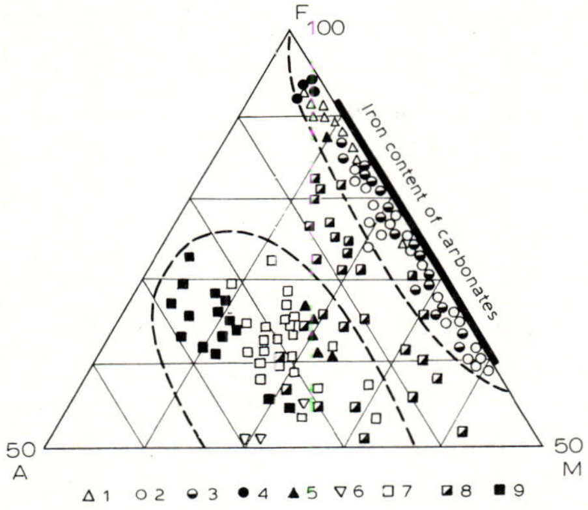

Fig. 16. Relationship of ratios of iron content (F), magnesium content (M), and aluminum content (A) in rock-forming silicates of the BIF and associated shales (schists): *1* = greenalite and minnesotaite (Biwabik, Minnesota; Gunflint, Ontario; Sokoman, Labrador Trough; Brockman and Roper River, Australia — Klein, 1974); *2* = amphibole (Ukrainian shield, Tobacco Root Mountains, Montana — Mel'nik, 1975; Immega and Klein, 1976); *3* = orthopyroxene (Ukrainian shield; Tobacco Root Mountains, Montana — Mel'nik, 1975; Immega and Klein, 1976); *4* = olivine (Ukrainian shield — Mel'nik, 1975); *5* = chamosite (Sokoman, Labrador Trough — Klein and Fink, 1976); *6* = ripidolite (Wishart Formation, Canada — Klein and Fink, 1976); *7* = chlorite (Ukrainian shield — Mel'nik, 1975); *8* = biotite and stilpnomelane (Ukrainian shield; Sokoman, Labrador Trough — Mel'nik, 1965; Klein and Fink, 1976); *9* = garnet (Labrador, Canada; Ukrainian shield — Klein, 1966; Mel'nik, 1975). Limits of iron content of carbonates designated by heavy line on FM side.

excess of silica in the form of quartz.

Comparison of the data given leaves no doubt as to the extreme similarity of the chemistry of the carbonates and silicates of typical low-alumina iron-formations regardless of the area of occurrence, age, rank of metamorphism, and formation to which the BIF belong.

On an AFM diagram (Fig. 16) all the minerals from iron-formations proper fall into a narrow and clearly defined field, while the rock-forming minerals of the associated iron-rich shales from BIF of Superior type (chamosite, ripidolite and other chlorites, garnets and some biotites) form another broader and less clear-cut field. These two fields do not overlap, which indicates the isochemical character of metamorphism and the limited mobility of the main rock-forming minerals, at least within individual layers and bands of the iron-formations and shales. The chemical composition of stilpnomelane occupies an intermediate position both in iron content and in alumina content. However, a high content of potassium (up to 2.2% K_2O), an element not typical of most BIF, is necessary for the formation of this mica.

Thus, study of the physicochemical conditions of formation of typical banded iron-formations at all stages of their geologic history should represent an analysis of the equilibria and mechanism of mineral formation in the system:

$$Fe_2O_3—FeO—(MgO, MnO, CaO)—SiO_2—C—H_2—O_2$$

and in the idealized version:

$$Fe_2O_3—FeO—SiO_2—C—H_2—O_2.$$

A solution of the analogous problem iron-rich shales containing various amounts of clastic material (associated with BIF of Superior type) is a separate and rather independent problem requiring special investigation.

Present ideas on the origin of the banded iron-formations

Let us examine present ideas on the reasons for the accumulation of iron in the Precambrian, on the sources and forms of transport of the iron and silica, and on the method of deposition and mechanism of formation of the typical banded structure.

An enormous number of published works, the greater part of them in English and Russian, have dealt with questions of the regularities of distribution and formation of the BIF. Recently several regional and international conferences devoted to discussion of this problem have been held. In particular, in 1966 an All-Union conference on problems of formation of Precambrian cherty iron-formations was held in Krivoy Rog in the U.S.S.R., where the geologic and physicochemical conditions of formation of these rocks were discussed and future directions for their study were pointed out. In 1970 an International Symposium on the Geology and Genesis of the Precambrian Iron and Manganese Formations of the World, organized by UNESCO, was held in Kiev. Its proceedings were published in Kiev (1972) and in Paris (1973). Finally, an issue of the journal "Economic Geology" (no. 7, 1973), devoted wholly to the problem of the origin of the BIF, was translated into Russian in 1975 by the "Mir" publishing house.

At the present time several versions of two main hypotheses of the formation of the BIF—clastic-sedimentary and volcanogenic-sedimentary—have been worked out in a fair amount of detail.

At different stages of study of the cherty iron-formations, completely different points of view, differing in principle from the "sedimentary" hypotheses, have emerged periodically. The scientific literature in Russian is particularly full of them; Alexandrov (1973) has given a review of them. Of new works, the book by Gusel'nikov (1972), should be mentioned; in it basaltic magma, differentiated into cleanly separated silicate and iron-ore parts, is considered to be the source of the ore-forming material; the formation of the oxide iron-formations was accomplished under high-temperature conditions "...near centers of volcanism, where direct segregation of the ore minerals (magnetite, iron mica) from highly concentrated iron-silica melts—solutions—could take place". Tochilin (1969) also assumes formation of jaspilites from iron-silica melts. Ideas on the metasomatic formation of some varieties of iron-formations have been expressed by Kozlov (1969), and in a somewhat modified version by Knyazev et al. (1969). Zhdanov and Malkova (1974) revised the traditional sedimentary-volcanogenic hypothesis of formation of the BIF of the Kola peninsula in the direction of a metamorphic-metasomatic origin. It is believed that the iron-formations and the skarns, microclinites, and skarnoids associated with them are the products of regional infiltrational-diffusional metasomatic basification. The accumulation of these products is related to migration of the basic front into a higher part of the Earth's crust in the course of development of powerful Early Proterozoic granitization. The process of formation of the BIF is subdivided into two stages—an early high-temperature stage, when barren skarns were formed, and a later lower-temperature stage with intensive iron metasomatism, and the thin banding is explained by segregation of the larger molecules of SiO_2 and Fe_2O_3 in "conditions of production of supramolar structures of polymeric solutions". Kondakov and Perunov (1973) assign the BIF of Tanzania to contact metasomatic products. However, no rigorous proof which would provide grounds for giving preference to a magmatic or metasomatic origin of typical BIF and rejecting the traditional "sedimentary" hypotheses is given in the works cited. A critical examination of these views (Belevtsev and Mel'nik, 1974) demonstrated the very poor argumentation of the main points of the metasomatic and particularly of the magmatic hypothesis. The subsequent reply by Gusel'nikov (1976) introduced practically nothing new and is not very convincing.

It is very typical that the different versions of the metasomatic or magmatic hypotheses arise, as a rule, in studying the highly metamorphosed BIF, which often have been altered by subsequent processes. At the same time, investigators of the unmetamorphosed or slightly metamorphosed BIF of Lake Superior and other areas (Bayley and James, 1973; French, 1973; Appel, 1974) have no doubts as to their primary sedimentary formation. It is quite obvious that in any genetic formulations it is impossible to ignore the vast factual material and its interpretation in the detailed investigations of Belevtsev, Goodwin, Gross, Plaksenko, Semenenko, and Chernov. In many recent publications based on study of the structural-textural, mineralogical, and geochemical particulars of the BIF of different areas, the primary-sedimentary nature of these rocks is demonstrated (Lepp and Goldich, 1959; Jones, 1963; Sastry, 1967; Gershoyg, 1968; Liddy, 1968; Fedorchenko, 1969; Turobova, 1969; Khodyush, 1969, 1971; Troshchenko, 1970). Isotopic investigations made by us on the ore minerals of the iron-formations of the Krivoy Rog and other districts of the Ukrainian shield and KMA also are in good agreement with ideas that the iron-formations of oxide-carbonate composition were deposited in an aqueous environment (Mel'nik, 1973). We note that there is no physicochemical basis at all for the magmatic and metasomatic formation of the BIF, as the authors of those hypotheses themselves admit.

Present ideas on the origin of iron-formations were analyzed in detail by Kaukin (1969), who classified the hypotheses and examined the main controversial questions that are answered differently by adherents of clastic-sedimentary and volcanogenic-sedimentary leanings.

The *clastic-sedimentary hypothesis* was developed in greatest detail by Plaksenko, Strakhov, Gershoyg, James, Sakamoto, and others. In the "pure" classic version of this hypothesis it is suggested that the source of the iron and silica was rocks weathered on an ancient landmass. Plaksenko (1969) believes that an unequivocal solution of the problem of the type and scale of the sedimentary basins is hardly possible. Some investigators postulate that they were narrow, deep trough-like depressions (Belevtsev, 1957; Markov, 1959a), others that the basins were fairly wide (Strakhov, 1962; Kalyayev, 1965), but restricted. The products of continental weathering were strongly differentiated during transportation to the region of deposition, and the iron and silica, migrating in the form of colloidal solutions, were deposited outside the zone of accumulation of clastic sediments under the influence of electrolytes in the sea water, or due to mutual coagulation. In connection with the poor stability of colloidal solutions, the existence of terrestrial vegetation in the Precambrian is sometimes postulated (Martynenko, 1950; Gruner, 1922); it would supply organic compounds of humic acid type, which have stabilizing properties (Strakhov, 1962). Data from study of the physicochemical properties of colloidal systems obliged geologists to modify the clastic-sedimentary hypothesis and in a newer work (Govett, 1966), we already find the conclusion that marine conditions are unfavorable for the primary accumulation of iron in substantial amount in sediments, and that the cherty iron-formations were formed in lake or near-shore (bay, lagoon) conditions, the distance of chemical transport being very short. That suggestion was made earlier by Hough (1958).

Dorr (1973) proposed that the Cauê Itabirite BIF (Brazil) was deposited in stable near-shore or miogeosynclinal conditions and was interbedded with thick layers of dolomite, and that in the upper part of the section of the Minas series, where carbonate and discontinuous oxide facies were deposited, the conditions were eugeosynclinal. In Dorr's opinion, the colossal scale of concentration of iron is evidence against volcanic emanations as a possible source of the Fe and SiO_2. It is proposed that the iron was supplied from ancient rocks in the course of prolonged peneplanation, weathering, and erosion. The climate was arid. The phyllites with very high aluminum content, underlying and overlying the BIF, were the result of weathering.

Garrels et al. (1973) believe that the BIF must have been formed in restricted basins in semi-enclosed water bodies, periodically communicating with the ocean via channels or over bars. Deposition of silica occurred mainly during evaporation, but deposition of iron was complex and is explained both by oxidation (hematite facies) and by evaporation (silicate and carbonate facies) and sulfate reduction (sulfide facies). It is suggested that the spatial distribution of the sedimentary facies of the BIF will correspond to the well-known scheme of James (1954), but to explain the similarity of banding in the face of different causes of precipitation of the iron raises difficulties.

Button (1976) considers the BIF of the Transvaal Supergroup, South Africa, to be the final member of a carbonate sedimentary cycle, when anomalous concentrations of iron, silica, and other elements were produced in restricted basins.

French (1973) also believes that BIF are the result of chemical precipitation in vast stable basins, analogous to the younger evaporite lakes. The alternation of layers rich in iron and silica probably reflects seasonal fluctuations or other cyclic climatic phenomena.

The leaning toward limitation and isolation of the sedimentary basins found its fullest expression in the original genetic model of Eugster and Chou (1973), based on investigations of an inorganic process of formation of cherty layers, hydrogeology, and chemical evolution of waters and brines under conditions of evaporation, and also on mineral formation and sedimentation in salt-marsh and shallow-lake conditions. In this model a mechanism of deposition of the cherty layers via an intermediate stage of magadiite in alkaline environments was successfully worked out. Silica arrived in continental water bodies from near-shore springs of percolating ground water and was concentrated in the waters as a result of evaporation. The matter of accumulation, and particularly the cyclic precipitation of iron, was a more complicated matter—the authors has to resort to several arbitrary constraints in formulating the model (such as "delaying" the precipitation of greenalite or siderite). The final conclusion of Eugster and Chou, that the BIF consist of different rocks and can hardly be explained by a single depositional model (p. 1167), does not contribute to a solution of the problem. At the same time the proposed model is in need of more detailed critical examination, especially its physicochemical aspect.

In explaining the banding of the BIF, which as is known presents a special problem, the experiments of Moore and Maynard (1929) are usually cited; in these, banded sediments were obtained due to different rates of coagulation and deposition of iron hydroxides and silica from dilute solutions. In transferring the experimental results to natural geologic processes, periodic arrival in the sedimentary basin of mixed iron-silica solutions, each batch of which yielded two bands—ferruginous and cherty—is presumed (Liddy, 1968). The reasons for such cyclicity are difficult to explain, therefore in some later works the formation of the banding is connected with separate migration of iron and silica. Sakamoto (1950) suggested that during weathering in the conditions of a tropical monsoon climate the iron migrated in acid waters at the time of the rainy season, and the silica in alkaline ground waters at the time of the dry season. On the basis of experiments on the leaching of iron oxides and silica, Alexandrov (1955) suggested that the alternation was caused mainly by selective weathering of the Precambrian soil. Due to seasonal variations in temperature and fluctuations in the pH caused by the formation and decomposition of humic acids, the soil solutions removed silica in the warm season and iron in the cold period of the year. Hough (1958) limited himself to referring to the usual seasonal fluctuations of temperature during deposition of cherty iron sediments in the conditions of a subtropical humid climate.

Trendall (1973c), citing his own previous work, believes that the microbanding (0.1 mm or less) in the BIF of Hamersley, Western Australia, is related to deposition of material in the form of bands reflecting annual seasonal fluctuations in the chemical environment of the basin, and that it is not very likely that the microbands represent varieties of Liesegang banding, which arises in a gelatinous colloid sediment even after its deposition. The macrobanding corresponds to cyclicity with a period of 23.3 years and is close to Hale's 22-year double sunspot cycle, which is related to periodicity of the planetary orbits.

Kaukin (1969) believes that seasonal pH fluctuations would hardly be large enough to cause separate migration; it is more plausible to presume that the silica and iron migrated together in the same solutions and continuously arrived in the basin, where they were separated into distinct bands in the course of sedimentation. The deposition of iron was controlled by the redox potential of the environment, which was determined by the amount of oxygen released into the sea water by photosynthetic organisms. The deposition of silica, which did not depend on the presence of oxygen, went on continuously, and periodic (seasonal) variation of the redox potential produced alternate deposition of bands of iron hydroxides with silica impurities and bands of silica with iron hydroxide impurities. The formation of the finest bands of iron oxides (second-order bands) and quartz is explained by mutual precipitation of iron and silica gels carrying different charges.

The factual geologic material which serves as the basis of this group of hypotheses in most cases is BIF of Superior type, which contain practically no volcanic rocks in their make-up.

The *volcanogenic-sedimentary hypothesis* also has been developed in several versions. Ideas that the iron and silica came from submarine volcanic activity are most widespread. Semenenko et al. (1959, 1967) believe that it is the outpouring of hydrothermal solutions and arrival of gaseous emanations in the sea water that served as the source of deposition of cherty iron, cherty manganese, and possibly some sulfide deposits. In the primary sediments, iron carbonates and silicates and silica gels were deposited directly from sea water mineralized by volcanic emanations. Along with iron silicates and carbonates, iron oxides were also deposited in places where the *Eh* of the water was high, under conditions of the siderite and chamosite facies. It is this succession of a regime of oxygen and of CO_2 that produced the rhythmic alternation of chert-siderite layers on the one hand and iron-silicate layers with chert bands on the other.

In a work by Tochilin (1963) the main ore minerals of the jaspilites—magnetite and hematite—are believed to be primary formations which were carried in hydrothermal solutions and precipitated by submarine fumarolic activity.

Goodwin (1961, 1973b) considers the BIF of the Canadian shield to be the products of chemical precipitation in submarine volcanic regions as a result of extensive development of hot springs and fumarolic activity. A typical feature of the Archean crust was the production of volcano-tectonic basins with exhalative-clastic-effusive filling. Universal emission of basic effusives, rhythmic eruption of acid pyroclastic material, and extensive development of exhalative activity were observed in their formation. After the ore components arrived in the sea water they were deposited mainly near springs, but substantial amounts of SiO_2 and part of the Fe were widely disseminated.

Banerji (1977) believes that the BIF of the Singhbhum region of eastern India represent the exhalative phase of volcanism along a fissure zone, where iron and silica were carried to an interior sea and deposited, yielding alternating bands.

In some works (Parak, 1975) even the apatite-iron ores of Kiruna in Sweden, which are believed to be intrusive-magmatic, are attributed to exhalative-sedimentary deposits.

The participation of volcanic material in iron-ore accumulation is confirmed by the distribution of trace elements, in particular germanium (Grigor'yev, 1971). The explanation of the formation of banding in the case of a volcanic source of the iron and silica and sedimentary mode of deposition presents considerable difficulties. Usually, pulsating volcanic activity, and as a result of this, periodic entry of iron and silica into solution, is suggested. Direct geologic or experimental substantiation of such pulsations can hardly be obtained. Moreover, the reasons for the diversity of the mineralogic composition of the iron-rich sediments and widespread development of magnetite and hematite are unclear. Domarev (1969) points out that the exceptionally extensive development of iron-formations in the Precambrian has not yet been explained from the standpoint of a volcanogenic origin. Post-Proterozoic effusives also occur over vast territories, but no interrelationship between them and the presence of effusive-sedimentary ores has been established.

In some works a tendency toward convergence of the volcanogenic-sedimentary and clastic-sedimentary hypotheses is noted. Belevtsev et al. (1966), who considered mainly the clastic-sedimentary hypothesis, postulate the extensive occurrence of acid waters in the Precambrian hydrosphere as the result of intensive volcanic activity. Tyapkin and Fomenko (1969) believe that the main source of iron and silica in the Precambrian was the basic rocks which were the chief constituent of the Earth's crust at that time, but that some was also derived from basaltic rocks erupted along abyssal faults and other products of basic volcanism. In this case it is impossible to deny the possibility that part of the iron and silica was supplied to the sea basins along with products of volcanic activity. In this scheme the role of volcanic activity in the formation of the BIF comes down chiefly to the creation of acid environments which promoted the leaching of iron compounds from basic rocks and its transport and subsequent accumulation. The primary banding is explained by periodic revival and extinction of volcanic activity, as a result of which the pH of the water basin varied, which ultimately led to deposition of iron or cherty sediments in turn. The periodicity of those cycles might have been of the order of several hundred years.

In this brief review, far from all the details of the numerous versions of the two main hypotheses of the origin of the BIF have been considered. At present nearly all possible ideas have been offered, but the factual data on which the various hypotheses are based are still too few for a strict substantiation. What is more, the very same observations are often regarded as proof of quite opposite points of view. The distinction of several types of cherty iron-formations, based mainly on study of the paragenetic associations of rocks, far from simplifies the solutions of such controversial problems as the formation of the banding or origin of the main iron-ore minerals. If it is believed that BIF very similar in texture, structure, and mineralogic and chemical composition arose in the very same geological epoch, but as a result of completely different processes, then it will be still more difficult to explain the geochemical and physicochemical mechanism of formation.

The solution to the complex problem of accumulation of huge masses of iron and silica in the Precambrian should be sought in understanding the regularities of sedimentation in the early stages of development of the Earth's crust, in analysis of the evolution of the atmosphere and hydrosphere, and in critical assessment of geologic observations on the basis of modern geochemical and isotopic data.

Analysis of these new data have led to intensive development of a version of the hypothesis of formation of the BIF which can be called the "biogeochemical accumulation" version. In the U.S.S.R., besides in the works by Khodyush (1969) and Kaukin (1969), this hypothesis has been developed in recent years by Mel'nik (1973, 1975) and Belevtsev and Mel'nik (1976) on the basis of geologic, geochemical, and physicochemical investigations, including thermodynamic calculations and experimental modeling. The factual data, their interpretation, and possible models of deposition of the BIF will be examined in more detail in this book.

Simultaneously and independently a similar hypothesis, in various modifications, has been developed in the U.S.A. (La Berge, 1973; Cloud, 1973; Garrels et al., 1973; Holland, 1973; Perry et al., 1973; Drever, 1974). The main point in this hypothesis is the assumption of an oxygen-free reducing atmosphere and hydrosphere, beginning from the moment of their formation and separation until ~2 billion years ago. The result of this is the possibility of long-term accumulation of substantial amounts of ferrous iron in solution, supplied to the water basins from various sources (terrestrial and submarine leaching, volcanic activity). An irreversible change in the reduction-oxidation conditions of the hydrosphere as a result of geochemical evolution was needed for precipitation of the iron and formation of the BIF. Possible mechanisms of circulation of the water masses which carried the iron from the deeper-water parts with reducing conditions to the near-surface region of intensive activity of photosynthetic microorganisms have been examined in detail.

As will be shown later, not only the formation of the "clastic-sedimentary" BIF of Superior type, but also that of the older "volcanogenic-sedimentary" BIF of Algoma type can be explained by that approach.

Chapter 2

Conditions of deposition in the Precambrian

Introduction

Lithologic investigations of the oldest metasedimentary rocks have shown that the conditions of sedimentation in the Precambrian were very unique and changed irreversibly, reflecting linearity in the geologic evolution of the Earth's crust.

The regularities of Precambrian lithogenesis have been examined on the basis of generalization of geologic and geochemical data in the works of Garrels and Mackenzie (1974), Laz'ko (1969), Ronov (1964), Sidorenko et al. (1969, 1973), Strakhov (1947, 1962), Tugarinov (1956), and other investigators. Analysis of these works and their comparison with the review of the genetic particulars of the Precambrian BIF given in the preceding chapter show that the physicochemical conditions for the accumulation of chemogenic sediments should be examined taking into account the specifics of the Precambrian, the role of volcanic and sedimentary processes at that time, the evolution of the atmosphere and ocean, variations in the reduction-oxidation setting, and the appearance and development of the first forms of life.

Evolution of sedimentary processes

Modern ideas on the evolution of processes of sedimentation are developed in detail in the monograph by Tugarinov and Voytkevich (1966).

The Early Precambrian was characterized by accumulation chiefly of effusive (later converted to amphibolites) and clastic tuffogenic rocks, which gradually were succeeded by chemogenic sediments, mainly iron cherts. Typical clastogenic sediments (now represented by feldspar-mica gneisses) did not appear until the end of this ancient stage (Fig. 17), indicating the beginning of differentiation of matter in the water basins. The middle Precambrian was a stage of more profound differentiation, which caused the accumulation of thick piles of clastic rocks: arkoses, conglomerates, graywackes, clays, and then of chemogenic iron-formations. The last were formed at the expense of material obtained as a result of intensive weathering of the Early Precambrian basic effusives. At the end of the period graphite-bearing shales appeared, reflecting the development of organic life.

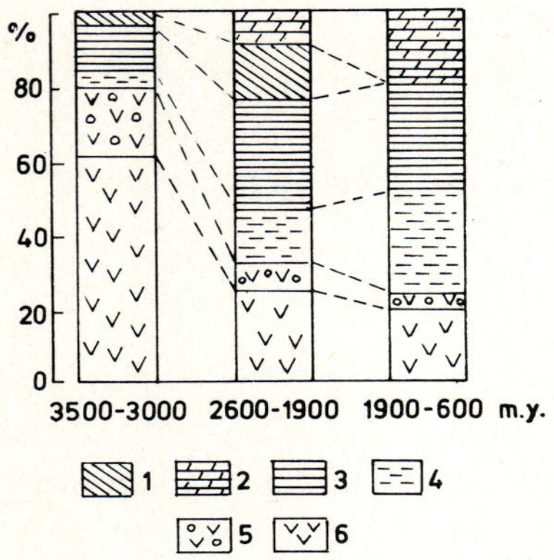

Fig. 17. Evolution of processes of sedimentation in the Precambrian (after Tugarinov and Voytkevich): *1* = iron-formations (iron-ore); *2* = carbonate strata; *3* = clays, mica-amphibole schists; *4* = sands, conglomerates, quartzites; *5* = graywackes; *6* = volcanics.

The Late Precambrian was characterized by an increase in the role of chemogenic and particularly of biogenic sedimentation. Deposition of iron-rich sediments was sharply curtailed, but thick piles of dolomite and limestone appeared, and the carbon content increased appreciably.

The change in character of sedimentation at the beginning of the Late Precambrian is explained by the fact that by this time a thick mantle of Middle Precambrian (Lower Proterozoic) sediments had arisen on the ancient continents. Subsequent denudation, erosion, and redeposition of these rocks, granitized to a considerable extent, contributed to further profound differentiation of matter. An increase in the role of the life activity of various organisms in sedimentation also was an important fact.

The general tendency in the change in character of the sediments, from volcanogenic to clastic and then biogenic with a changing amount and composition of chemogenic deposits, is traced clearly on Ronov's scheme (1964) (Fig. 18). Ronov (1959) established interesting regularities in a study of the distribution of carbonate rocks. For post-Precambrian time the amount of sediment deposited in a certain period is directly proportional to the intensity of volcanic activity (Fig. 19) and to the area of distribution of the intracontinental seas. The relatively limited distribution of carbonate rocks in the Archean cannot be taken as evidence that carbonate sediments were not deposited on a large scale at that time. As Sidorenko et al. (1969) believe, the question of the time of appearance of carbonate sediments in the Precambrian is important in principle for understanding the evolution of the Earth's substance in the outer shell, atmosphere, and hydrosphere and for reconstructing the conditions of evolution of life on Earth. Although quantitative data on the role of carbonate sediments in Precambrian stratigraphic

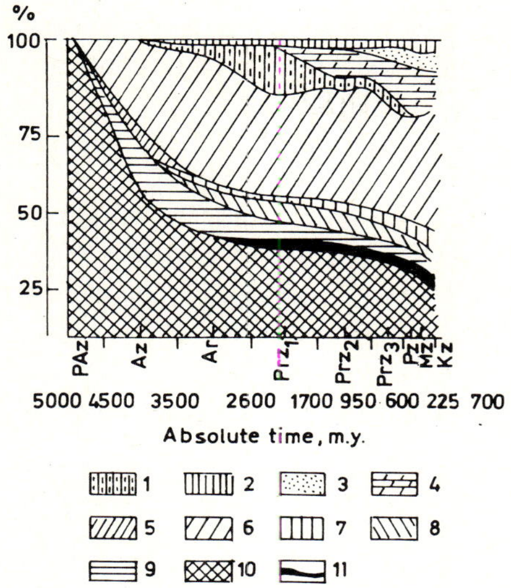

Fig. 18. Scheme of evolution of the composition of sedimentary rocks (after Ronov, 1964): *1* = jaspilites and their analogs; *2* = evaporites; *3* = limestones; *4* = dolomites; *5* = secondary quartzites; *6* = clays and their metamorphic equivalents; *7* = quartz sands; *8* = arkoses; *9* = graywackes; *10* = submarine volcanics, chiefly basic; *11* = terrestrial volcanics.

sections are limited and cannot be considered reliable because of strong metamorphism, individual estimates suggest that chemogenic carbonates were substantially abundant. The thick piles of dolomite in the section of the Upper suite of the Saksagan group of the Krivoy Rog, the piles of dolomite underlying the BIF in eastern Transvaal (Button, 1976), the dolomitic itabirites of Minas Gerais in Brazil (Dorr, 1973), and many others are examples of purely carbonate sediments. The content of primary iron and

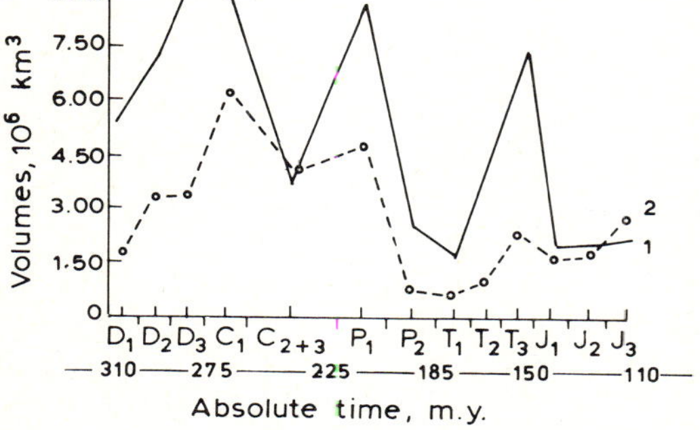

Fig. 19. Relationship between volumes of submarine and terrestrial volcanic effusions (*1*) and volumes of CO_2 (*2*) buried in carbonate rocks, in the area of the present continents (after Ronov).

magnesian-iron carbonates in the composition of the cherty iron sediments could also have been very considerable. It is not ruled out that the regularity inferred by Ronov for the Precambrian is valid, if the probable break in time between the volcanic activity and deposition of purely carbonate rocks is taken into account.

The question of the distribution of free carbon in Precambrian sedimentary rocks also needs special consideration. Ronov and Migdisov (1970) show that the C_{org} content in Lower and Upper Proterozoic complexes remained at the same level, but at the boundary of the Proterozoic and Paleozoic there is observed a two-fold increase in the carbon content. However, quantitative data on the carbon content in different types of rocks, including the BIF, are very few.

Geochemical history of the atmosphere and ocean in the Precambrian

According to present ideas on the chemical evolution of the Earth (Rubey, 1957; Vinogradov, 1959, 1967; Marshall and George, 1969), the formation of the original atmosphere and hydrosphere is explained by intensive degassing of mantle material and condensation of vapor on the cooling surface of the planet. This process went on most actively in the initial periods of the life of the Earth (Strakhov, 1960), which caused the formation of the main volume of the ocean as early as the Early Precambrian (Tugarinov and Voytkevich, 1966). It is usually presumed that the ocean waters originally were characterized by low pH values due to the presence of dissolved "acid fumes" (HF, HCl, etc.). However, these acid waters were neutralized fairly rapidly by interaction with the surrounding rocks. The evolution of the ocean must be considered in relation to the evolution of the atmosphere.

The composition of the original atmosphere— the gas phase remaining in the form of an independent shell after condensation of water vapor and

TABLE I
Scheme of main stages in the evolution of the atmosphere (after Vinogradov)

Temperature of Earth's surface, °C	Composition of the atmosphere	
	Main components	Secondary components
100	H_2O	N_2, NH_3, $B(OH)_3$, CO, CO_2, CH_4, HCl, HF, inert gases etc.
100	N_2	CO_2, CO, CH_4, O_2, inert gases
10–20	O_2, N_2	Ar, CO_2, etc.

formation of the hydrosphere—can be reconstructed only by means of a series of postulates and assumptions. At the present time there is hardly any doubt as to the reducing character of the original atmosphere, which contained practically no free oxygen (Rankama, 1955, 1957; Urey, 1956; Vinogradov, 1959, 1964; Berkner and Marshall, 1965; Tugarinov and Voytkevich, 1966; Miyake, 1969; Mason, 1971). However, the proportions of the main components at different stages of the evolution of the Earth remain unclear.

Vinogradov (1964) divides the history of the atmosphere into three phases (Table I): ancient (water vapor), transitional (nitrogen atmosphere), and present (oxysphere). Apparently the Precambrian BIF were deposited at the boundary of the transitional atmosphere and the oxysphere. Therefore, it is of particular interest to examine the evolution of the nitrogen atmosphere and its individual components, mainly nitrogen and carbon.

The basis of the transitional atmosphere—gaseous nitrogen—could have been formed from ammonia of juvenile origin, occurring in the ancient atmosphere and dissolved in the waters of the hydrosphere. When NH_3 is oxidized N_2 is produced, and the nitrogen atmosphere proper appeared at the same time as the appearance of photosynthetic oxygen (Vinogradov, 1964). According to Urey's data (1956), even in a reducing atmosphere nitrogen occurs in the form of N_2 rather than as NH_3. Mason (1971), who also distinguishes three stages in the evolution of the atmosphere, postulates the existence of a nitrogen atmosphere with an admixture of CO_2 and H_2O in the second stage. Ammonia is believed to have been a minor component ($10^{-2} > P > 10^{-4}$ atm) of the atmosphere of the first stage, which consisted of methane and possibly hydrogen. However, Miyake (1969), comparing the compositions of the atmospheres of the planets, mentions that on no other planet except Earth are there free oxygen and nitrogen in the atmosphere. Venus and Mars are blanketed by atmospheres consisting mainly of CO_2; CH_4 and NH_3 are characteristic of the outer planets with low density, which confirms the secondary formation of N_2.

The existence of the methane phase also remains in dispute, although arrival of carbon in the form of CH_4 from the interior of the Earth is quite probable (Vinogradov, 1964; Miyake, 1969). It follows from Urey's calculations (1956) that with some excess of hydrogen, carbon existed mainly in the form of methane.

Garrels and Mackenzie (1974) report that as the crust solidified and magmatic minerals crystallized, volatile gases were liberated, forming the atmosphere. Below 100°C all the H_2O would have condensed and all acid gases would have reacted with the primordial crust. In the first version, at 600°C the atmosphere would have had a $P_{H_2O} \approx 300$ atm (it corresponds to the mass of water on the present oceans), $P_{CO_2} \approx 45$ atm (it corresponds to the mass of carbon dioxide contained in carbonate rocks and organic matter), and $P_{HCl} \approx 10$ atm. Such a system is extremely active chemically and in the course of condensation would have reacted intensively with the rocks of the Earth's surface, extracting practically all the cations. A slow and continuous course of reactions during cooling is more likely, but this model of the accumulation of dissolved matter in the ocean by its reaction with the primordial crust can be useful for solving the problem of accumulation of the BIF.

Thermodynamic data

To describe the composition of the gases liberated as the planet cooled, Garrels and Mackenzie give French's diagram (1966), calculated for $P_f = 1000$ bar, with a graphite oxygen buffer and constant ratio of hydrogen to oxygen (2:1).

Urey's calculations (1956) also are based on several arbitrary assumptions: it is postulated that $P_{H_2} = 10^{-3}$ atm, all the carbon in the Earth's crust was bound in CH_4 ($P_{CH_4} = 7.94$ atm), and it is assumed that there was no graphite at all.

The idea that free carbon might already have existed on the Earth's surface before the generation of life and the appearance of photosynthetic oxygen, in the form of the dynamically stable phase—graphite—seems more sound. In that case gaseous equilibrium in the atmosphere was controlled by a graphite buffer (Mel'nik, 1972a).

The composition of the reducing fluid arriving from the mantle was controlled, under isobaric-isothermal conditions, by two independent factors —the presence of an excess of free carbon (graphite), which is an oxygen buffer together with iron minerals, and the fugacity of hydrogen. In modern geochemical models of the Earth, in particular in Semenenko's (1975) oxygen-hydrogen model, it is assumed that hydrogen is given off from the core to the mantle. In the pressure range up to 200 kbar there is an excess of hydrogen and carbon in the composition of mantle fluids with respect to oxygen, in connection with which native iron and other metals exist here along with silicates and oxides. The variations in composition of such a fluid can be estimated approximately for various moderate parameters—P up to 15 kbar and T up to 1200–1300°C, at which we recently obtained fairly reliable data on the fugacity of real gases (Mel'nik, 1978).

A calculation based on the presumed buffer reaction in which graphite participates:

$$Fe_2SiO_4 + 2\,C = 2\,Fe + SiO_2 + 2\,CO \qquad (2.1)$$

and the Boudouard reaction:

$$2\,CO = C + CO_2 \tag{2.2}$$

shows that at any fluid pressures the partial pressure of carbon monoxide does not exceed 250 bar. As pressure increases to 10–15 kbar, P_{CO} decreases to 30–50 bar, as a result of which at high pressures the CO content in the fluid is not more than 0.2–0.3% (by volume). The CO_2 content decreases still more sharply, and at pressures above 4–5 kbar there is practically no carbon dioxide in the fluid.

The partial pressures of CH_4 and H_2O, like the overall fluid pressure (P_f), depend on the fugacity of hydrogen arriving from deep sources under a certain pressure. The quantitative relationships between these gases are determined by the reactions:

$$C + 2\,H_2 = CH_4 \tag{2.3}$$

$$CO_2 + 2\,H_2 = 2\,H_2O + C \tag{2.4}$$

The calculated composition of a fluid in equilibrium with the association fayalite + metallic iron + graphite at different P_f and P_{H_2} is given in Fig. 20 (the relative amounts of the gases are shown in molar percentages). It is typical that the atomic ratio H/C = 5 hardly varies in a wide pressure range (from 15 to 2–3 kbar). At these pressures the (H + C)/O ratio likewise varies little. Only when the pressure decreases to 0.5–1.5 kbar do both these ratios

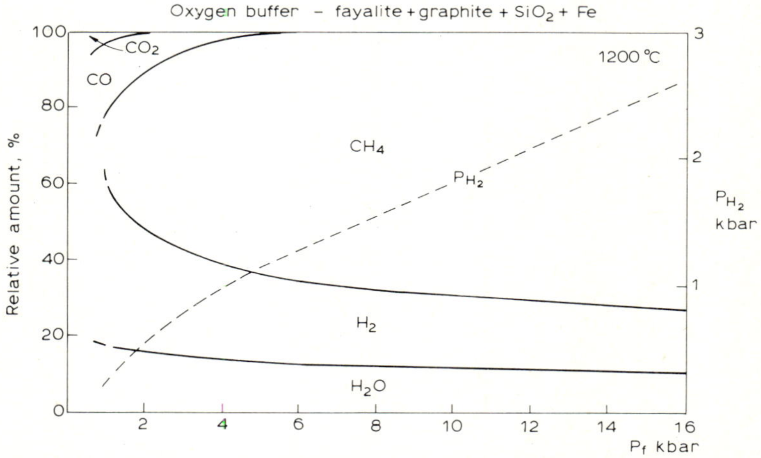

Fig. 20. Evolution of the composition of reducing fluid arriving from the mantle, as pressure drops under isothermal conditions (relative amounts are expressed in molar percentages). Dashed line shows partial pressure of incoming hydrogen.

decrease. Thus, a supply of both carbon and oxygen is necessary to maintain equilibrium in the face of isothermal expansion of a deep reducing fluid. Possible sources of these elements ("solution" of graphite, metallization of silicates, introduction of CO_2 from decomposing carbonates) remain indefinite.

As the temperature falls the composition of the fluid in equilibrium with graphite also undergoes essential changes. Several variants of the thermodynamic model of the evolution of the dense primordial atmosphere can be presented for the case of early ("catastrophic") degassing of the Earth. One such variant, based on the assumption that the mass of hydrogen was constant during isobaric cooling, is given in Fig. 21. In this case it is postulated that $P_f = 500$ bar, and that the total mass of H_2 is approximately equal to the total amount of hydrogen fixed in the present hydrosphere and sedimentary rocks. The picture obtained graphically demonstrates the possibility of transformation of a deep reducing fluid consisting of H_2 and CO into a dense atmosphere consisting of H_2O and CO_2. However, in this case too the problem of the source of the oxygen needed to oxidize the reducing gases remains unsolved. Another complication is to explain the geochemical cycles of carbon during the cooling of the fluid—first introduction, and then removal of large amounts of carbon from the gas phase in the form of graphite or carbonates.

If we turn to the geologic data, in particular to analysis of the conditions of formation of ancient sedimentary rocks including the BIF, then the

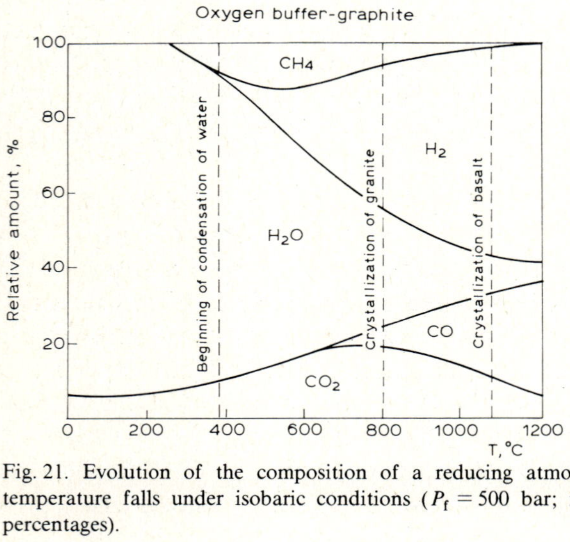

Fig. 21. Evolution of the composition of a reducing atmosphere in equilibrium with graphite, as temperature falls under isobaric conditions ($P_f = 500$ bar; relative amounts are expressed in molar percentages).

TABLE II
Probable chemical composition of a methane-nitrogen atmosphere in equilibrium with graphite for different ratios of carbon and nitrogen, in vol.% ($T = 25°C$, $P_f = 1$ bar)

Components	Atmosphere		
	mainly methane C:N = 100	intermediate C:N = 1	mainly nitrogen C:N = 0.01
CH_4	95.90	48.43	0.96
N_2	0.96	48.43	95.84
H_2	$3.6 \cdot 10^{-3}$	$2.6 \cdot 10^{3}$	$3.6 \cdot 10^{-4}$
NH_3	$1.7 \cdot 10^{-3}$	$6.9 \cdot 10^{-3}$	$5.2 \cdot 10^{-4}$
CO_2	$7.1 \cdot 10^{-4}$	$1.4 \cdot 10^{-3}$	$7.1 \cdot 10^{-2}$
CO	$1.3 \cdot 10^{-11}$	$1.9 \cdot 10^{-11}$	$1.3 \cdot 10^{-10}$
O_2	$5.8 \cdot 10^{-73}$	$1.1 \cdot 10^{-72}$	$5.8 \cdot 10^{-71}$
H_2O	3.13	3.13	3.13

hypothesis of gradual degassing, especially with respect to methane, carbon dioxide, and "acid" gases, seems to be more workable. Before the appearance of free oxygen, neutral free nitrogen was sharply predominant in that atmosphere, and the pressure after condensation of water possibly was somewhat lower than at present.

Table II gives the results of thermodynamic calculations of the composition of the equilibrium gas phase in the system C–H–N–O for different ratios of carbon and nitrogen (the inert gases were not taken into account). The following reactions were analyzed:

$$CH_4 = C + 2\,H_2; \quad \log K_p = -8.86; \tag{2.5}$$

$$2\,NH_3 = N_2 + 3\,H_2; \quad \log K_p = -5.78; \tag{2.6}$$

$$(2\,H_2O)_l = 2\,H_2 + O_2; \quad \log K_p = -83.12; \tag{2.7}$$

$$(H_2O)_l = (H_2O)_g; \quad \log K_p = -1.50 \tag{2.8}$$

$$CO_2 = C + O_2; \quad \log K_p = -69.09 \tag{2.9}$$

$$2\,CO = 2\,C + O_2; \log K_p = -48.05 \tag{2.10}$$

From Table II it is seen that regardless of the ratio of carbon and nitrogen N_2 and CH_4 are the predominant components, the NH_3 content is not more than thousandths of a percent, there is no free oxygen at all, and the CO content is low. The reduction in the content of free hydrogen and the marked increase in the content of carbon dioxide gas upon transition from a

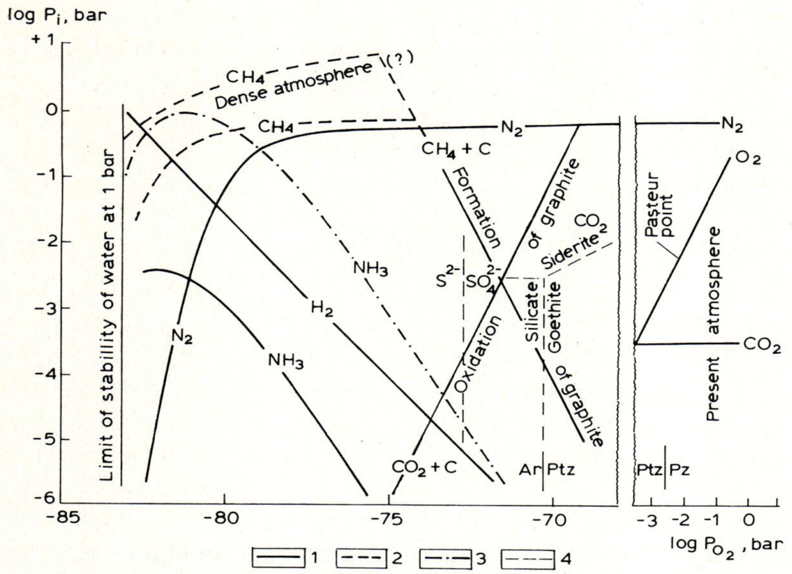

Fig. 22. Chemical evolution of the atmosphere: *1* = calculated equilibrium partial pressures of gases at atmospheric pressure; *2* = presumed partial pressures of gases; *3* = arbitrary partial pressure of ammonia (not counting NH_3 dissolved in water); *4* = phase boundaries on stability diagrams of iron minerals and sulfur ions.

methane to a nitrogen atmosphere is striking. Thus the content of primary juvenile methane determined the oxidation-reduction properties of the transitional atmosphere.

The chemical evolution of the atmosphere can be considered from the thermodynamic standpoint in a broader scheme, including the early stages when the environment was so sharply reducing that free carbon could not yet exist stably. Calculations show that in the presence of liquid water at 25°C a hydrogen-ammonia atmosphere, which is stable when $\log P_{O_2} = -83.0$ to -80.0, is highly reducing (Fig. 22).

When carbon in any form gets into this atmosphere, the irreversible process of formation of methane according to reaction (2.5) takes place. The irreversibility of the process is proved by the fact that at such low P_{O_2} the equilibrium partial pressure of CH_4 is very high ($10^{+8.7}$–$10^{+5.7}$ bar) and the equilibrium is shifted sharply to the left. Thus in the primordial hydrogen-ammonia atmosphere, if such existed in the history of the Earth, carbon occurred entirely in the gas phase in the form of methane and other hydrocarbons in equilibrium with CH_4. From this it follows that either the supply of juvenile carbon was limited at that time, or a dense methane atmosphere with a pressure much higher than one atmosphere existed (Urey has suggested $P_{CH_4} \approx 8$ bar). Besides methane, H_2, NH_3, and N_2 occurred in minor amounts in this atmosphere ($n \cdot 10^{-1}$–$n \cdot 10^{-2}$ bar). Figure 22 shows two schematic curves of the evolution of the methane content: for a normal ($P = 1$ bar) and for a dense atmosphere.

We note that methane could have been the only gas in the history of the Earth capable of forming a dense atmosphere at low temperatures in the presence of liquid water. Ammonia could not have been such a gas in connection with high solubility and lack of an appreciable amount of fixed nitrogen in sedimentary rocks. The last also pertains to free nitrogen. Sharply reducing conditions, not very likely on the Earth's surface in the presence of iron silicates, which contain a certain amount of Fe^{3+} in addition to Fe^{2+}, are needed for the existence of a dense hydrogen atmosphere. Moreover, constant dissipation (loss to the atmosphere) of light H_2 molecules was possible (Vinogradov, 1959). The partial pressure of carbon dioxide gas, as will be shown later, could not have been more than $n \cdot 10^{-1}$–$n \cdot 10^{-2}$ bar in connection with the formation of chemogenic carbonates in association with silicates.

Free oxygen could not have existed stably in an atmosphere containing CH_4, NH_3, and H_2, therefore the appearance of even small amounts of O_2 led to irreversible oxidation reactions. First of all hydrogen would be oxidized according to reaction (2.7) to form water, then ammonia according to the reaction:

$$4\,NH_4 + 3\,O_2 = 6\,H_2O + 2\,N_2 \qquad (2.11)$$

to form water and free nitrogen. In the opinion of many modern geochemists, the nitrogen of the atmosphere was formed in just that way (Vinogradov, 1959; Miyake, 1969). As a result of these reactions a hydrogen-ammonia atmosphere with an undetermined methane content was transformed into a nitrogen atmosphere, stable when $\log P_{O_2} > -79.0$. In the reducing region ($\log P_{O_2} = -79.0$ to -75.0) this atmosphere still contained an admixture of ammonia and hydrogen and all the carbon, accumulated exclusively in the gas phase in the form of methane.

Oxidation of methane to form water and free carbon:

$$CH_4 + O_2 = C + 2\,H_2O \qquad (2.12)$$

began after practically total oxidation of hydrogen ($P_{H_2} = 10^{-5}$–10^{-4}) and ammonia ($P_{NH_4} = 10^{-4}$–10^{-3}). Thermodynamic calculations show that the "graphite phase" could have been of great geochemical importance in the history of the primordial atmosphere, explaining the appearance of the appreciable amount of disseminated carbon in ancient sedimentary rocks.

The circumstance that further oxidation of graphite to carbon dioxide according to reaction (2.9) begins only after the oxidation of a substantial part of the methane is very important.

This scheme of the evolution of the original atmosphere is valid only if an oxidant, mainly oxygen, was constantly introduced. According to the data of Vinogradov (1964), oxygen was formed continuously in very small amounts as a result of photochemical processes in the upper layers of the atmosphere. When water vapor decomposes photochemically O_2 and H_2 are produced; the hydrogen goes off into outer space (dissipation) but the oxygen is used in oxidizing the atmosphere. The concentration of O_2 could not have been more than 10^{-3} times the present level (Berkner and Marshall, 1965), but this is quite enough for the irreversible reactions of oxidation of CH_4, NH_3, C, etc. to have occurred. The amount of oxygen produced was not great—a maximum of $2 \cdot 10^{12}$ g per year (Vinogradov, 1964). Calculations show that even if all of this oxygen was used to oxidize methane according to reaction (2.12), it would take 5 b.y. in the case of initial $P_{CH_4} = 1$ bar, and the oxidation of methane to carbon dioxide requires a still longer time. Thus, only photosynthesis could have yielded oxygen in an amount sufficient to oxidize the primordial atmosphere (an annual production of O_2 of up to $1.5 \cdot 10^{17}$ g) during Precambrian time. This conclusion leads to recognition of several implications, the main one of which is the appearance of very simple plants (bluegreen algae) as early as the Early Archean in an oxygen-free atmosphere containing CH_4 and NH_3.

As has already been mentioned, oxidation of methane led to the formation of free carbon which apparently was involved in biochemical cycles. Therefore the oldest sedimentary rocks could have contained both "juvenile" carbon, produced in chemical oxidation of methane, and biogenic carbon, formed in the decomposition of organisms.

Vinogradov has pointed out that with the appearance of the biosphere somewhere on the verge of $3 \cdot 10^9$ yr ago, there was a major upheaval in the evolution of the Earth. Oxidizing processes were abruptly accelerated, a nitrogen atmosphere arose in which carbon dioxide predominated over methane, and free carbon was oxidized to CO_2. After the carbon was oxidized or at the same time as that process, there began oxidation of divalent iron (at $P_{O_2} = 10^{-70}$), which led to subsequent wholesale deposition of the sediments of the Precambrian BIF. Free carbon in equilibrium with the atmosphere appeared only after complete oxidation of ferrous iron compounds in the hydrosphere and on the land surface.

Let us note that in a consideration of the evolution of the oxidizing-reducing properties of the atmosphere (see Fig. 22), the concept "partial pressure" of oxygen (P_{O_2} or log P_{O_2}) is purely illustrative and has no immediate physical meaning, inasmuch as at $P_i < 10^{-19}$–10^{-21} bar the laws of thermodynamics do not apply (Zakgeym, 1966). In fact, the properties of the primordial atmosphere were determined by the partial pressures of the gases (CH_4, NH_3, CO_2, etc.) that actually took part in reactions of the type of reaction (2.6).

The next stage in the development of the atmosphere began when formation of oxygen exceeded its consumption (Mason, 1971). This occurred in the period between 1.7 and 1.2 b.y. ago; at about 1 b.y., at the boundary of the Proterozoic and Paleozoic, the amounts of O_2 reached the level of the Pasteur point (~ 1% of the present oxygen content), as a result of which more complex organisms could appear in the biosphere. The CO_2 content at this stage was controlled by carbonate-silicate equilibria in the hydrosphere.

In the present ocean the pH is controlled chiefly by silicate buffer systems (Sillen, 1961; Holland, 1965, 1972; Kramer, 1965; Mackenzie and Garrels, 1966), including the reactions of solution and cation exchange. Knowing the cation concentration in the water, it proves possible to calculate equilibrium values of the pH which are in good agreement with the natural values. However, such a calculation cannot be made for the Precambrian inasmuch as the cation concentrations in the ancient oceans are unknown, and data on the stability of sedimentary minerals are limited. Vinogradov (1967) mentions that the first batches of water of the hydrosphere were acid and basic anions were the product of degassing of the mantle, as can be seen from the balance (Table III), whereas the cations of ocean water were the products of

TABLE III
Balance of matter in weathering and transport to the ocean* (after Vinogradov)

Element	Average content in original rocks of Earth's crust (g/kg)	Supplied to ocean (per 600 g of originally destroyed rocks), (g)	Present content in sea water (g/kg)	Content in sea water, % with respect to supply
Cl	0.17	0.10	19.350	19350
Br	0.002	0.0012	0.0659	5492
F	0.66	0.396	0.0013	0.3
S	0.37	0.222	0.900	400
B	0.01	0.006	0.0045	75
Na	25	15.0	10.764	64
K	25	15.0	0.388	2.6
Mg	18.7	11.2	1.297	11.5
Ca	33	19.0	0.4080	2.1
Sr	0.34	0.2	0.008	4.0

* Balance calculated from Na content in sea water. After Goldschmidt, it is assumed that for each kilogram of sea water, 600 g of original rock were destroyed in the weathering process.

weathering of the rocks of the Earth's crust—the pH of the solution possibly was lower than today, approximately seven. At the present time the uniformist point of view prevails in the geochemical literature, according to which the composition of the oceans has not changed essentially in the course of geologic time. However, it is not ruled out that in the Archean the volume of the hydrosphere was only a part of the total present volume of the ocean (see the review of the hypotheses of Conway, Twenhofel, and Walter (Mason, 1971)), and the content of most cations in the slightly acid waters was substantially higher.

The appearance of free carbon dioxide in the atmosphere denotes a change to carbonate or carbonate-silicate equilibria in the ocean, which played an exceptional role in the Precambrian. Rough estimates of P_{CO_2} in the ancient atmospheres are known, based on thermodynamic calculations of the equilibrium of calcite with wollastonite and magnetite with entastite (Rubey, 1957); they yielded a value of $P_{CO_2} \approx 10^{-8}$, and estimates made for the equilibria of silicates with dolomite (Holland, 1965) gave $P_{CO_2} \approx 10^{-4}$. However, in most of the works metastable equilibria involving anhydrous silicates that are unstable at low temperatures were assessed, and the thermodynamic constants of the minerals were not consistent with the experimental equilibrium data. Moreover, in the Archean and beginning of the Proterozoic, magnesian-iron carbonates and silicates predominated in the sediments, calcite was not deposited at all, and dolomite appeared after deposition of the BIF in most regions. Therefore to reconstruct the carbon dioxide regime in the Precambrian it is necessary to analyze the equilibrium of the Fe-Mg carbonates with stable hydrous silicates.

For the further development of the thermodynamic model of the geochemical evolution of the ocean and analysis of chemogenic sedimentation in the Precambrian, the results of the investigations of Garrels and Mackenzie, summarized in the well known monograph "Evolution of Sedimentary Rocks" (1974), can be used. In this work the early ocean is considered to have been the product of leaching (terrestrial and submarine) of basaltic and andesitic rocks under anaerobic conditions. HCl predominated among the volatile acid gases dissolved in the hydrosphere. In the course of leaching of the rocks, which was accompanied by liberation and deposition of silica, a chloride ocean was gradually formed, containing dissolved Na, K, Ca, Mg, Al, and Fe in the same ratios as in the original rocks. After complete neutralization of the solution, precipitation of Al became possible, with the formation of aluminosilicates, then of Fe and Mg with the formation of magnesian-iron silicates and aluminosilicates. In the resulting solution, Cl would be balanced by Na and Ca in approximately equal amounts with small additions of Mg and K. The general picture of sedimentation at all stages except the very earliest was complicated by solution and deposition of atmospheric CO_2 in the form of carbonates.

We also note that at the moment deposition of Al began, congruent solution of the original rocks would have been replaced by incongruent selective solution, manifested more completely in submarine conditions. As the "acid" ocean was neutralized, the residual products were enriched first in SiO_2, then in Al, and in neutral environments Fe and part of Mg were added.

If the scheme of a steady-state ocean is adopted, in which volatile components (mainly HCl and CO_2) occurred from the beginning in the same proportions in which they are now found in the lithosphere and hydrosphere, then the original solution would have been hydrochloric acid saturated with CO_2 at high pressure. On the basis of calculations of mass made by Garrels and Mackenzie, the mass of the present ocean is $13.5 \cdot 10^{23}$ g, the mass of CO_2 (total amount of CO_2 occurring in all the sediments of the Earth's crust, hydrosphere, and biosphere) is $2.25 \cdot 10^{23}$ g, and the amount of neutralized HCl now contained in the hydrosphere and evaporite deposits is $\sim 0.56 \cdot 10^{23}$ g.

It is easy to calculate that if solution of acid gases in the water is not taken into account, the values of partial pressures would be: $P_{CO_2} = 45$ bar and $P_{HCl} = 10$ bar. After solution of gases in the water and reduction of temperature to normal, the concentration of HCl in the water would have been ~ 40 g/l or ~ 1.0 g·mol/l, and the pH of the solution would have been close to zero. Part of the CO_2 would have gone into solution, forming undissociated carbonic acid:

$$(CO_2)_g + (H_2O)_l = H_2CO_3; \quad K = 10^{-1.46} \tag{2.13}$$

Taking into account the mass proportions of H_2O and CO_2 we find the equilibrium value to be $P_{CO_2} \approx 32$ bar and the carbonate content of the ocean to be ~ 1.1 g·ion/l. As Garrels and Mackenzie point out, the chemical activity of such an ocean-atmosphere system would have been frightful.

Another version of the model of the evolution of the primordial ocean in which volatiles, including water, arrive constantly from the mantle in approximately the same proportions as in the steady-state model, leads to an analogous conclusion.

Only if it is presumed that the mass of the hydrosphere remained more or less constant in geologic history, and CO_2 and HCl continually arrived from the mantle and gradually accumulated in the hydrosphere and atmosphere, will the characteristics of the primordial ocean be different. Taking the initial masses of CO_2 and HCl to have been ten times less than at present ($m_{CO_2} \approx 250 \cdot 10^{20}$ g and $P_{CO_2} = 5$ bar; $m_{HCl} = 50 \cdot 10^{20}$ g and $P_{HCl} = 1$ bar) with the mass of H_2O being constant ($P_{H_2O} = 13.5 \cdot 10^{23}$ g) we obtain a pH of the ocean of ~1.0 for a carbonate content of ~0.1 g·ion/l and an equilibrium P_{CO_2} in the atmosphere of ~3.5 bar. However, even with such constraint the primordial ocean is much too acid to have existed for any appreciable period in geologic history, and as a result of the high P_{CO_2} and "greenhouse" effect of the carbon dioxide atmosphere the temperature of the Earth's surface would have been so high that liquid water would have been converted to steam.

Rapid neutralization of the acid gases coming from the mantle and dissolved in the hydrosphere, upon interaction with the rocks, seems to be more plausible. Regardless of the actual course of this process, an attempt can be made to determine the conditions of the individual stages or steps by means of thermodynamic calculations.

Let us assume that the original amount of HCl was neutralized in the initial stage of the process as a result of congruent solution (not taking into account SiO_2 and Al) of an original rock which corresponded in composition to an average basalt with the proportion (in atomic numbers)

$$\mathrm{Fe} : \mathrm{Mg} : \mathrm{Ca} : (\mathrm{Na} + \mathrm{K}) = 1 : 1 : 1 : 0.5$$

The equation for electroneutrality of such a solution can be written in the following form:

$$2a_{Fe^{2+}} + 2a_{Mg^{2+}} + 2a_{Ca^{2+}} + a_{Na^+} + a_{H^+} = a_{Cl^-} + a_{OH^-} + a_{HCO_3^-} + 2a_{CO_3^{2-}} \qquad (2.14)$$

In moderately acid environments ($3 < \mathrm{pH} < 6$) chiefly cations of the metals and the chlorine ion are in solution and the values of activities of H^+, OH^-, and carbonate ions are relatively small.

Table IV gives the composition of the Precambrian ocean after neutralization of HCl, calculated under these assumptions. The pH values correspond to the end of "congruent" solution of the original rocks of the Earth's crust and beginning of deposition of the carbonate and silicate facies of the BIF (amorphous sediments). The values of P_{CO_2} correspond to the beginning of deposition of $FeCO_3$ (first value) and the beginning of deposition of $Fe_3Si_2O_5(OH)_4$ (second value). The values of the sum of carbonate ions in solution ($\Sigma CO_{2_{p\text{-}p}}$) are consistent with the values of P_{CO_2} in the atmosphere. Questions of the proportion of the masses of CO_2 in the atmosphere, hydrosphere, and carbonate sediments will be considered in more detail later.

From Table IV it is seen that the total mass of dissolved iron accumulated in the ocean as a result of "congruent" solution of basalt was ~10^{15} tons.

TABLE IV
Chemical composition of primordial Precambrian ocean after completion of neutralization

Components	Precambrian ocean						Present ocean	
	Mass of HCl equal to 10% of present			Mass of HCl equal to present			c	m
	a (g·ion/l)	c (g/l)	m ($g\times10^{20}$)	a (g·ion/l)	c (g/l)	m ($g\times10^{20}$)	(g/l)	($g\times10^{20}$)
Fe^{2+}	0.015	0.86	11.6	0.175	9.8	132.3	–	–
Mg^{2+}	0.015	0.37	5.0	0.175	4.3	58.0	1.3	17.6
Ca^{2+}	0.015	0.60	8.1	0.175	7.0	94.5	0.4	5.4
$Na^{+}+K^{+}$	0.007	0.16	2.2	0.087	2.0	27.0	10.9	147.1
Al^{3+}	10^{-8}	10^{-7}	–	10^{-6}	10^{-5}	–	–	–
SiO_{2ag}	0.002	0.12	1.6	0.002	0.1	1.6	0.01	0.2
Cl^{-}	0.100	3.60	50.0	1.14	40.5	547.0	19.0	256.5
SO_4^{2-}		not considered					2.6	35.1
CO_{2ag}	0.020	0.88	11.9	0.20	9.0	130.0	0.3	4.0
Total salinity (g/l)		6.6			72.7		34.5	
pH		5.2–6.0			4.6–6.0		8.0–8.5	
P_{CO_2} (bar)		0.15–0.0075			1.8–0.0075		0.0003	

This value coincides with estimates of the mass of iron (10^{14}, and perhaps 10^{15} tons) included in the BIF of the whole world (James and Sims, 1973).

If we come back to the model of a steady-state ocean with constant masses of HCl and H_2O, then the value of activity of Fe^{2+} after the end of neutralization due to "congruent" solution of basalt is $\sim$ 10 g/l (Table IV). In the primordial ocean more than $13 \cdot 10^{15}$ tons of iron will accumulate, and the total salinity of the solution will be unusually high. In comparing the amount of iron which theoretically can accumulate in solution and the iron occurring in chemogenic sediments, the fact that we now actually are finding and investigating the remains of very thick Precambrian BIF which have survived after 2 b.y. of weathering, granitization, and other destructive processes must not be forgotten. Therefore even order-of-magnitude agreement indicates the advisability of further consideration of the model.

Before the beginning of deposition of iron the original sediments were redeposited siliceous-aluminous products of terrestrial or submarine "leaching" of basalts.

The next stage would have been chemogenic deposition of iron in the form of very finely divided carbonates:

$$Fe^{2+} + (CO_2)_g + (H_2O)_l = FeCO_3 + 2\,H^+ \qquad (2.15)$$

$$pH = 4.58 - \tfrac{1}{2}\log P_{CO_2} - \tfrac{1}{2}\log a_{Fe^{2+}} \text{ (amorphous sediments)}$$

$$pH = 3.85 - \tfrac{1}{2}\log P_{CO_2} - \tfrac{1}{2}\log a_{Fe^{2+}} \text{ (crystalline minerals)}$$

or silicates:

$$3\,Fe^{2+} + 2(SiO_2)_{am} + 5\,H_2O_l = Fe_3Si_2O_5(OH)_4 + 6\,H^+ \qquad (2.16)$$

$$pH = 5.17 - \tfrac{1}{2}\log a_{Fe^{2+}} \text{ (amorphous sediments)}$$

$$pH = 4.91 - \tfrac{1}{2}\log a_{Fe^{2+}} \text{ (crystalline minerals)}$$

the interactions between which are governed by the reaction:

$$3\,FeCO_3 + 2(SiO_2)_{am} + 2(H_2O)_l = Fe_3Si_2O_5(OH)_4 + 3\,CO_2 \qquad (2.17)$$

$$\log P_{CO_2} = -1.19;\ P_{CO_2} = 6.4 \cdot 10^{-2} \text{ bar (amorphous sediments)}$$

$$\log P_{CO_2} = -2.13;\ P_{CO_2} = 7.5 \cdot 10^{-3} \text{ bar (crystalline minerals)}$$

This was the first buffered reaction which could have controlled the partial

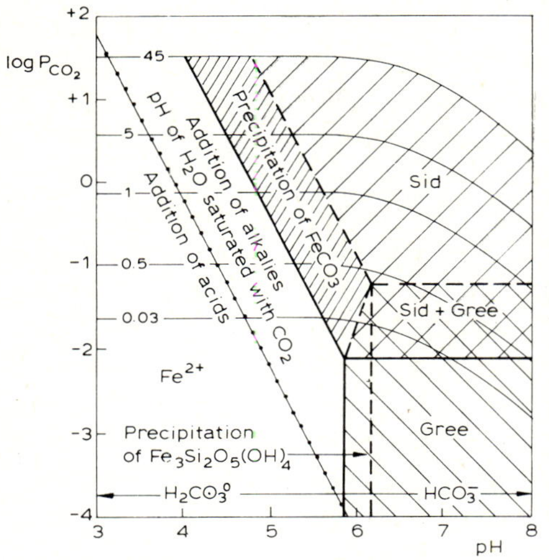

Fig. 23. Diagram of carbonate and carbonate-silicate equilibria in deposition of BIF: solid lines bound field of deposition of crystalline minerals; dashed lines—field of deposition of amorphous sediments; thin lines—isobars of P_{CO_2} in primordial atmosphere. *Sid* = siderite; *Gree* = greenalite.

pressure of carbon dioxide in the Precambrian atmosphere in conjunction with deposition of substantial amounts of carbonate and silicate iron-formations.

Figure 23 is a graph in coordinates of pH–log P_{CO_2} which illustrates the conditions of deposition of the original iron deposits in association with silica. Inasmuch as the degree of crystallinity of the original sedimentary minerals might have varied within wide limits, the fields of their deposition situated between corresponding limits of deposition of crystalline and amorphous solid phases are shown in this graph. The line for the pH of pure water saturated with CO_2 is entered on the same diagram. To create a lower pH in the region to the left of this line, it is necessary to add a certain amount of acid to the saturated CO_2 solution (HCl in this model). Analogously, to create a higher pH in the region to the right of the line it is necessary to add an alkali, neutralizing the solution.

In addition, several lines of equal log $P_{CO_2} = f(\mathrm{pH})$ are entered on the diagram, characterizing the partial pressure of CO_2 over the ocean for different initial P_{CO_2} in the original atmosphere. The figures on these lines designate the P_{CO_2} which was attained if all the carbon dioxide dissolved in the ocean was converted to the gaseous phase and mixed with atmospheric CO_2. The lines of equal $P_{CO_2(\mathrm{original})} = 45$ bar correspond to the model of a steady-state ocean with constant masses of H_2O and CO_2. The lines of equal $P_{CO_2(\mathrm{original})} = 0.032$ bar describe the "evolution" of the present ocean—such partial pressure is obtained if the CO_2 dissolved in the ocean is converted to the gaseous phase (0.00265 mol/l, which corresponds to a mass of $1.574 \cdot 10^{20}$ g) and mixed with the CO_2 of the atmosphere ($P_{CO_2} = 3 \cdot 10^{-4}$ bar, which corresponds to a mass of $0.015 \cdot 10^{20}$ g).The other isolines are intermediate.

Let us examine the more general diagram of carbonate silicate equilibria in the hydrosphere which reflects the evolution of chemogenic sedimentation after the end of the stage of accumulation of cations due to "congruent" solution of the original rocks. A typical feature of "incongruent" solution is its selectivity—in an idealized model deposition of Fe^{2+} would have been accompanied by the solution of compensating amounts of more soluble Mg, Ca, and alkali metals. Correspondingly, deposition of Mg^{2+} was accompanied by an increase in concentrations of Ca^{2+}, Na^+ and K^+, etc. These processes occur when the pH is increased, first due to neutralization of HCl and then (beginning at pH > 6) also of carbonic acid.

Inasmuch as the condition of electroneutrality of the solution must necessarily be fulfilled, knowing the amounts of HCl and H_2CO_3 the activities of the cations at each step can then be calculated. If the mineralogical and geochemical characteristics of the BIF of the Krivoy Rog basin and an original composition of the ocean with a mass of HCl equal to 10% of the present (Table IV) are taken as the basis, then deposition should have begun

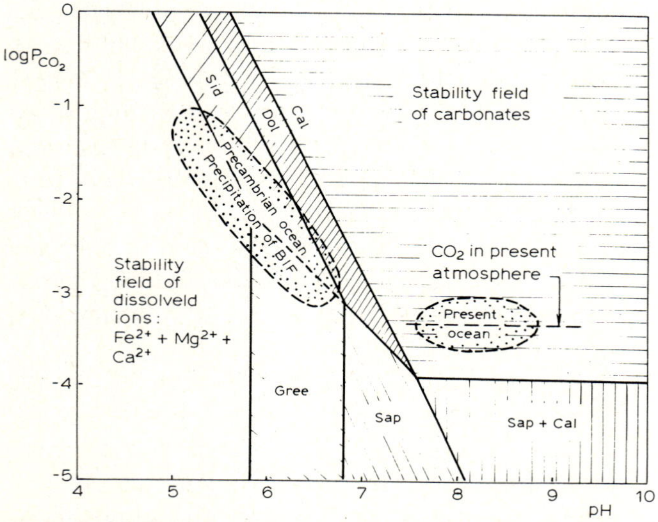

Fig. 24. Conditions of precipitation of carbonates and silicates of iron, magnesium, and calcium in the presence of excess silica. Graphic representation of model with initial: $a_{Fe^{2+}} = a_{Mg^{2+}} = a_{Ca^{2+}} = 0.015$ g·ion/l and $a_{Cl^-} = 0.10$ g·ion/l. *Cal* = calcite; *Dol* = dolomite; *Gree* = greenalite; *Sap* = saponite; *Sid* = siderite.

at pH $\leqslant 6$ and $P_{CO_2} \geqslant 0.075$ bar. The original sediment apparently was highly dispersed crystalline magnesian-iron carbonate alternating with amorphous silica. However, inasmuch as there are no reliable thermodynamic constants for amorphous or highly dispersed carbonates and silicates of magnesium and calcium, the diagram given (Fig. 24) reflects the conditions of deposition of crystalline minerals. In the case of amorphous sediments the sequence of deposition will not change, but all the phase boundaries will be shifted toward lower pH values, reflecting the increase in solubility with the increase in degree of dispersion.

The primary sedimentary-metamorphic sideroplesites and magnesian-iron silicates of the BIF of the Krivoy Rog basin contain on the average 25–30% magnesian and 70–75% iron components and less than 2% calcium—in molar percentages (Mel'nik, 1975). The $FeCO_3$ content in the dolomites of the upper suite is low and can be neglected in calculations. Similar ratios of Mg and Fe are observed in the BIF of other areas of the world, although in individual deposits the Ca contents are high and in some other cases, $CaO \geqslant MgO$.

After deposition of nearly all the iron, part of the magnesium, and a small amount of calcium, there remained in solution ~75% of the original amount of Mg^{2+}, ~98% of the original amount of Ca^{2+}, and all the alkali cations. The solution was saturated with respect to amorphous silica and supersaturated with respect to crystalline quartz. As has already been mentioned, to maintain electroneutrality an equivalent amount of Mg^{2+}, Ca^{2+}, and $(Na^+ + K^+)$ must be introduced into the solution to compensate for the deposited iron and magnesium. Assuming that "congruence" of solution of Mg, Ca, and Na (proportion 1:1:0.5) is maintained in submarine leaching and that not only Si and Al but also Fe accumulate in the residual products, we find the composition of the solution before the beginning of deposition of dolomite to be (in g · ion/l):

$$Mg^{2+} = 0.0195;\ Ca^{2+} = 0.0229;\ (Na^+ + K^+) = 0.0152$$

The deposition of dolomite according to the reaction:

$$Ca^{2+} + Mg^{2+} + 2(CO_2)_g + 2(H_2O)_l = CaMg(CO_3)_2 + 4\,H^+ \qquad (2.18)$$

$$pH = 4.41 - \tfrac{1}{2}\log P_{CO_2} - \tfrac{1}{4}\log a_{Mg^{2+}} - \tfrac{1}{4}\log a_{Ca^{2+}}$$

is expressed by the equation:

$$pH = 5.25 - \tfrac{1}{2}\log P_{CO_2}$$

Equal amounts of Mg^{2+} and Ca^{2+} will be removed from solution, compensating for the introduction of calcium and alkalis into solution. At the moment deposition of dolomite begins a certain amount of iron still remains in solution ($\sim 10\%$ of the original, for $a_{Fe^{2+}} = 1.6 \cdot 10^{-3}$ g · ion/l or ~ 90 mg/l). That is, some of the Fe-Mg sideroplesites and Fe-Mg-Ca carbonates of ankerite type will be deposited along with dolomite. Those mineral associations are observed in the oxide facies of the BIF, in particular in the jaspilites of the Krivoy Rog basin.

After a substantial part of the Mg^{2+} is precipitated in the form of dolomite, deposition of calcite begins as a result of the relative and absolute enrichment of the solution in Ca^{2+}:

$$Ca^{2+} + (CO_2)_g + (H_2O)_l = CaCO_3 + 2\,H^+$$

$$pH = 4.82 - \tfrac{1}{2}\log P_{CO_2} - \tfrac{1}{2}\log a_{Ca^{2+}} \qquad (2.19)$$

Taking into account the equilibrium activities of the ions:

$$a_{Mg^{2+}} = 0.0007;\ a_{Ca^{2+}} = 0.0292;\ a_{Na^+} = 0.04042$$

we obtain the equation:

$$pH = 5.59 - \tfrac{1}{2}\log P_{CO_2}$$

By the beginning of deposition of calcite, a substantial part of the magnesium (nearly 95% of the original amount) and iron ($\sim 98\%$ of the original amount) will have come out of solution.

Theoretically this sequence should be observed in the deposition of purely carbonate sediments. In the actual conditions of nature, including deposition of the BIF, a substantial part of the Fe and Mg went into the composition of primary silicates of greenalite, minnesotaite, and iron chlorite type. Taking ferrous greenalite — $Fe_3Si_2O_5(OH)_4$) — and magnesian saponite — $Mg_3Si_4O_{10}(OH)_2$ — as the stable end members, one can find the interrelationships between carbonates and silicates.

The deposition of greenalite and its interrelationships with siderite are controlled by reactions (2.16) and (2.17), considered earlier. The line separating the field of deposition of Sid and Gree (Fig. 24) in the first approximation characterizes the conditions of joint deposition of Fe-Mg carbonates and silicates of variable composition. From the diagram it is seen that as the content of Fe^{2+} in solution decreases and, apparently, the magnesium content of the minerals deposited increases somewhat, the buffer value of P_{CO_2} also decreases several times: from $\sim 7 \cdot 10^{-3}$ to $\sim (1\text{–}2) \cdot 10^{-3}$ bar. Inasmuch as in most Precambrian BIF magnesian silicates occur relatively rarely (the finding of talc in the unoxidized oxide facies of the BIF of the

Krivoy Rog basin can be given as an example–Yaroshchuk et al., 1976) and dolomite likewise does not occur very widely except in the later overlying piles of molasse formation, the physicochemical characteristics of the Precambrian ocean at the time of chemogenic deposition of the BIF are fairly precisely determined (Fig. 24). The general trend of evolution was a gradual decrease in P_{CO_2} from 0.10–0.15 bar (these evidently are the maximum possible values before the beginning of deposition of the carbonate facies of the BIF) to 0.010–0.001 bar (end of deposition of the carbonate-silicate and silicate facies, periodic deposition of dolomite). The slightly acid reaction of the solution gradually became neutral.

After deposition of the BIF, dolomite was deposited in a neutral environment and P_{CO_2} gradually came down to the present level. The actual picture undoubtedly was complicated by oxidation-reduction reactions and periods of clastic sedimentation, which will be considered in more detail in formulating the general model of the genesis of BIF.

To complete the picture, let us also examine the conditions of deposition of magnesian silicate in neutral and slightly alkaline environments at low P_{CO_2}, although these processes, theoretically possible after deposition of the BIF, would hardly have been extensive in the Precambrian, yielding intensive accumulation of carbonate strata as in the Phanerozoic.

With the decrease in P_{CO_2} as carbon dioxide was consumed in the formation of carbonates, incongruent solution of dolomite was possible, with the formation of saponite (or talc) as a result of reaction with silica:

$$3\,CaMg(CO_3)_2 + 4\,SiO_2 + 6\,H^+$$

$$= Mg_3Si_4O_{10}(OH)_2 + 3\,Ca^{2+} + 2\,H_2O + 6\,CO_2 \qquad (2.20)$$

$$pH = 2.87 - \log P_{CO_2} - \tfrac{1}{2}\log a_{Ca^{2+}}$$

at the point where deposition of dolomite began ($a_{Ca^{2+}} = 0.023$ g · ion/l):

$$pH = 3.96 - \log P_{CO_2}$$

and at the point where deposition of calcite began ($a_{Ca^{2+}} = 0.029$ g · ion/l):

$$pH = 3.64 - \log P_{CO_2}$$

In connection with variation in the activity of calcium the boundary between the fields of deposition of dolomite and saponite is somewhat crooked.

Outside the region of carbonate stability, direct deposition of magnesium in the form of saponite (or sepiolite) is possible according to the reaction:

$$3\ Mg^{2+} + 4\ SiO_2 + 4\ H_2O = Mg_2Si_4O_{10}(OH)_2 + 6\ H^+ \quad (2.21)$$

$$pH = 5.96 - \tfrac{1}{2}\log a_{Mg^{2+}}$$

And finally, the interrelationships between dolomite and saponite are determined by the buffer reaction:

$$3\ CaMg(CO_3)_2 + 4\ SiO_2 + H_2O = Mg_3Si_4O_{10}(OH)_2 + 3\ CaCO_3 + 3\ CO_2 \quad (2.22)$$

$$\log P_{CO_2} = -3.67;\ P_{CO_2} = 2.1 \cdot 10^{-3}\ \text{bar}$$

The data given indicate that at no stage in geologic history did the partial pressure of carbon dioxide gas in the atmosphere in equilibrium with the ocean fall below 10^{-3} bar.

In the primordial ocean sulfur was represented mainly by sulfide formations (HS^-, S^{2-}), which is confirmed by thermodynamic calculations of ionic equilibria:

$$S^{2-} + 2\ O_2 = SO_4^{2-};\ \log K_p = +145.36 \quad (2.23)$$

$$4\ HS^- + 5\ O_2 = 2\ SO_4^{2-} + 2\ S + 2\ H_2O;\ \log K_p = +352.25 \quad (2.24)$$

In the case of equality of concentration of sulfide and sulfate ions the equilibrium values of partial pressure of oxygen are $10^{-72.68}$ and $10^{-70.45}$ bar, respectively. These values fall in the stability field of siderite and ferrous-iron silicate (see Fig. 22). In the stability field of siderite plus goethite or hematite, the sulfate ion already predominates. Thus sedimentary sulfides in ancient rocks (Archean, or Azoic) evidently contain juvenile sulfur, introduced in the form of hydrogen sulfide. The sulfides in iron-formations could have been formed both by way of chemogenic deposition, and by way of diagenetic reduction of sulfates. In the presence of free oxygen only sulfate ions are stable.

The relatively low abundance of the sulfide facies of BIF and its spatial and genetic restriction to volcanogenic sequences indicates that sulfur concentrations were low in the Precambrian ocean. In reducing conditions practically all the sulfur was fixed in pyrite.

To solve the problem of the possibility that a dense atmosphere with a high CO_2 content existed in the early stages of formation of the world ocean, we can also resort to rough calculation.

Deposition of all the iron, magnesium, and calcium that accumulated during the period of accumulation (see Table IV) in the form of carbonates requires $\sim 27 \cdot 10^{20}$ g of CO_2 (for $m_{HCl} = 0.1$ of the present mass), which corresponds to an original partial pressure of CO_2 in the atmosphere, in equilibrium with the ocean, of ~ 0.4 bar. After deposition of all the iron in the form of $FeCO_3$, P_{CO_2} decreases to ~ 0.25 bar. It is quite obvious (see Figs. 23 and 24) that at such high values of P_{CO_2} the formation of Fe-Mg silicates is impossible and the oldest Archean BIF would originally have been represented exclusively by carbonate facies, which is not supported by the geologic data. If part of the iron was deposited in the form of magnetite or hematite, P_{CO_2} turns out to be still higher, and the formation of silicates becomes still less likely. Similar conclusions are reached if the original assumptions of the model are changed, for instance, the mass of the ocean or mass of HCl. Deposition of BIF in the form in which we now know them is possible only if the original values of P_{CO_2} were not more than 0.10–0.15 bar and then as a result of deposition of the carbonate facies of the BIF were reduced to 0.05–0.001 bar. If that is so, then constant (steady or, more likely, pulsed) introduction of CO_2 from deep sources was required for deposition of the carbonate (mainly dolomitic) sequences of the Proterozoic and then Phanerozoic. From this standpoint, the relationship of periods of carbonate deposition to intensity of volcanic activity, mentioned earlier, is quite explainable.

Thus the hypothesis of gradual degassing and gradual accumulation of volatiles (CO_2, HCl, SO_2, etc.) in the ocean and then in chemogenic deposits at relatively low partial pressures of "oxide" gases seems most logical.

However, the problem of the relationship of the mass of the hydrosphere in the Precambrian and Phanerozoic cannot be solved by developing thermodynamic models. The data of isotope geochemistry may be more promising for a solution of the problem of a "steady-state" or "growing" ocean.

Isotopic data

At the present time there are sufficiently reliable data on the distribution of the stable isotopes of oxygen, carbon, and sulfur in Precambrian sedimentary rocks. On the basis of analysis of these data, it is possible to obtain additional information on the geochemical history of the ocean and conditions of sedimentation.

Oxygen is the most common and most important chemical element of the Earth's crust, but study of its geochemical evolution did not become possible until the development of methods of isotopic analysis. Numerous determinations of $\delta^{18}O$, i.e., the relative content of the heavy oxygen isotope compared to a generally accepted standard, have indicated the existence of considerable variations, which carry geologic information. Substantial enrichment of the sediments in ^{18}O compared to the water of the basin has proved to be the most striking feature of chemogenic deposition. The value of isotopic fractionation ($\Delta\delta^{18}O$) between sedimentary carbonates or silicates and the water is a thermodynamic constant which depends only on temperature. As the temperature rises, $\Delta\delta^{18}O$ decreases and in the conditions of deep petrogenesis it approaches zero.

In study of the oxygen isotopic composition of sedimentary carbonates and cherts, the dependence of $\delta^{18}O$ on the age of the sediment has been established. Weber (1965) investigated the oxygen isotopic composition of ancient and recent marine limestones. It turned out that the $\delta^{18}O$ of Precambrian carbonates is much lower than that of Quaternary sediments (respectively −11.39 and −1.18‰, PDB standard).

Sidorenko and Borshchevskiy (1977), who analyzed the general trend in the evolution of the isotopic composition of carbonates in the Precambrian and Phanerozoic, summarized all published data (up to 1976) and depicted them graphically (Fig. 25). In their opinion, the age effect in the oxygen isotopic composition is the result of regional metamorphism, as a global process.

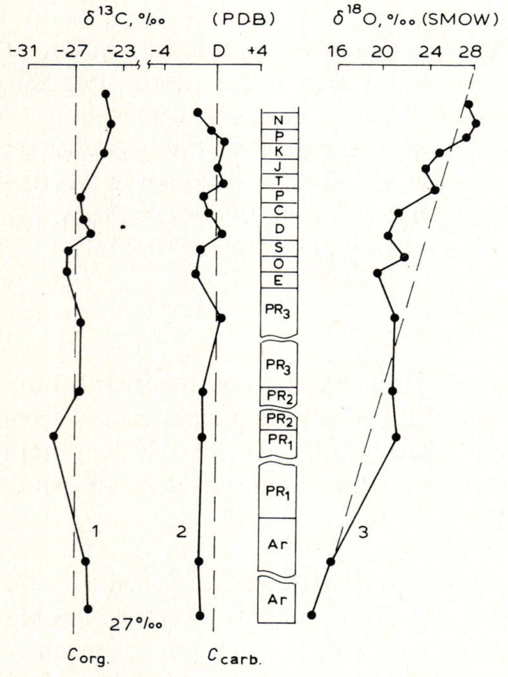

Fig. 25. Variation in average values of $\delta^{13}C$ of organic matter (*1*) and of carbonates (*2*) and in $\delta^{18}O$ of carbonates (*3*) during geologic time (world data for 1976 are summarized) (Sidorenko and Borshchevskiy, 1977).

For ancient chemogenic cherts, including cherts from the BIF of Gunflint, Ontario, and Biwabik, Minnesota, Perry (1967) established a functional relationship between the $\delta^{18}O$ of silica and the time of deposition. On this basis it was concluded that the waters of the ancient ocean were depleted in the heavy oxygen isotope, and that the $\delta^{18}O$ value was about −10 to −12‰ for the Proterozoic and up to −18‰ for the Archean (compared to the oxygen of recent ocean water, SMOW standard).

Isotopic data, first obtained by us (Mel'nik and Lugovaya, 1972) for syngenetic iron carbonates from the BIF of the Krivoy Rog basin, gave values of $\delta^{18}O$ in the range of +12.3 to +12.9‰, which corresponds to a $\delta^{18}O$ of ~ −14 to −15‰ for the water from which they were deposited at room temperature. However, these rocks were metamorphosed under greenschist facies conditions, which could have led to redistribution of ^{18}O between coexisting siderite, magnesite, quartz, and silicates and the pore solution.

Further investigations by Chase and Perry (1972), based on voluminous factual material (Perry and Tan, 1972), led to the development of a dynamic model of the isotopic evolution of the oceans. Making a number of assumptions, Chase and Perry came to the conclusion that an excess of ^{18}O in the crustal rocks is consistent with the widespread plate tectonics concept. Figure 26 presents calculated curves characterizing the evolution of $\delta^{18}O$ of ocean water for different initial assumptions. Curve *I* was plotted for an ocean with constant mass with an initial value of $\delta^{18}O = -12‰$ $3.3 \cdot 10^9$ yr ago. Curve *II* characterizes a gradual (linear) degassing with a corresponding increase in mass of the hydrosphere. In this case, with a doubling of the water mass in $3.3 \cdot 10^9$ yr, the initial value of $\delta^{18}O = -519‰$ seems

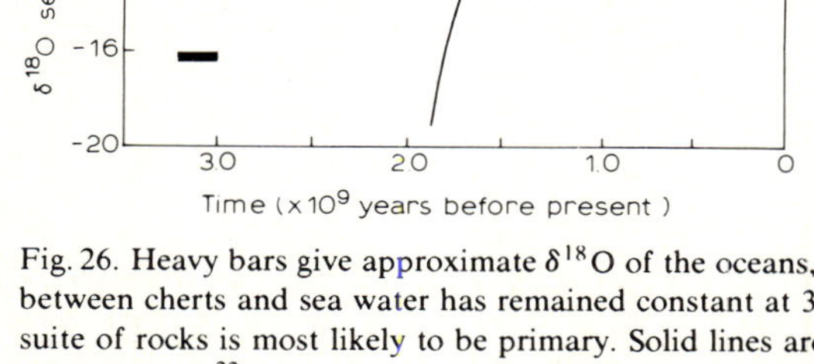

Fig. 26. Heavy bars give approximate $\delta^{18}O$ of the oceans, based on the assumptions that the isotopic shift between cherts and sea water has remained constant at 34‰, and that the highest value for $\delta^{18}O$ from a suite of rocks is most likely to be primary. Solid lines are three calculated models, all with present ocean mass $3.89 \cdot 10^{22}$ moles O_2, and $\delta = 0$ (Chase and Perry, 1972).

completely unrealistic. Curve *III* possibly represents the most realistic picture of the variation in $\delta^{18}O$ both with respect to its nonexponential form and with respect to its agreement with the $\delta^{18}O$ of cherts. In the end, Chase and Perry concluded that the mass of the ocean has not changed during geologic history and that $\delta^{18}O$ has varied widely due to the extensive development of cyclic circulation of water through the mantle.

At the same time, calculations based only on the mass balance of oxygen in the ocean and rocks, made earlier by Silverman (1954), led to quite opposite results.

Assuming that the sedimentary rocks were derived from rocks with an original $\delta^{18}O$ value of +7.0‰ (original basalts, meteorites), and that the total amount of oxygen in the ocean is $1.2 \cdot 10^{24}$ g and in sedimentary rocks $0.9 \cdot 10^{24}$ g, the value found for the $\delta^{18}O$ of the water of the primordial ocean is +12.0‰ (for the water of the present ocean, $\delta^{18}O$ was taken as 0.0‰—SMOW standard). If granitized rocks are taken into account, then the $\delta^{18}O$ of the ancient hydrosphere increases to +16.2‰. These calculations cannot be considered reliable, as Silverman himself remarks, inasmuch as the composition and amount of primary rocks of the Earth's crust remain uncertain; nor was juvenile water introduced from the mantle taken into account.

Other rough estimates of the $\delta^{18}O$ of the Precambrian ocean are known. Degens (1967), who assumed that in the course of Earth history there has been little change in the isotopic composition of sea water for 2500 m.y., arrived at a value of $\delta^{18}O = +2$‰ for the Precambrian. We (Mel'nik and Lugovaya, 1972) obtained a similar value (+2.8‰) for water in equilibrium with hematite and magnetite in the oxide facies of the BIF of the Krivoy Rog basin. To extrapolate these data to the temperature conditions of the environment of deposition, more reliable information on the isotopic fractionation of oxygen in the system magnetite-hematite-water is needed, and the effect of metamorphism on isotopic exchange with quartz must be taken into account.

Knauth and Epstein (1976) report that the isotopic composition of some Precambrian cherts does not obey the regularities observed by Perry. In reality, large fluctuations are observed, possibly caused by climatic temperature fluctuations. If this is so, then the temperature of the Earth's surface $1.3 \cdot 10^9$ yr ago might have reached 52°C, and $3 \cdot 10^9$ yr ago even 70°C. Becker and Clayton (1976), who investigated the BIF of the Hamersley area in Western Australia, believe that isotopic exchange between carbonates and quartz during metamorphism is substantial. In particular, the $\delta^{18}O$ of siderite associated with quartz is very constant (+20.12 to +21.20‰). In their opinion, the $\delta^{18}O$ of ocean water $2.2 \cdot 10^9$ yr ago almost certainly was greater than −11‰ and probably less than −3.5‰. Interpretation of data obtained from metamorphosed rocks and especially from individual minerals is attended by considerable uncertainty. At the same time metamorphism of uniform thick carbonate rocks or cherts, unaccompanied by phase transitions or mineral reactions, changed the original isotopic relationships to a minimal extent.

Therefore the model of Chase and Perry, in which the geologic and isotopic data can be reconciled and which gives a quantitative estimate of the evolution, seems to be rather sound. Naturally, periodic fluctuations of $\delta^{18}O$ caused by mass deposition of chemogenic carbonates, silicates, and silica in individual geologic epochs, and also by temperature variations, cannot be ruled out.

At the same time, the conclusion concerning the catastrophic early degassing of the Earth and early formation of an ocean whose volume was still limited in the Early Precambrian, can hardly be extended automatically to other volatiles, in particular CO_2, sulfur, and probably, HCl.

Carbon. Galimov (1968) pointed out two lines of evolution of carbon. The first is traced from gaseous CO_2 and CH_4 coming from the mantle and endogenetic minerals into the carbon of the carbonaceous matter of the mantle and later into the carbonaceous matter of meteorites (Fig. 27). By far the largest part of the carbon characterized by a heavy isotopic composition ($\delta^{13}C$ from -4 to -7%) arrived in the Earth's crust from the mantle by that route. The second line is related to carbide matter in the mantle and meteorites, which does not enter into the formation of gaseous compounds. Carbon of the first line of evolution predominates in the Earth's crust.

In studying the carbon isotopic composition of Precambrian carbonates Galimov (1968) established a correspondence of the $\delta^{13}C$ of Paleozoic and

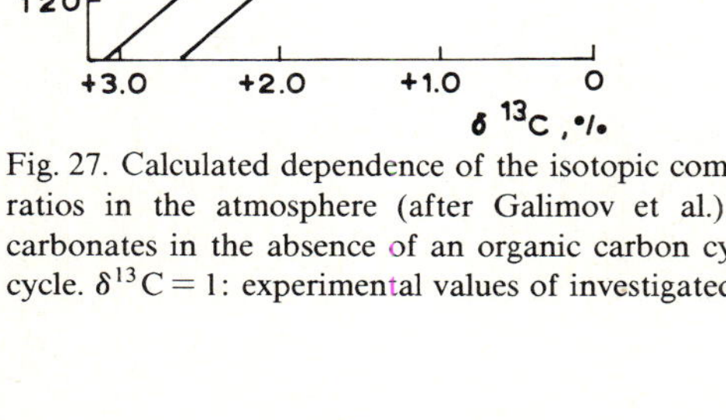

Fig. 27. Calculated dependence of the isotopic composition of sedimentary carbonates on CO_2 and CH_4 ratios in the atmosphere (after Galimov et al.): A = calculated values of isotopic composition of carbonates in the absence of an organic carbon cycle; B = same, taking account of the organic carbon cycle. $\delta^{13}C = 1$: experimental values of investigated Precambrian carbonates.

later carbonates, which indicates the predominance of CO_2 among the gaseous carbon compounds in the ancient atmosphere. Among the samples investigated, a group was distinguished with anomalous isotopic composition ($\delta^{13}C$ from +6.3 to +9.1‰), which is not typical of the sedimentary carbonates of the post-Precambrian period. These samples reflect the period when there was no organic carbon cycle and carbonates were deposited chemogenically (Fig. 24) in a carbon dioxide atmosphere. Although no clear-cut correlation is observed between the carbon isotopic composition of the carbonates and their age, the lowest $\delta^{13}C$ values belong to the Late Proterozoic and the highest to the Early Archean.

According to the data of Perry and Tan (1973), most of the carbonates from the oldest rocks (from $3 \cdot 10^9$ to $1 \cdot 10^9$ yr) of South America and the Canadian shield have $\delta^{13}C$ within ±2‰. Determinations by Hoering (1967) also showed that the carbon isotopic composition of Precambrian carbonates falls within the values established for normal Phanerozoic marine sediments.

A value of $\delta^{13}C = -33.2$‰ has been established for the free carbon from the slates of the Biwabik formation (Perry and Tan, 1973), and for the slates of the Thompson and Gunflint formations, −30.5 and −30.3, respectively (Hoering, 1967). These values are reported for the carbon of the schists of cherty iron-formations: −34.9 and −31.4‰ (graphite schists of the Upper suite of the Krivoy Rog group) and −23.3‰ (carbonaceous-biotite-sericite schist of the Kursk group, Belgorod district, KMA). These $\delta^{13}C$ values fall within those characteristic of the carbon of organic matter.

Sidorenko and Bortschevskiy (1977) report that much factual material indicates that the average isotopic composition of two forms of carbon—carbonate ($\delta^{13}C_{av} = 0.0$‰) and organic ($\delta^{13}C_{av} = -27$‰)—was constant not only for the Phanerozoic, but also for the whole Precambrian (Fig. 25). In their opinion the value $\Delta(C_{org} - C_{carb})$ is an important planetary biogeochemical constant which indicates that $3.5 \cdot 10^9$ yr ago there was a stable geochemical carbon cycle in the Earth's crust. This conclusion is based on numerous measurements of $\delta^{13}C$ of coexisting carbonates and organic matter in various parts of the world. Among the works of recent years, the investigations of Eichmann and Schidlowski (1975) and Schidlowski et al. (1976) should be mentioned. In the first of these works it is shown, on the basis of measurements of $\delta^{13}C_{org}$ and $\delta^{13}C_{carb}$ of 58 coexisting pairs from Precambrian sediments (mainly from South Africa) that the value of isotopic fractionation $\Delta\delta^{13}C \cong 25$‰ has not varied for $3.3 \cdot 10^9$ years, but was somewhat higher (≅28‰) in the Early Precambrian. In the second work analogous data are given for carbon isolated from argillite and carbonate sediments from the Rio das Velhas and Minas group of Brazil. The $\delta^{13}C_{org} = -24.3$‰ values indicate the very early evolution of photosynthetic microbiota and the extensive participation of biogenic agents in the process of formation of the BIF. These data are supported by $\delta^{13}C$ determinations on the massive siderites from the Archean BIF of Michipicoten and Woman River (Canadian shield), which gave a value of +2.4‰ (Goodwin et al., 1976), while for graphite carbon the $\delta^{13}C$ ranges from −20.4% to −27.7‰ (average −24.7‰).

However, in many cases essential departures from the indicated regularity are observed, particularly in the rocks of the BIF. Often the carbonates from iron-formations have lower $\delta^{13}C$ values (to −18‰). This anomaly was first mentioned by Becker and Clayton (1970), who established that the $\delta^{13}C$ of the limestones and dolomites of the Dales Gorge strata ($1.9 \cdot 10^9$ yr) are analogous to those of Phanerozoic carbonates, whereas the iron-rich rocks of the Brockman BIF have $\delta^{13}C$ lower by 10–15‰. On the basis of these data it was concluded that these iron-formations of Western Australia were deposited in a restricted basin with active volcanic activity, where the effect of juvenile carbon from volcanic sources is not precluded. Very low values of $\delta^{13}C$ (−6 to −17‰) have been established in the iron carbonates associated with magnetite rocks of the Biwabik iron-formation (Lake Superior, U.S.A.). In the rocks containing no magnetite, the $\delta^{13}C$ of the carbonates usually is higher than 4‰. In the opinion of Perry and Tan, such isotopic distribution is explained by oxidation-reduction reactions in diagenesis and metamorphism.

In several cases considerable variations in $\delta^{13}C$ of free carbon have also been established. Snezhko and Berezovskiy (1975) obtained values from −14.2 to −31.3‰ for sedimentogenic carbonaceous matter from the schists of the Upper suite of the Krivoy Rog group. Barghoorn et al. (1977) studied the variations in $\delta^{13}C$ in organic matter (kerogen) from the very slightly metamorphosed rocks of the Gunflint BIF (Lake Superior, U.S.A.). It turned out that in the shallow-water algal chert facies the $\delta^{13}C$ values were more negative (−25 to −30‰) than in the deeper-water cherty carbonate and taconite facies (−15 to −20‰). Lenses of anthraxolite were most depleted in ^{13}C (δ to −39‰). It is quite obvious that in interpreting isotopic data it is necessary to take into account both the kind of organisms and their physiological particulars and living environment, and the possible presence of abiogenic carbon. In addition, the original isotopic ratios of carbon, as of oxygen, could have been disturbed during metamorphism.

Thus available data on the variations in the carbon isotopic composition in Precambrian carbonates, graphitites, and organic matter confirm the important role of organic processes in the formation of the BIF, especially in the Proterozoic. In the Archean, apparently, organic life was more limited and processes of chemical deposition of carbonates prevailed.

Sulfur. A kinetic isotopic effect, accompanying the conversion of sulfate

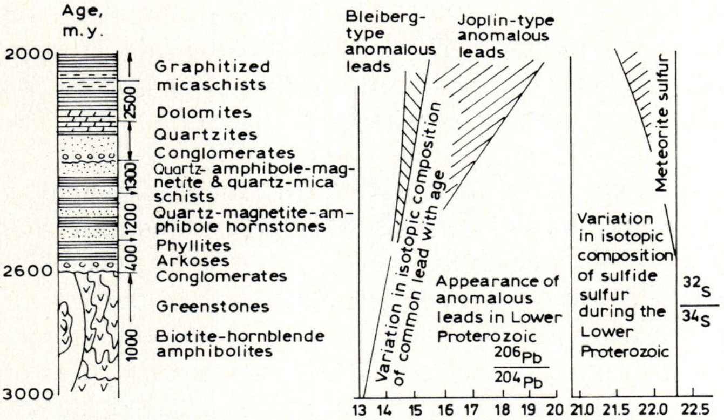

Fig. 28. Evolution of the isotopic composition of lead and sulfur in the Early and Middle Precambrian, as illustrated by a typical section in Central Ukraine (after Tugarinov and Voytkevich).

sulfur into sulfide sulfur, is produced in the biosphere by the life activity of anaerobic sulfate-reducing organisms. In the opinion of Chukhrov et al. (1969), these organisms could not have existed until after the appearance of a substantial amount of free oxygen in the atmosphere and the formation of sulfates, and organic matter was present.

According to the data of Tugarinov and Grinenko (1965), obtained for the Basal and Upper suites of the Lower Proterozoic of Krivoy Rog and the KMA, fractionation of sulfur isotopes was insignificant at the beginning of the Middle Precambrian (Fig. 28). This indicates a low sulfate content in sea water and an atmosphere with little free oxygen. By the end of the Middle Precambrian, fractionation had reached the present level. Subsequent detailed investigations on pyrite from the Lower Proterozoic and Archean rocks of the KMA and Krivoy Rog (Chukhrov et al., 1969) showed that the sulfides from the Mikhaylovka and Oboyan' groups of the Archean contain sulfur whose isotopic composition coincides with the composition of meteorite sulfur ($\delta^{34}S = +0.2$ to 1.0‰), whereas the sulfur of the rocks of the Kursk and Krivoy Rog groups shows signs of some isotopic fractionation ($\delta^{34}S = -6.1$ to $+3.6$‰), and high values of $\delta^{34}S$ (up to $+24.5$‰) are observed only in cross-cutting veinlets of pyrite. From these data it follows that isotopic fractionation of sulfur related to organic activity appeared in the Lower Proterozoic in the deposition of the sediments of the BIF of Krivoy Rog and the KMA.

Isotopic fractionation of sulfur ($\delta^{34}S = +8.2$ and $+3.1$‰) has also been established for some samples of sulfides from the graphite-rich ancient (~ 3 b.y.) shales of the Soudan formation (Lake Superior). Other determinations of the sulfur isotopic composition of Precambrian sedimentary rocks are known. A wide range of variations of $\delta^{34}S$ has been established in the cupriferous lagoonal-delta deposits of the Udokan group (1.7 to 2.1 b.y.) of the Olekma-Vitim hill country (Bogdanov and Golubchina, 1969), and somewhat less fractionation is observed in the phlogopite-iron ore deposits of the Aldan (Vinogradov et al., 1969).

Not long ago Goodwin et al. (1976) published new data on the distribution of sulfur isotopes in the volcanically enclosed Michipicoten and Woman River BIF (Canadian shield), $2.75 \cdot 10^9$ yr old. In the section of the Michipicoten series the average $\delta^{34}S$ value is +0.02‰, with variations from −10.5 to +10.1‰; in the samples from the Woman River series the average value is −1.3‰, with variations from −6.8 to +8.2‰. In the author's opinion, such large fluctuations over short distances indicate isotopic fractionation during biologic reduction of sulfates under anaerobic conditions in a restricted reservoir with low concentrations of sulfur. However, Chukhrov et al. (1977) believe that the question of biogenic fractionation of sulfur in the Early Precambrian remains open. In the oldest rocks, including the Onverwacht strata ($3.35 \cdot 10^9$ yr) and Fig Tree formation in Swaziland, the $\delta^{34}S$ of sulfates and sulfides do not differ greatly from mantle sulfur (from +3.1 to +3.8‰ in barite and from −0.1 to +1.1‰ in pyrite) (Perry's data). It is suggested that although photosynthesis already existed at that time ($> 3 \cdot 10^9$ yr), the total amount of organic matter in the sediments was insufficient for bacterial sulfate reduction to make oceanic sulfate isotopically heavier. With the development of life in the Precambrian seas and increase in the amount of organic matter in the muds, the prerequisites for the development of sulfate-reducing bacteria were created, but due to the insufficiency of O_2 the process of regeneration of sulfate from biogenic H_2S was only slightly manifested.

Thus the activity of sulfate-reducing bacteria was going on as early as the end of the Archean-beginning of the Proterozoic. However, it should be noted that it cannot be inferred directly from the isotopic data that a substantial amount of free oxygen appeared in the atmosphere at that time, although that is precisely the question raised by Chukhrov et al. (1977).

In the thermodynamic analysis it has already been mentioned that oxidation of the sulfide ion is possible at lower partial pressure of oxygen than in the case of oxidation of ferrous iron, and both these reactions are irreversible in the presence of even mere traces of free O_2. Therefore the suggestion of Cloud and Licari (1968) and Cloud (1969), that primitive organisms could have used "electron donor reactions" of the type:

$$Fe(OH)_2 + OH^- \rightarrow Fe(OH)_3 + e^- \qquad (2.25)$$

instead of reactions in which free oxygen participated, seems to be more sound. Under these same conditions green or purple sulfur bacteria oxidized

H_2S or S^{2-} to SO_4^{2-} in photosynthetic layers. As has already been mentioned, the sulfate content in the ancient ocean was lower, which also could have led to a reduction in isotopic fractionation of sulfur.

Evolution of life in the Precambrian

At the present time it can be considered generally accepted that organic life existed on the Earth at the time when the oldest sedimentary rocks were deposited. This conclusion was even formulated by Vernadskiy (1940, 1967): "A most important fact is that the biosphere existed during all the geologic periods, since their very oldest manifestations, since the Archean era". He is supported by numerous investigations devoted to the study of biogenic carbon in Precambrian rocks and the search for traces of ancient organisms. The problem of the distribution of free carbon in Precambrian sedimentary rocks has occupied many outstanding geochemists, the number of publications is very substantial, and the geography of the regions studied is continually being broadened. Therefore only some general information on the geochemistry of carbon in the Precambrian will be given below with a more detailed consideration of the BIF. Along with well known published data for the rocks of the BIF of the Ukrainian shield and particularly for the Krivoy Rog region, systematized data on the distribution of free and carbonate carbon are given for the first time.

Carbon in Precambrian sedimentary rocks

Numerous examples of the finding of graphite-bearing sedimentary rocks in the make-up of the Precambrian sequences are usually given in the latest geochemical review works.

Rankama (1957) wrote that finely divided carbon in crystalline schists of mudstone origin is much more widespread in Precambrian rocks than previously believed. The origin of this carbon probably is organic.

Mason (1971) points out that even in very old Archean formations slates are found which contain free carbon of possibly organic origin. In particular, such slates occur in the Gunflint BIF (Lake Superior), with an age of about 1.9 b.y.

An analysis of sedimentation in the Precambrian made by Tugarinov and Voytkevich (1966) on the basis of a summary of recent geochemical data gives a clear picture of the directionality and irreversibility of the geologic evolution of the Earth's crust. Whereas in the Early Precambrian the composition of the sedimentary-metamorphic complexes is characterized by the predominant development of basic and ultrabasic effusives, altered diabases, and spilites, in the Middle Precambrian an appreciable amount of organic matter appears in the thick piles of sedimentary rocks (Ukrainian shield, Canada, South Africa, etc.). The role of organic matter increases toward the end of the Middle Precambrian, when graphite-bearing schists appear. The importance of biogenic factors in sedimentation increases in the Upper Precambrian.

Detailed investigations of carbon in Precambrian rocks made by Sidorenko and Sidorenko (1968, 1970, 1971) showed that in all the rocks studied (Keyvy suite, the shungite-bearing slates of the Trans-Onega area, graphite-bearing gneisses and schists of the Krivoy Rog group of the Ukrainian shield and the Pri-Azov crystalline massif, etc.) the carbonaceous component was a sedimentogenic and synchronous inclusion in the rocks. Very high contents of free elementary carbon, undoubtedly representing metamorphosed organic matter, are characteristic of many originally sedimentary rocks. The carbonaceous matter typically is associated with argillaceous schistose rocks, gneisses, rarely with carbonate rocks (marbles and dolomites), and the graphite and graphite-like accumulations observed in the metasedimentary rocks are the remains of carbon which constituted only part of the carbon of organic matter present in the sediment from the beginning.

However, the carbon content observed in metamorphosed Precambrian rocks is appreciably lower than in post-Precambrian strata. In the opinion of Garrels et al. (1973) the reserves of organic matter $\sim 3.5 \cdot 10^9$ yr ago were about half as much as at present. According to estimates by Ronov and Migdisov (1970) the C_{org} content in the Lower and Upper Proterozoic complexes still remained at a low level, slightly increasing from the former to the latter, whereas at the Proterozoic–Paleozoic boundary there is observed a more than twofold enrichment in organic matter of all types of sedimentary rocks. Differences in the extent of metamorphic reworking of the rocks could hardly have led to the removal of so much carbon from the Precambrian rocks, inasmuch as numerous examples are known where concentrations of carbon were preserved during metamorphism, forming deposits of crystalline graphite in gneisses.

On the Ukrainian shield accumulations of graphite are confined to certain stratigraphic horizons of metasedimentary rocks, which indicates an originally sedimentary origin of the carbon. Levenshteyn and Polunovskiy (1971) believe that intensive enrichment of sandy-shaly-marly sediments in organic matter occurred under the conditions of a near-shore marine and lagoonal setting. In the graphite- and graphite-bearing gneisses formed from these sediments the carbon content reaches 10–30% or more, and the thickness of individual members and horizons in the section of the gneissic group of the Pri-Azov area reaches 100 m. Dinisenko and Kanigin (1970) regard the graphite gneisses as regular members of the Precambrian metamorphic sequences and believe that the distribution of carbon in the quartzites, gneisses, and marbles corresponds to the distribution of carbon in the sedimentary rocks.

Carbon in Precambrian BIF

Nearly all the metasedimentary rocks entering into the make-up of the BIF contain free carbonaceous matter, and its typical association with the argillaceous (shaly) component of the rocks is observed. Let us examine some typical examples.

Krivoy Rog district

Carbonaceous matter occurs in the rocks of all three suites of the sedimentary cherty iron-formation.

In the Lower suite (K_1) a small amount of finely dispersed carbonaceous matter is sometimes observed in the sandstones, but general incorporation of carbonaceous matter is typical of the layer of phyllites conformably overlying the sandstones (quartzites or arkoses) and conglomerates. The phyllites are quartz-sericite and quartz-biotite-sericite schists, very fine-grained, with a carbonaceous pigment. Sometimes beds enriched in carbonaceous matter, called graphite (graphitite)-quartz-sericite schists, occur within the phyllite sequence. The content of C_{org} in most of the phyllites is not high and falls within 0.1–1.0‰

In the *Middle suite (K_2)* carbonaceous matter is characteristic of clastic shaly horizons, layers, and individual bands within the cherty iron-formation. The chemogenic rocks (oxide and carbonate facies) are practically devoid of carbonaceous matter and the C_{free} content in them usually is not more than a few hundredths of a percent.

In the unoxidized rocks of the Middle (iron-ore) suite of the Saksagan district of the Krivoy Rog the content of C_{free} is 0.29% and of CO_2, 4.10%. These figures were obtained from 109 analyses of composite samples (the total number of original samples was more than 1000) of core from eleven boreholes. In the calculations it was assumed that the ratio of the masses of shaly and iron horizons was 2:1 and the total thickness of the middle suite, 1000 m.

The distribution of C_{free} and CO_2 in the stratigraphic section is shown in Fig. 29. The average contents of C_{free} and CO_2 in the iron horizons are 0.11 and 4.06% respectively, and in the shaly, 0.38 and 4.13%. Here it should be kept in mind that the iron horizons, together with the chemogenic iron-formations, always contain layers and intercalations of schist, and the shaly horizons are characterized by extensive development of barren and low-grade cherts and carbonate-chert rocks. The lack of correlation between the

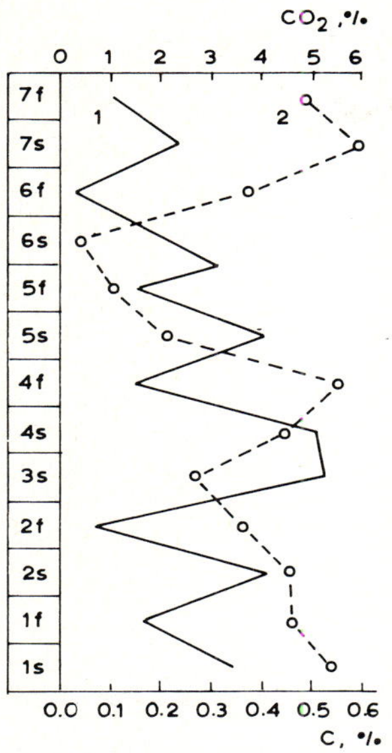

Fig. 29. Distribution of carbon (*1*) and CO_2 (*2*) in the stratigraphic section of the Middle suite of the Saksagan group of the Krivoy Rog. *f* = iron-rich horizons; *s* = slaty or shaly horizons.

contents of C_{free} and CO_2 in all the rocks is noteworthy, and the total amount of carbonates (CO_2) in the iron and shaly horizons is almost the same.

Free carbon occurs in the rocks studied in the form of spot accumulations a few microns in size with indistinct crystal outlines, perceptible only under the electron microscope.

X-ray and electron microscope investigations of pure fractions of the carbonaceous matter separated by means of chemical enrichment showed (Mel'nik et al., 1975) that it is graphite with a relatively ordered structure— from fully ordered graphite in the gneisses to graphites d_1 and d_{1A} in the schists (according to Landis' classification, 1971). The completely ordered graphite is characterized by an intensive sharp and symmetrical peak—the (002) peak— on the diffractogram at 3.35–3.36 Å; in the less ordered structures the peak is wider and asymmetrical. Disordered matter gives a very broad and faint peak corresponding to an interplanar spacing of 3.50–3.75 Å.

All the graphites studied are characterized by a rather distinct reflection at 3.35–3.37 Å, but both the interplane distance and especially the values of half-widths of the reflection appear to be dependent on the extent of metamorphism (Fig. 30). The least ordered graphite (half-width of reflection up to 32′) occurs in the slightly metamorphosed chlorite schists of the

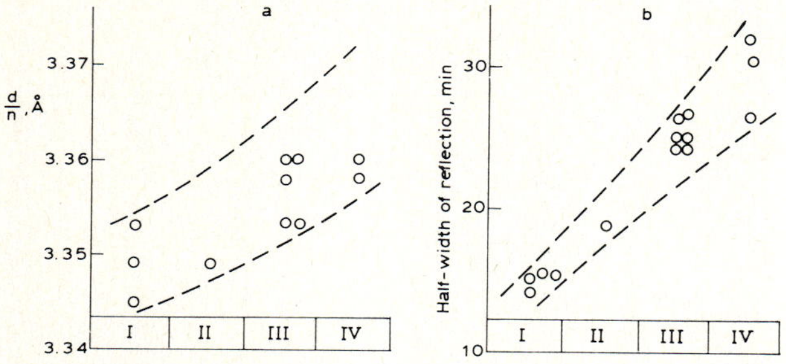

Fig. 30. Dependence of interplanar spacings (*a*) and sharpness (half-width) of peak (*b*) of the (002) reflection of graphite on the extent of metamorphism of the enclosing rocks in the Krivoy Rog basin: *I* = biotite gneiss (Petrovo); *II* = staurolite-biotite schist (Ingulets district); *III* = garnet-biotite and garnet -amphibole-biotite schists (Ingulets and Saksagan districts); *IV* = biotite-chlorite schist (South Mining and Beneficiating Combine and Saksagan district). After Mel'nik et al. (1975).

Saksagan district, the graphite of the garnetiferous schists of the Ingulets district is more ordered, and the structure of the graphite from the gneisses is completely ordered. A systematic change in the properties of graphite during metamorphism has been reported by other investigators (French, 1968; Diezsel and Offler, 1973).

Such dependence is still another proof of the originally amorphous nature of the carbonaceous matter and of its organic origin.

In the *Upper suite* (K_3) rocks with a high carbon content are widespread; according to the data of Kalyayev (1965) these rocks are part of the carbonaceous(graphitic)-clastic, carbonaceous-carbonate-lagoonal, and quartzite-conglomerate-molassic formations, with a total thickness of up to 4000–5000 m.

A horizon of carbonaceous-graphitic biotite and sericite schists is distinguished within the carbonaceous-clastic formation, and they occur in the lagoonal formation along with carbonate quartz-biotite schists. In addition to rocks in which graphitite is of minor importance, there are rocks in these sequences which consist almost totally of carbonaceous matter or graphitite, associated with thin veinlets of pyrite, quartz, and carbonate. In these schists the basic fabric of the rock is formed of fine-grained quartz and carbonaceous matter. Besides these schists, there occur carbonaceous-graphitite-quartz-carbonate rocks that consist of 60–70% carbonate which is intergrown with graphitite, and the carbonate is primary-sedimentary, deposited together with organic matter in lagoonal conditions. Kalyayev also studied carbonaceous matter in schists (Table V) in which the residual carbon content is more than 20%. It was established that the main mass of residual carbon in the samples probably was not graphite, but carbonaceous matter apparently of organic origin.

TABLE V
Results of analyses of carbonaceous-graphitic quartz-biotite schists (after Kalyayev)

Sample	Borehole	Amount of bitumen		Hydrogen content	Residual carbon content
		g	wt.%	wt.%	wt.%
755	9318	0.0931	0.09	0.49	7.59
910	8118	0.752	0.07	0.66	3.18
947	8123	0.0985	0.19	0.10	23.04
986	8527	0.1116	0.11	0.16	12.98

In the composition of the molassic formation a small amount of graphitic matter occurs in the cement of conglomerates.

Snezhko and Berezovskiy (1975) established that in the graphite-bearing rocks the carbonaceous matter is of sedimentary origin and synchronous with the rest of the rock-forming minerals. In the course of metamorphism not all of the matter was converted to the end product, graphite; part of it (up to 20–45%) was preserved in the form of hydrocarbons, about 30% of the samples studied showing bitumoids in luminescence-bitumen analysis. Ivantsív and Nenchuk (1974) attributed the carbonaceous matter from the K_3 rocks to graphitoids, which metamorphically are close to cryptocrystalline graphite with the following properties: $\rho = 1.90$–2.00 g/cm^3; specific electrical resistance 2.4–2.6 $\cdot$ 10^{-2} ohm-cm; C:H = 90–150; liberation of volatiles less than 1%; temperature of thermal effects—520–600°, 850–870°C (exoeffects) and 720–750° (endoeffect). According to these parameters the carbonaceous matter is identified as a material intermediate in character between meta-anthracite and subgraphite.

These data confirm the suggestion made earlier (Kalyayav, 1965) that although the carbonaceous(graphitic)-clastic formation cannot be considered to be a full-fledged analog of the caustobiolithic formations of the Hercynian and Alpine mobile zones, the carbon in these rocks probably was originally of organic origin, and the presence of bitumens and high pyrite content indicate formation in water bodies of lagoonal character, where conditions were suitable for the burial of very simple aquatic-plant organisms, mainly blue-green algae.

Kremenchug district

According to the data of Bordunov (1964), the BIF of the Kremenchug belt correlate with those of Krivoy Rog. A certain regularity in the distribution of carbon is observed, caused by cyclicity of sedimentation. Variations in the content of carbonaceous matter are related to the values of the oxidation ratio of the rocks and to zoning in the distribution of the iron minerals. The sandstones and light-colored schists that occur at the base of the oxide-facies member contain almost no carbonaceous matter, the oxidation ratio is high, and the iron minerals are represented by hematite and magnetite. In the dark two-mica and chlorite-biotite schists above them the C_{free} content increases to 0.84–1.19%, and the iron is completely fixed in silicates, and the oxidation ratio decreases. Higher in the section come ferruginous-silicate biotite-chlorite schists with intercalations of siderite-chlorite-quartz rocks; in this sequence the content of carbonaceous matter is maximal (1.65%), and the iron occurs in the form of silicates and carbonates. Then there appear chlorite-magnetite-quartz rocks; toward the center the number of schist intercalations is reduced, which in itself leads to a decrease in the content of alumina and of carbonaceous impurities. Above the iron-rich layers this mineralogical zoning is repeated in reverse order.

In the composition of the Ingulets suite, which is correlated with the upper suite of the Krivoy Rog district, carbonaceous-argillaceous schists which in places contain up to 7–12% of carbonaceous-graphitic matter are extensively developed.

Kursk Magnetic Anomaly

The regularities in the distribution of free carbon in the rocks of the BIF of the KMA have been studied in detail by Plaksenko (1966). The data obtained were fundamental for reconstructing the processes of sedimentation in the Precambrian (Strakhov, 1960) and the formation of ore mineral parageneses.

The carbon contents in the schists underlying the BIF and in the overlying schists (analogs of the schists of the lower and upper suites of the Saksagan group) are about the same as in the BIF of the Ukrainian shield.

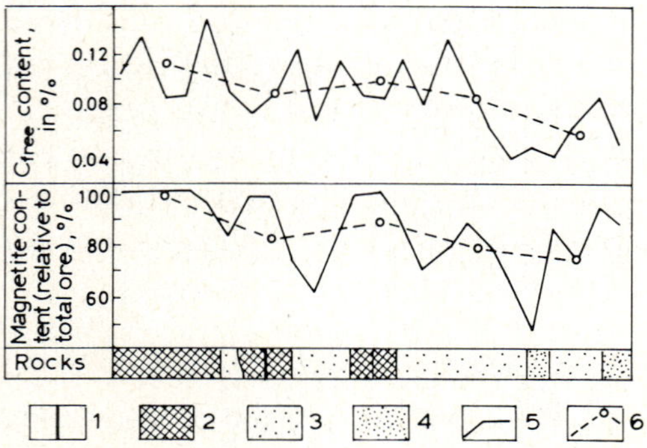

Fig. 31. Relationship between content of C_{org} and magnetite-hematite ratio in BIF of the KMA (Korobovskoye deposit, after Plaksenko): *1* = schist intercalations; *2* = dolomite-magnetite rocks; *3* = iron mica-magnetite rocks; *4* = magnetite-iron mica rocks. Sample data: *5* = seriate; *6* = combined, consisting of seriate.

TABLE VI
Free carbon content in various rocks of the facies profile of the BIF of the KMA (after Plaksenko)

Component	Iron-rich rocks					
	Underlying rocks	Carbonate-Mgt ores w. sulfides	Barren or low grade	Cum-Mgt	Do-Mgt	Hem-Mgt
C_{free}	0.447	0.82	0.22	0.19	0.14	0.086
Number of analyses	32	4	48	42	92	51

In addition to conglomerates and quartzites the underlying rocks contain a sequence of two-mica and muscovite (sericite) schists with a small admixture of free carbon (0.2–0.4%). The sericite-biotite phyllitic schists immediately beneath the iron-ore sequence often are carbon-bearing (of carbonaceous schist type). The average carbon content in the underlying schists is 0.45%.

The BIF. Plaksenko (1966) showed that the amount of free residual carbon in the BIF correlated with the relative content of magnetite and hematite (Fig. 31), the minimal concentrations of C being established in essentially hematitic rocks although the variations of individual determinations were rather considerable. A gradual increase in carbon content on passing from the deeper-water hematite facies to less deep-water facies (Table VI) is more clearly established, with a distinct maximum in the schists

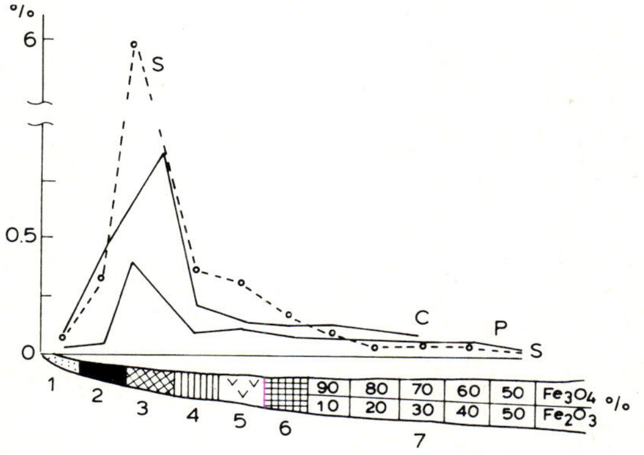

Fig. 32. Distribution of sulfur (*S*), phosphorus (*P*), and organic carbon (*C*) in the facies profile of the BIF of the KMA (after Plaksenko): *1* = arkosic metasandstones, barren blastopsammitic quartzites; *2* = clay shales, phyllites, crystalline schists; *3* = carbonate-magnetite rocks with sulfide ores; *4* = non-ore and low-grade iron-formations with carbonates and silicates; *5* = silicate-(cummingtonite)-magnetite rocks with various proportions of magnetite and hematite; *6* = dolomite-magnetite iron-formations; *7* = hematite-magnetite rocks with various proportions of magnetite and hematite.

of the barren facies (fine argillaceous sediments), if the special and not very widespread ores of the sulfide-carbonate-magnetite facies are not taken into account (Fig. 32). The regular distribution of C_{org} according to the grain size of the sediments indicates that it accumulated at the expense of plankton.

The *overlying sequence* consists of a complex of rocks similar to the rocks of the Krivoy Rog Upper suite. The carbonaceous-biotite schists are richest in free carbon (up to 8.5%). Its average content is 4.6%, i.e. 10 times greater than in the schists of the underlying sequence, which are similar in conditions of formation, grain size, and chemical composition.

Karelia

In the sections of the schist-leptite BIF of Karelia, sequences of metamorphosed aluminosilicate sedimentary rocks containing layers of sulfide-bearing graphitic rocks are distinguished, among which there occur quartz-biotite and biotite-quartz graphitic schists, feldspar-biotite-quartz gneissic schists with graphite, and sericite graphitic schists (Chernov et al., 1970).

In the quartz-biotite schists the content of sparsely disseminated graphite is between 2 and 5%. It occurs in layers; usually the dark gray, essentially graphitic bands are interbedded with lighter quartz and quartz-biotite bands.

The biotite-feldspar-quartz gneissic schists are highly recrystallized and contain somewhat less graphitic material (1 – 3%, in isolated cases up to 5%). The graphite forms flocculent particles 0.01 – 0.05 mm in size, sometimes concentrated in indistinct patches and bands. On the basis of thermographic and X-ray investigations the carbonaceous matter is attributed to a transitional variety of shungite-graphite.

Canadian shield districts

Cloud et al. (1965) call the Soudan formation, which contains carbonaceous layers, one of the oldest carbonaceous deposits on Earth (at least 3 b.y. old). Individual layers and lenses in this sequence are so rich in carbon that they resemble pure graphitized coal in appearance. Chemical, spectrophotometric, and X-ray fluorescence investigations suggest a possible organic origin of the carbon. Under the electron microscope a kerogen-like material is found which can be interpreted either as the remains of tissue or as a crystal formation. Gross (1973) mentions that black carbonaceous slates are typical of both the clastic-sedimentary BIF of Lake Superior type and BIF of Algoma type, associated with volcanic rocks.

In the Mesabi, Gunflint, and Cuyuna districts (Minnesota, Lake Superior) bands of graphititic argillites occur in the section of the BIF (Morey, 1973).

According to Goodwin's data (1973), thin zones of black carbonaceous sulfide-bearing cherts, up to 30 feet thick and several hundred feet long, are also observed in the Helen district (Michipicoten, Lake Superior).

Other areas of the world

The BIF of the greenstone belts of Southern Africa are very old (older than $3.3 \cdot 10^9$ yr). Beukes (1973) mentions that in the lower part of the section of the Hooggenoeg formation of this belt, carbonaceous cherts completely devoid of iron appear in a thick sequence of metamorphosed tholeiitic basalts with intercalations of ultrabasic and acid lavas. The first signs of iron appear in the upper part of the section in a banded chert directly underlying the BIF. In this chert, small segregations of iron-rich material are closely associated with black bands of carbonaceous chert or replace them.

In a new summary work by Dimroth and Kimberley (1976) devoted to a comparison of the distribution of carbon and some other elements in Precambrian and Phanerozoic sedimentary rocks, it is mentioned that in most Precambrian sequences the normal negative correlation between grain size and C_{org} concentrations is established regardless of the depositional environment. That regularity indicates the great importance of the role of plankton in the accumulation of C_{org}. Garrels et al. (1973) came to the same conclusion.

Organic compounds in Precambrian sedimentary rocks

In recent years some data have been obtained on the content of organic compounds (hydrocarbons, carbohydrates, paraffins and aromatic compounds, alcohols, etc.) in Precambrian sedimentary rocks, including the rocks of the cherty iron-formations.

Sidorenko and Sidorenko (1970) made a detailed study of carbon-containing rocks, including the graphitic gneisses of the Pri-Azov area and Krivoy Rog (Table VII). It was established that despite the profound

TABLE VII
Characteristics of bitumens from graphite gneisses of the Ukrainian shield (after A. and S. Sidorenko)

District	C_{org} (%)	Content of bituminous matter (%)	Element analysis of chloroform extract		Brief interpretation of infrared spectra of bituminous matter of chloroform extract
			C	H	
Pri-Azov area	10.3	0.004	70.16	10.52	Paraffin and aromatic structures (about equally abundant). Main part—non-carbon structures
Pri-Azov area	10.8	0.003	71.57	10.15	Naphthene-aromatic structures with short paraffin chains. Of nonhydrocarbons—complex ester and alcohol groups
Krivoy Rog	13.7	0.005	not	det.	

metamorphism, the Precambrian rocks contain a certain amount of bitumens and hydrocarbons which are the products of metamorphism of original sedimentary organic matter. Bituminous substances soluble in organic solvents were present in all the rocks investigated, in amounts from hundredths to whole percents. Hydrocarbons also were always found in sorbed gases. Methane and ethane have been found in the gases from samples from Krivoy Rog ($C_{org} = 0.1\%$) (Sidorenko, 1971). In the carbonaceous schists of the Upper suite of Krivoy Rog, which sometimes contain more than 20% carbon, the amount of bituminous matter is about 0.1%, but the nature of the bitumens is not adequately known. Kalyayev (1965) studied the hydrocarbon content is some schists and petroleums from 3 to 2.7 b.y. old. By means of gas chromatography and mass spectrometry, C_{18}, C_{19}, C_{20}, and C_{21} isoprenoids-alkanes have been established in the slates of the Soudan formation in Minnesota (U.S.A.); their finding indicates that there were life processes in that period of time.

Swain et al. (1970) made a detailed study of carbohydrates from Precambrian rocks, including slates of the BIF. Carbohydrate remains were found in eleven samples from Early to Late Precambrian rocks and fossils. The authors believe that the acid-extractable (polymeric) sugars they found are indigenous to the rocks themselves, not introduced from outside. However, in two Lower Precambrian samples more than 2.8 b.y. old (from the Soudan and Coutchiching districts, U.S.A.) no starch remains showed up, although traces of cellulose and laminarin were found in the sample from Coutchiching. In the authors' opinion this might indicate that structural polysaccharides of cellulose type and sugars of laminarin type already existed in the Early Precambrian, but there were not yet compounds of starch type. In the Cuyuna BIF from the Huronian Animikie system, 2.2 b.y. old, several samples of the upper slate series (argillites) were studied; a similar assemblage of compounds was found. Samples from the Biwabik belonging to the same system contained acid-hydrolyzable monosaccharides. It is typical that the ratios of acid-hydrolyzable sugars found in Precambrian rocks are close to the ratios established for various marine and nonmarine algae (including Recent ones).

Let us mention a few other works of recent years. Nagy et al. (1975) presented results of a study of traces of ancient manifestations of the life activity of organisms. In the rocks of the Onverwacht series (3355 m.y. old) in South Africa a condensed aromatic substance was found; in the rocks of the Fig Tree group, aliphatic and aromatic hydrocarbons; and in the stromatolite formations of Transvaal, kerogen residual products probably including alkaloids and also complex organic compounds of the furan, benzonitrile, and aliphatic nitrile groups, which could have been formed by decomposition of proteins and chlorophyll.

Ozonolysis of insoluble organic matter from the cherts of the Onverwacht group showed (Brooks et al., 1973) that it is related to sporopollenin, a polymer of biologic origin. Data on the finding of organic compounds of possible biogenic origin in Precambrian rocks, including BIF, are also given in other works (Harrington and Cilliers, 1963; Fiebiger, 1975; Voytkevich et al., 1975; Schidlowski et al., 1976; Kolesnikov and Yegorov, 1977).

The development of analytical methods has led to a sharp increase in the number of publications in the field of organic geochemistry of ancient sediments. However, it must be kept in mind that there are no clear-cut criteria for the recognition of organic matter of biogenic origin. In some cases, as McKirdy (1974) mentions in his review work, an abiogenic origin of both the carbon and the "organic" matter cannot be ruled out.

Traces of organisms in the rocks of the cherty iron-formations

In a review by Voytkevich and Belokrys (1960) it was concluded that on the basis of present data on Precambrian life it can be said that fossil remains of relatively highly organized animals are of great antiquity and remains of plants of still greater antiquity. It is presumed that traces of the existence of living organisms can be found even in the oldest rocks, with an age of 3000–3500 m.y. No animal remains have been reliably established in the Precambrian BIF, but traces of ancient plants and bacteria have been described repeatedly. In particular, the finding of accumulations of traces of autotrophic iron bacteria in iron-formation from the Proterozoic Greben suite on the Angara River (Vologdin, 1969), remains of bluegreen algae in the Huronian deposits of Michigan and Minnesota (Gruner, 1923), and also in the cherts of the Gunflint formation in Canada (Tyler and Barghoorn, 1954) has been reported.

Detailed petrographic investigations have shown (La Berge, 1967) that structures of possible organic origin are extremely common in many Precambrian BIF. These structures are chiefly spheroids, ranging in diameter from 5 to 40 μm. Specific types of structures usually are of similar size and occur in regions far apart. The structures in the Lake Superior district have been studied in most detail; the first finds of possible macro- and microfossils were known more than 60 yr ago. Even in early works (Leith, 1903; Grout and Broderick, 1919) it was suggested that some structures in the Biwabik iron-formation might be organic. Hypotheses of the possible relationship between the deposition of the cherty iron-formation and the activity of various primitive organisms were offered more than once after that(Moore, 1918; Harder, 1919; Gruner, 1922, 1923, 1925), but few proofs were given of the occurrence of definite fossils.

In 1953 Tyler found extremely well preserved organic structures in the

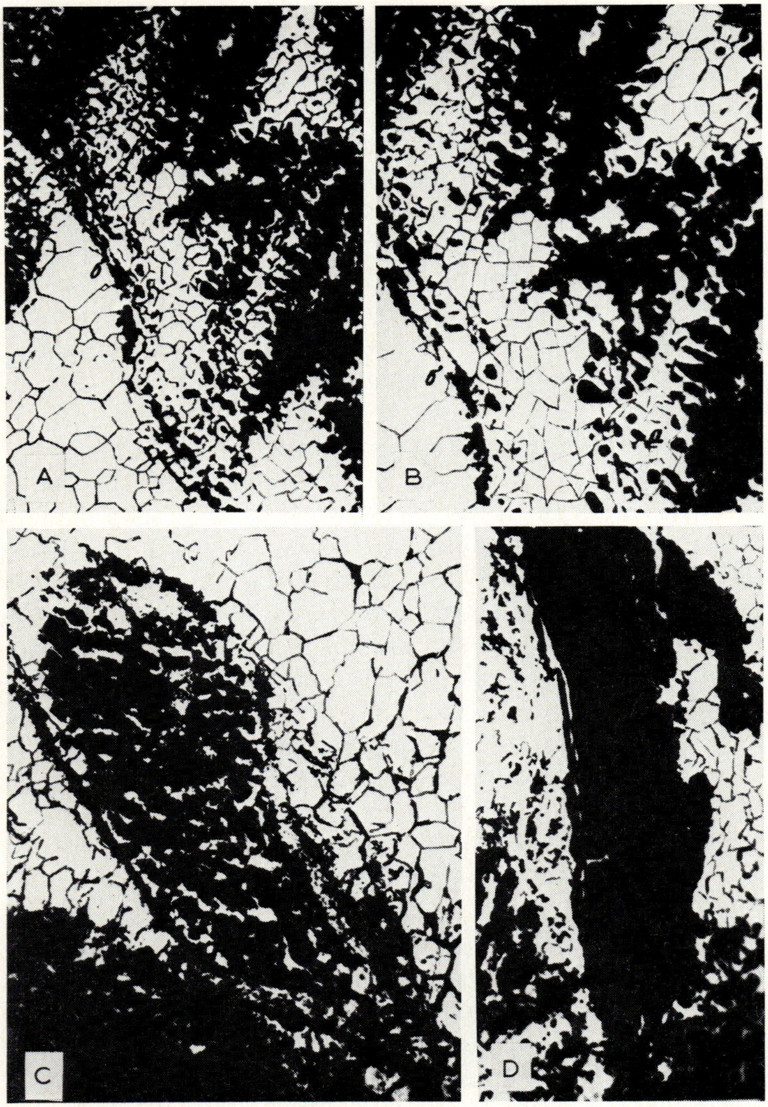

Fig. 33. A. Longitudinal section of remains of the alga *Protospira strygini* Vologdin, gen. et sp. n., in thin-section, ×60; B. as A., ×120; C. longitudinal section of corallite of *Aseptalia ukrainika* Vologdin, gen. et sp. n., internal space filled with remains of single-celled planktonic algae; thin-section, ×120; D. longitudinal section of *Aseptalia* sp. with signs of fibrous structure of internal layer of the corallite, thin-section, ×60 (Vologdin and Strygin, 1969).

eastern part of the Gunflint district in Ontario. In the paper on his discovery (Tyler and Barghoorn, 1954) and in a more detailed work (Barghoorn and Tyler, 1965) the great diversity and wide distribution of fossil remains in the basal algal cherts of the iron-formation were mentioned. As a consequence of these works a careful search for organic structures in relatively slightly metamorphosed iron-formations, shales, and carbonate rocks began all over the world (Moorhouse and Beales, 1962; Cloud, 1965a, b, 1968, 1969). Edgell (1964) described possible algal structures in the Archean Soudan formation in Minnesota. In particular, microstructures which in shape and size might represent bacteria or bluegreen algae were observed in pyrite from the carbonaceous shales. These structures, and also geochemical and isotopic investigations and the presence of molecules of biogenic hydrocarbons in the carbonaceous rocks, make it possible to say that the oldest autotrophic organisms and possible photosynthetic activity appeared more than 2.7 b.y. ago.

In the rocks of the Krivoy Rog group Besedin (1958) observed some "problematic" remains of organisms. In the oxidized martite ores remains of spores, shreds of plant tissue, and iron bacteria were found in substantial amount, the spores constituting an assemblage characteristic of Devonian sediments (Ishchenko, 1957). However, a control analysis of the carbonaceous schists showed that the spores were washed into the older rocks along cracks (Shatskiy, 1957). Investigations of phyllitic material from the Middle and Upper suites of the Krivoy Rog group (Timofeyev, 1966) showed that the spore-like formations (*Protosphaeridium laccatum* etc.) found in the rocks belong to the Upper Proterozoic.

In 1965 Strygin found well preserved remains of organisms in conglomerates (Upper suite of the Krivoy Rog group), which were described in 1969 by Vologdin and Strygin (1969). In particular, remains of bluegreen algae in association with remains of coral-like representatives of invertebrate animals were identified in a pebble of marmorized limestone (Fig. 33).

Most of the described remains of ancient organisms were not found in the iron-formations, but in the surrounding rocks (phyllites, carbonaceous shales) or even in the rocks enclosing cherty iron-formation proper (carbonate rocks, conglomerates, shales). Therefore the descriptions of microscopic structures found in iron-formations that have appeared in recent years are of special interest.

Results of study of Precambrian BIF in the U.S.A., Canada, Australia, and South America are summarized in a work by La Berge (1967). Numerous spherical structures, easily perceptible if there is pigment in them, have been found in cherts. The author distinguishes seven types (varieties) of structures.

The most common are structures of type A, which are spheroids 15–40 μm in diameter (Fig. 34). In well preserved samples the surface consists of

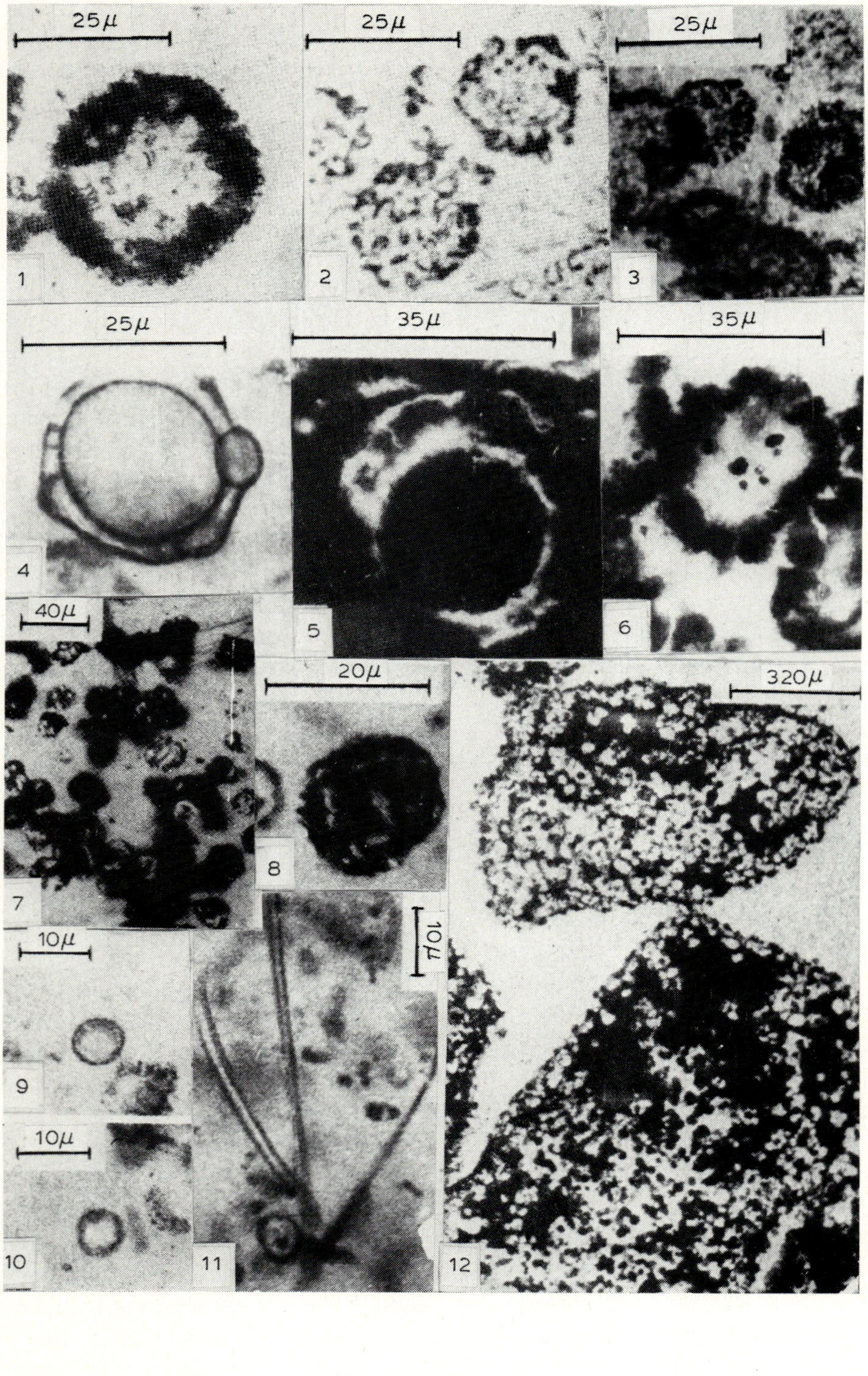
25μ
25μ
25μ
1
2
3
25μ
35μ
35μ
4
5
6
40μ
20μ
320μ
7
8
10μ
10μ
9
10μ
10
11
12

extremely fine worm-like fibers less than 0.5 μm in cross-section, curved or looped. The structures usually are pigmented with black X-ray-amorphous carbonaceous material. In less well preserved structures the pigment is bleached out (oxidized); sometimes replacement of the organic pigment by extremely fine silicate (greenalite-like) or hematitic material is observed. Nearly all the structures of this type that have been described were found in the silicate (greenalite or minnesotaite) facies of iron-formations of Archean age.

Type B structures, which usually occur in cherts (jaspers) of different regions, are interesting. The structures consist of spherical central masses of more or less pure chert about 15–25 μm in diameter, surrounded by a variable number of randomly arranged ellipsoidal bodies (up to 10) 5–8 μm in diameter. The satellite ellipsoids usually are pigmented with a variable amount of extremely fine hematite. These structures, found especially frequently in bright red jaspers, may be *Eosphaera tyleri* (Barghoorn and Tyler, 1965). The replacement of the walls by hematite probably was accomplished in several ways, and the whole structure may be replaced or preserved, or separate fragments of it. Complete replacement of a spheroid by finely divided hematite leads to very dense, almost opaque formations (see Fig. 34).

Spheroidal bodies preserved by carbonaceous matter are assigned to types C and D.

The smaller spheroidal and ellipsoidal bodies of type E, 4–8 μm in size and preserved due to impregnation with carbonaceous material, fine-grained hematite, or greenalite, differ from the structures that have been described. These structures are found in black cherts, red jasper bands, and granule-containing greenalite cherts (see Fig. 34). The algal structures at the base of the Biwabik formation also contain unique filiform structures of type F; their diameter is 1–2, sometimes up to 6 μm.

Fig. 34. Relict organic structures in Precambrian cherty iron-formations (after La Berge). *1* = type A structure, well preserved due to carbonaceous matter in chert matrix; note fibrous surface and clear interior (lower chert unit of Biwabik Iron Formation, Pilotac mine, Minnesota); *2* = type A structure, largely replaced by greenalite; *3* = type A structure, preserved due to unidentified brown substance (hematite?) in chert matrix (Belcher Islands iron-formation); *4* = type B structure (*Eosphaera tyleri*), well preserved due to carbonaceous matter in chert matrix (Gunflint district); *5* = type B structure in which most of the organic matter has been replaced by extremely fine hematite (Vicar mine, Gogebic district); *6* = type B structure in which only satellite ellipsoids have been replaced by fine hematite (same locality); *7* = type C and D structures in various states of preservation; some are highly pigmented, others contain only relicts of carbonaceous pigment, sometimes with stilpnomelane impurities; chert matrix (Brockman Iron Formation); *8* = type D structure, densely pigmented by carbonaceous matter, in chert (same locality); *9* = type E spheroidal body, preserved by carbonaceous matter, in chert matrix (Gunflint district); *10* = type E spheroidal body preserved by finely divided hematite (Erie mine, Mesabi district); *11* = type F filiform and spheroidal structures, preserved by carbonaceous matter, in chert matrix (Gunflint district); *12* = greenalite granules full of 20- to 30-μ spheroidal structures. Dark material is mainly greenalite, white is chert (Gunflint district).

The data of Cloud and Licari (1968) support La Berge's conclusions. Spherical structures 4–10 μm in size, similar to "*Huroniospora*", are found even in the very old rocks of the Soudan BIF. In the Proterozoic formations of Gunflint and Biwabik (Lake Superior) and in the Pretoria series (South Africa), spherical forms which probably were pelagic organisms living in the water layer near the bottom of the photozone are extensively developed in the banded iron-formations.

Let us mention a few other finds of presumably organogenic forms in ancient rocks.

Nagy et al. (1975) established that there are no identifiable microfossils in the 3800-m.y.-old metaquartzites from Greenland, but globular graphite formations are found and they contain methane. It is not ruled out that the methane is of inorganic origin and that the globules were formed as a result of reactions between the methane and iron oxides. Round and ellipsoidal bodies in the rocks of the Onverwacht group of Southern Africa are probably of inorganic origin, although aromatic hydrocarbons and n-alkanes also occur in them. However, in the opinion of Brooks et al. (1973), the spheroids (7–10 μm) and threads (diameter 4–8 μm, length up to 20 μm) found in the cherts of this series (age 3300–3400 m.y.) undoubtedly represent the remains of living organisms. The presence of bacterial forms of life, traces of filiform and coccoid bluegreen algae, and remains of decomposed cellulose have been definitely established in rocks of the Witwatersrand (age 2600 m.y.) in South Africa (Nagy et al., 1975).

Study of the distribution of the structures has shown that minute spheroidal and ellipsoidal bodies are characteristic of almost all BIF, both Archean and Proterozoic, but there are none in rocks that have undergone appreciable metamorphism. These structures are observed in all types and facies of BIF; they have been identified in cherts which contain hardly any iron minerals and in ore jaspilites. Many jaspers apparently actually consist of spherical structures, clearly outlined by hematite dust. Only in rocks where hematite forms larger flakes are no spherical structures encountered. Thus organic life was very abundant at the time and place of deposition of the BIF, and the iron-formations themselves were deposited as the result of biological processes. It is supposed that these very simple organisms could have reacted with an oxygen acceptor—Fe^{2+}—in solution, and the compounds of trivalent or trivalent-divalent iron that were formed were deposited along with a rain of dead and sometimes impregnated phytoplankton. All these components were interbedded with chemically formed silica gel.

Stages of development of life in the Precambrian

Modern ideas on the evolution of life on Earth are intimately related to ideas on the evolution of the ocean and atmosphere. In many works it is postulated that the qualitative jump at the Precambrian–Cambrian boundary was caused by the appearance of traces of abundant animal life in the marine sediments of the whole world (Rutten, 1962). Berkner and Marshall (1965) have suggested that the beginning of Cambrian time marks a threshold in the evolution of the atmosphere—the attainment of about 1% of the present content of oxygen or P_{O_2} (Pasteur point), at which the level of respiration of oxygen became suitable for the maintenance of animal life. The next qualitative jump in the evolution of the organic life, which took place in Devonian and Silurian time, marks the stage when the level of P_{O_2} in the atmosphere rose to 10% of the present level and because of this, dangerous ultraviolet radiation decreased to a tolerable level at the Earth's surface, which led to colonization of the land by organisms (Fisher, 1965).

During the greater part of the Precambrian, a dangerous amount of ultraviolet radiation penetrated a layer of water up to 10 m deep, due to the absence of oxygen and of the ozone layer in the atmosphere. In the Late Precambrian–Early Cambrian the thickness of this layer might have decreased to a few centimeters, and tolerable subaerial levels of radiation were reached in the Silurian. However, it is not ruled out that at that time adaptive organisms might have existed in the danger zone. As a result, it is presumed that only marine organic life existed in the Precambrian. Until the Silurian only isolated zones in the soil probably were inhabited (so-called colonization of sheltered oases), later the coastal belt flooded by the tide was settled, and in the Silurian the first land plants and animals appeared.

Photosynthesis apparently began more than 2.7 b.y. ago (Korde, 1974; Sokolov, 1975; Timofeyev, 1977), but the greater part of the oxygen was used to oxidize inorganic matter in the atmosphere and on the Earth's surface, and also went into supporting the life activity of the plants themselves.

Cloud (1973) points out that the predominant types of organisms were prokaryotes (very simple bacteria without nuclei and bluegreen algae), which could easily withstand the harmful action of ultraviolet radiation. They also are the only organisms which as a rule live in an anaerobic environment without aerobic hosts and are capable of fixing nitrogen. An anaerobic environment is an important condition for the emergence of life, as free oxygen is toxic to all forms of life in the absence of appropriate enzymes for its assimilation, and life could not have arisen in the presence of oxygen. According to Timofeyev (1977), in the Precambrian microplankton proliferated in basins with normal salinity, in the shelf zones. Inasmuch as bluegreen algae disappear at a pH of less than 5 (Brock, 1973; Shapiro, 1973), the environment was close to normal, which is in good agreement with the thermodynamic model (Fig. 24). The upper temperature limit for the life of photosynthetic prokaryotes is 73–75°C (Brock, 1967), thus the tempera-

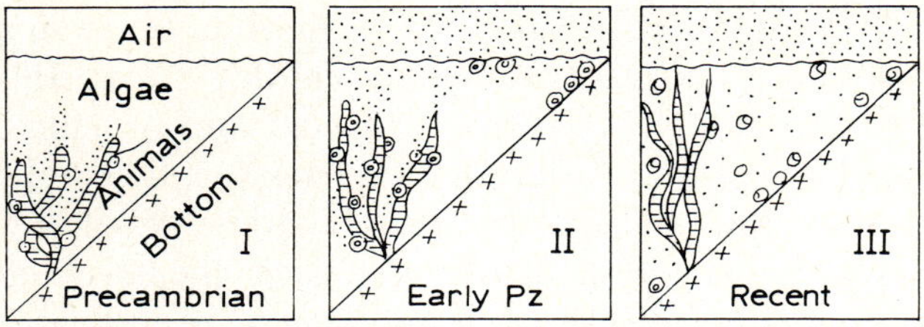

Fig. 35. Stages in the evolution of animals (after Fisher): I = animals live in total respiratory dependence on the number of plants; the atmosphere is practically devoid of free oxygen. II = the atmosphere reaches the "Pasteur point", animals can leave plants but seek water-air interface. III = the atmosphere and water contain a large amount of oxygen, animals are widespread.

ture range of life activity would have been fairly wide. Before the pasteur point was reached (according to Cloud, between 1.2 and 0.6 b.y. ago) animal life was very primitive; aerobic bacteria had developed, but the first Protozoa evolved as a function of the number of plants. These very simple organisms were active when oxygen was plentiful, and abated or curtailed enzyme activity when there were no plants. In the opinion of Fisher (1965), primitive types of Metazoa evolved in similar conditions, including the ancestral sponges, turbellarians, and coelenterates; these organisms were small and unprotected, they lived outside the ultraviolet radiation danger zone, and it is very unlikely that remains of these organisms were preserved in rocks. Figure 35 schematically shows the stages in the evolution of animals. Fisher proposed a general scheme of evolution of organic life,

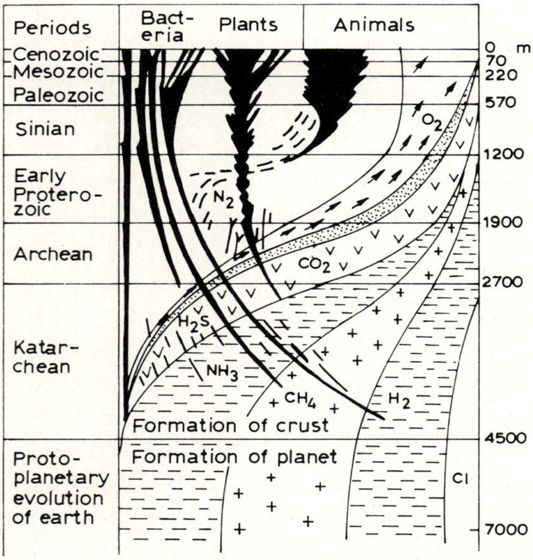

Fig. 36. Evolution of the environment and of life on earth (after Vologdin).

drawn up on a model of the evolution of the atmosphere. A similar scheme, but in a more general and more schematic form, is also given by Vologdin (1969) (Fig. 36).

On the basis of the data of the Fourth International Conference on the Origin of Life (1973), Lopuchin (1974) gives the following more precise scheme of the evolution of organisms (in billions of years ago):

(1) Origin of prokaryotes (mainly of heterotrophic life)—more than 3.3.
(2) Origin of autotrophs (mainly bacterial)—3.2 ± 0.2.
(3) Origin of photoautotrophic bluegreen algae—3.0 ± 0.2.
(4) Formation of an atmosphere with a small oxygen content—2.1 ± 0.3.
(5) Origin of mitotic (asexual) eukaryotic algae—1.6 ± 0.2.
(6) Origin of mitotic eukaryotic algae (with separation of sexes)—1.1 ± 0.2.
(7) Origin of diploid benthic mega-algae—0.75 ± 0.05.
(8) Origin of primitive (chiefly pelagic) eumetazoa—0.70 ± 0.05.

Commoner (1965) suggests that the first organisms probably were anaerobic heterotrophs which drew energy from partial oxidation of geochemically produced organic matter (substrate). The transition to organisms in which light absorption intensified metabolism should be considered the next stage, and then photosynthesis, a process which goes on with the assimilation of carbon. The primitive anaerobic photosynthetic organisms probably were the ancestors of organisms capable of photolysizing water to form molecular oxygen. As a result, the primitive aerobes determined the metabolic conditions for terrestrial and aquatic organisms which prevail in the present biosphere.

The general scheme of gradual development was complicated by periodic "flare-ups" and "extinctions" of biologic activity, possibly caused by cosmic factors. In the opinion of Korde (1974), photosynthetic activity of bluegreen algae began 2.8–2.6 b.y. ago; at the Precambrian–Cambrian boundary the bluegreen algae were superseded by red algae. The Vendian–Lower Cambrian flare-up of life is explained not only by the gradual accumulation of oxygen in the atmosphere of the Earth (as in the concept of Berkner and Marshall), but also by a sharp reduction in intensity of cosmic radiation in the period from 900 to 300 m.y. ago. A new intensification of radiation at the 300-m.y. boundary no longer had such effect on the animal world of the planet, inasmuch as by that time the "ozone shield" already existed.

Chapter 3

Conditions of migration and deposition of iron and silica

Introduction

At the present time fairly complete and reliable information exists concerning the state of iron and silica in solution, forms of transport, conditions of deposition, and formation of ore concentrations. This information mainly includes data from geochemical study of recent waters and sediments and the results of experimental laboratory investigations. Comparison of the geologic and physicochemical data makes it possible to obtain a finished picture of the processes that occurred in the not very distant geologic past. The classic works of Strakhov (1947, 1960, 1963), based on a profound understanding of the geochemistry of sedimentation, and the works of Garrels (1962) and Garrels and Christ (1968), who continued along the path via the precise laws of thermodynamics to a theory of lithogenesis and ore formation, are most noteworthy at the present stage of development of geologic science. It is obvious that these directions must be made to converge further, primarily by making the idealized thermodynamic formulations, usually formal and simplified, approach as closely as possible to actual natural phenomena and objects. The extrapolation of conclusions obtained in physicochemical analysis of recent processes compared to the Precambrian, with its special conditions of sedimentation and different composition of the atmosphere and hydrosphere, is of considerable complexity. By keeping in mind the special features of the Precambrian, constraints can be placed on genetic hypotheses and the conditions of accumulation of the BIF can be represented more definitely.

In this chapter the experimental physicochemical and the theoretical thermodynamic data enabling us to obtain an overall picture of the relationships between the ions and molecules in the solutions and the various sediments are examined. All the thermodynamic calculations were made on the basis of the systems of consistent constants first worked out by us (Mel'nik, 1972b), including the oxides, carbonates, and silicates of iron. As new experimental data have been obtained, the constants have been refined and supplemented (Tardy and Garrels, 1976; Klein and Bricker, 1977, Mel'nik and Radchuk, 1977b). The consistent thermodynamic constants used in this work are given in the Appendix. The plotting of the "stability" diagrams of the sediments also is essentially different, corresponding more closely to natural sedimentation than stability diagrams of crystalline minerals.

The results obtained in studying colloidal forms of transport are less certain and not always reproducible. Therefore special experimental investigations of the stability of colloidal solutions of iron, silica, and mixtures of the two were conducted in conditions as close as possible to nature, in addition to summarizing and critically examining published data.

In a separate section information on the geochemistry of iron and silica in recent waters (rivers, seas, oceans) and on the conditions of their deposition and the formation of concentrations is summarized. Special attention is paid to the role of organic compounds in the migration of iron, and volcanogenic-sedimentary processes are also analyzed thoroughly.

In conclusion, experimental and theoretical thermodynamic and geochemical data are compared and possible factors causing the formation of concentrations of iron and silica sediments in various conditions, including the specific setting of the Precambrian, are examined. The main purpose of this was not to create any universal hypothesis of the formation of the Precambrian BIF, but to determine the conditions which, when fulfilled, make the ore-forming process possible.

Physicochemical data on the forms of transport and conditions of deposition of iron and silica

In this section, the conditions of migration of iron and silica first in ionic and then in colloidal solutions are examined. Molecular forms—uncharged and unpolymerized particles existing stably in solution—also belong to the ionic forms.

Migration and deposition of iron and silica from ionic solutions

Forms of transport of iron

In aqueous solutions iron can occur in two valency states, Fe^{2+} and Fe^{3+} (Latimer, 1954). In the presence of anions, iron forms numerous complex ions. Preliminary assessment of the role of sulfate and chloride complexes in the transport of iron has shown that these compounds are stable only in very acid environments. Carbonate and bicarbonate complexes are unstable (Singer and Stimm, 1970), and phosphate complexes are of no importance in connection with the low concentrations of phosphoric acids in natural waters. Thus the main role in the migration of iron is played by simple and hydrated ions (Kalinin and Kashik, 1968). Unfortunately the stability of

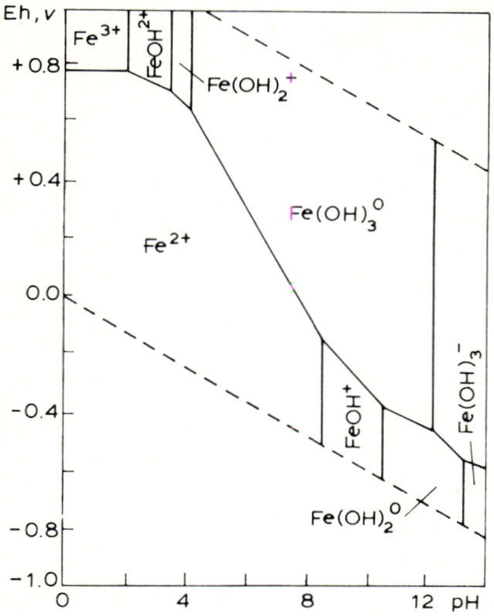

Fig. 37. Stable ions of iron in the system Fe–H_2O (*Eh*-pH diagram): dashed lines—boundaries of field of stability of water at $P = 1$ bar (on all diagrams of ionic equilibrium).

complexes of iron with natural organic acids (of humic type) and of metallo-organic compounds has not yet been studied.

Thermodynamic calculations using the system of consistent constants and taking into account the formation of hydrocomplexes show that migration of iron in the form of the simple Fe^{3+} ion is limited to highly acid environments (pH < 2.2) with a redox potential higher than +0.77 V (Fig. 37). In moderately acid waters (pH = 2.2–4.1) with *Eh* values above +0.62 V ferric oxide occurs in solution in the form of the hydroxy-complexes $FeOH^{2+}$ and $Fe(OH)^{2+}$. From the diagram it is seen that undissociated molecules of $Fe(OH)_3^0$, the stability field of which occupies the region in the pH range of 4.18 to 12.35, are of special importance. The minimum *Eh* value for which this form is stable is −0.4 V in alkaline environments. However, it must be kept in mind that the molecular forms cannot be considered completely stable due to their tendency toward aggregation and condensation, with the subsequent formation of hydrosols. In highly alkaline environments (pH > 12.35) iron occurs in the form of negatively charged hydroxy-complex $Fe(OH)_4^-$.

In the case of high contents of ferric iron, polymeric complex compounds can appear—binuclear $Fe_2(OH)_2^{4+}$, trinuclear $Fe_2(OH)_2^{5+}$ and more complicated complexes, formed in the moderately acid region (Fig. 38). In this case the concentration of iron in solution would be so high (more than 10^{-2} mol/l) that the hydroxide precipitates out—$Fe(OH)_3$. Thus polymeric compounds evidently are metastable forms transitional to colloidal, and there-

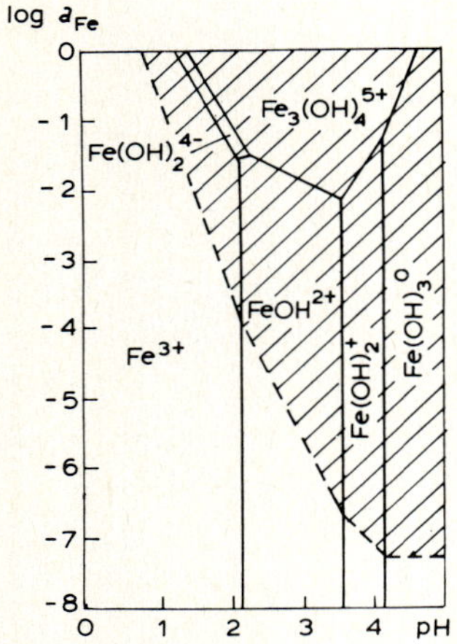

Fig. 38. Monomeric and polymeric ions of trivalent iron (log a_{Fe}–pH diagram). Field of precipitation of hydroxides—$Fe(OH)_3$—is hatched.

fore they do not play an independent role in the migration of iron.

Divalent (ferrous) iron is stable in a reducing environment in solutions containing virtually no free oxygen ($P_{O_2} < 10^{-20}$ bar). The main form of transport is the simple Fe^{2+} ion (pH from 0 to 8); in alkaline environments the hydroxy-complexes $FeOH^+$, $Fe(OH)_2^0$ and $Fe(OH)_3^-$ are stable.

Compounds of Fe with organic matter also are possible. Shcherbina (1956) has suggested that complex anions of iron are formed with humic acids (by analogy with complexes with citric acid), with alkalis playing the part of the cation. Humate complexes can be destroyed as a result of oxidation by the oxygen in the air. Mokriyevskaya (1959) also explains the high contents of Fe in swamp waters by the formation of a complex with humic acids. In the opinion of Antipov-Karatayev and Tsyuryupa (1961), in podzol soils iron migrates in the form of chelates, which is confirmed by the experiments of Kaurichev et al. (1958).

Levashkevich (1968) experimentally studied the ability of specific humic acids in several types of soils to form chelated compounds with ionic forms of iron (Fe^{3+}). It was established that in the reactions of chelate formation Fe_2O_3 is bound with humic acids into compounds in which the iron loses the properties of an independent ion, and fulvic acids have higher complex-forming capability than humic. The maximum amount of Fe that can be bound with fulvic acids reaches 650–711 mg per 1 g of organic matter, with humic acids, 119–152 mg. Chelated compounds of Fe_2O_3 with fulvic acids are considerably mobile, but solution of humate complexes occurs only in alkaline environments (at pH = 7.3–7.7). In acid environments these compounds decompose, the iron going into solution as ionic forms and the humic acids remaining in the residue.

Senesi et al. (1977), using the methods of electron spin resonance and Mössbauer spectroscopy in conjunction with chemical methods, established that at least two and possibly three forms of binding of Fe^{3+} occur in humic materials. Ferric iron is firmly bound and protected in tetrahedral or octahedral coordination; this form of binding of iron is resistant to chemical complexing and reduction. Fe^{3+} adsorbed on the outer surfaces of humic materials is less firmly bound. The iron-fulvic acid complexes studied contain from 5.5 to 50.1% Fe, but a large part of the iron is bound to the surficial octahedral position.

Conditions of deposition of iron

The following are taken to be the original compounds constituting the primary cherty iron sediments: amorphous hydroxides of trivalent and divalent iron—$Fe(OH)_3$ and $Fe(OH)_2$; amorphous silica—$SiO_{2(a)}$; finely dispersed crystalline magnetite—$Fe_3O_{4(d)}$; siderite—$FeCO_{3(d)}$; greenalite—$Fe_3Si_2O_5(OH)_{4(d)}$; pyrite and pyrrhotite—$FeS_{2(d)}$ and $FeS_{(d)}$. The new thermodynamic constants of the amorphous and dispersed metastable phases were used in the calculations.

In choosing the constants of the original Fe^{3+} hydroxides we proceeded from the fact that deposition of amorphous compounds on the floor of a sedimentary basin usually occurs some time after they are formed in the water layer. This time depends on many factors, in particular on the depth of the basin, the size of the aggregates, the rate of their settling to the floor, etc. Therefore, out of the several varieties of iron hydroxide with different aging time, distinguished earlier, the hydroxide $Fe(OH)_3$-II, which forms in a few hundred hours, was chosen. For magnetite, the recalculated consistent value ΔG^0_{298} of the finely dispersed sediment containing water was also used.

The formation of other primary iron mineral compounds is determined by the presence in solution of active forms of silica, carbon, and sulfur. The sediments can be subdivided provisionally into four groups: oxide, silicate, carbonate, and sulfide, formally corresponding to the sedimentary facies of the iron-formation, according to James (1954). For each group of sediments the relationships between iron compounds were determined by calculation, in a wide range of pH (from 0 to 14) and *Eh* (from +1.0 to −1.0 V).

Iron oxide sediments are formed in the absence of or at low concentrations of active forms of silicic acid, carbonic acid, and sulfur in the waters. According to the data of investigations of the composition of recent ferric iron sediments, they consist mainly of amorphous hydroxides of Fe^{3+} (Strakhov, 1947, 1960; Ostroumov, 1955; Volkov and Tikhomirova, 1966;

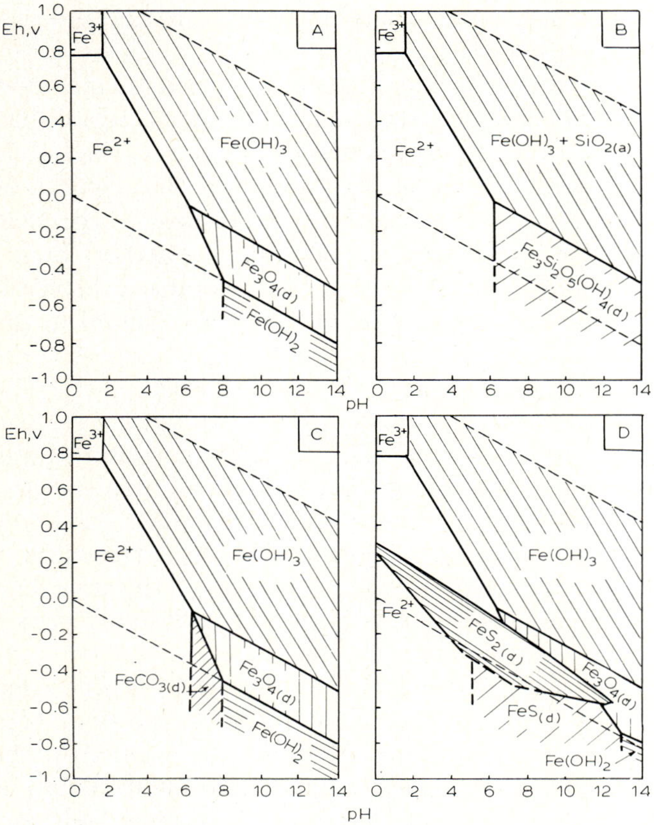

Fig. 39. Relationships between iron compounds in various primary sediments for activity of $a_{Fe} = 10^{-2}$ g·ion/l: A = oxide; B = silicate; C = carbonate; D = sulfide ($a_S = 10^{-3}$ g·ion/l).

Sevast'yanov, 1968); in some cases finely dispersed magnetite has been found (Strakhov, 1960; Stashchuk et al., 1964; Harrison and Peterson, 1965), and very rarely, in highly reducing conditions—for instance in the sediments of swamps and peat bogs—amorphous hydroxides of Fe^{2+} are encountered (Astapova et al., 1969; Kovalev and Generalova, 1969; Stashchuk, 1969).

Thermodynamic analysis of the relationships between the primary iron sediments confirms the geologic data. On the *Eh*-pH diagram (Fig. 39) it is seen that in oxidizing environments iron is deposited in the form of amorphous Fe^{3+} hydroxides, and in reducing environments, finely divided magnetite is formed. Hydroxides of divalent iron are deposited only in a narrow field of highly reducing environments near the limit of stability of water at atmospheric pressure.

The oxidation-reduction conditions of formation of iron compounds can also be characterized by values of partial pressure of oxygen in the atmosphere, in equilibrium with the waters of the sedimentary basin. For the $Fe(OH)_3$–Fe_3O_4 and Fe_2O_3–$Fe(OH)_2$ interfaces the values of P_{O_2} are $10^{-61.8}$ and 10^{-84}, respectively. These values mean that when the first magnetite and Fe^{2+} hydroxides were formed, free oxygen was totally absent and in that case the oxidation-reduction processes would be characterized by the relationships between such components as H_2O and H_2, CO_2 and CO, CH_4, etc. CH_4, etc.

From the diagram it is seen that in the presence of free oxygen in the waters, iron can be transported only in the form of Fe^{3+} or other ferric ions (Fig. 39). However, the migrational capacity of Fe^{3+} is not great even in acid environments, for instance, at pH ~ 1.5 about 10^{-2} g·ion/l of iron are in equilibrium with $Fe(OH)_3$ in solution. These figures are very approximate, inasmuch as they characterize the thermodynamic concentration or activity. To calculate the real concentrations it is necessary to take into account the ionic force of the solution, temperature, etc. These calculations are not given here, inasmuch as the main purpose of the work—to give a general analysis of the regularities of sedimentation—is achieved with such approximate formulations.

A small increase in pH (to 3.5–4.0) leads to practically total removal of iron from solution. In the pH range most typical of natural waters (4–10) the content of iron in the form of $Fe(OH)_3^0$ is of the order of 10^{-7}–10^{-8} g·ion/l ($5 \cdot 10^{-3}$–$5 \cdot 10^{-4}$ mg/l).

Thus the geochemical barrier which causes nearly total precipitation of ferric oxide from water is a change in the reaction of the environment from highly acid (pH = 0–2) to moderately acid (pH = 2–4). In the case where equilibrium is reached the amount of iron hydroxide precipitated from a saturated solution is inversely proportional to the pH, the dependence being nonlinear and the main mass of sediment being deposited from highly acid solutions.

In the absence of free oxygen the migrational capacity of iron increases sharply, which is expressed on the diagram by an increase in the stability field of the Fe^{2+} ion. Here the dependence on pH is ambiguous, inasmuch as deposition of the solid phases is determined to a substantial extent by the redox potential of the environment.

In analyzing the sequence of deposition of iron oxides, the question arises concerning the potential-determining systems that regulate the *Eh*. This question has been examined in detail by Stashchuk (1968) as it applies to the diagenesis of recent sediments with the formation of iron sulfides. In the general case the regulator of the *Eh* may be oxygen buffer systems consisting of two minerals of constant composition, for instance hematite and magnetite, or one compound in equilibrium with gas, for instance carbon at

P_{CO_2} = const, or any other external source that keeps the partial pressure of oxygen at a constant level (P_{O_2} = const). For recent equilibrium processes in the well aerated surface layers of water bodies the potential regulator is the atmosphere, with its vast reserves of free oxygen which specify $P_{O_2} = 0.21$ bar.

In the Precambrian atmosphere there were no substantial amounts of oxygen capable of maintaining P_{O_2} at the same level for a long time. The appearance of oxygen as a result of photosynthesis led to oxidation of Fe^{2+} and other reducing agents, as a result of which the *Eh* values fluctuated, reflecting irreversible oxidation-reduction processes. Therefore it is feasible to regard the evolution of the *Eh* of the sedimentary basins as the result of the general development of the atmosphere and hydrosphere, tending toward an increase in the content of ferric forms, and then leading to the appearance of free oxygen.

If it is considered that in the initial stages of evolution the atmosphere was sharply reducing and the *Eh* was determined by variations in CO_2, CO, H_2, CH_4 and other hydrocarbons in the atmosphere in equilibrium with carbon (see Fig. 22), then ferrous iron would have predominated in the solutions. The migrational capacity of Fe^{2+} is very great under these conditions: near the lower boundary of decomposition of water (see Fig. 39) the iron content can reach several tens of grams per liter even in neutral solutions (pH = 6–7).

In reducing conditions the main component of the sediment is metastable dispersed magnetite:

$$3\,Fe^{2+} + 4\,H_2O = Fe_3O_4 + 8\,H^+ + 2\,e^- \qquad (3.1)$$

$$Eh = +1.229 - 0.236\,\text{pH} - 0.088\log a_{Fe}$$

The magnetite phase would prevail in the sediments so long as the overall character of the atmosphere and hydrosphere are unchanged; temporary "flare-ups" of life activity of the first organisms and the local appearance of oxygen would not affect the composition of the solid phase. Even in those cases where such phenomena cause direct oxidation of Fe^{2+} to $Fe(OH)_3$:

$$Fe^{2+} + 3\,H_2O = Fe(OH)_3 + 3\,H^+ + e^- \qquad (3.2)$$

$$Eh = +0.941 - 0.177\,\text{pH} - 0.059\log a_{Fe}$$

the hydroxide that is formed is unstable and will irreversibly convert to magnetite:

$$3\,Fe(OH)_3 + H^+ + e^- = Fe_3O_4 + 5\,H_2O \qquad (3.3)$$

$$Eh = +0.366 - 0.059\,\text{pH}$$

However, for this process to be realized in nature a combination of several conditions is necessary. Kinetic factors—high rate of crystallization of the oxide compared to silicates or carbonates—would favor the deposition of iron in the form of magnetite. For $Fe(OH)_3$ to convert to Fe_3O_4, the presence of a sufficient amount of a reducing agent, for instance organic matter, in the lower water layer or directly in the sediment is necessary, but in that case the final product more likely will be siderite or greenalite:

$$Fe(OH)_3 + C_{am} = FeCO_3 + 3\,H^+ + 3\,e^- \qquad (3.4)$$

$$Eh = +0.078 - 0.059\,pH$$

The appearance of stable hydroxides of Fe^{3+} in the sediments occupies the next stage in the evolution of the atmosphere and hydrosphere, when the oxidation-reduction setting is determined by reaction (3.3) of coexisting Fe_3O_4 and $Fe(OH)_3$. With large enough amounts of these oxides the sediment itself becomes the potential-determining system, and the equilibrium concentration of iron in solution depends only on the pH. Let us give some figures (activity of Fe^{2+} in g/l): 167 at $pH = 5$; 1.67 at $pH = 6$; 0.02 at $pH = 7$. As a result, in the pH range of 6–7 practically all the iron precipitates and less than 0.01% of the original content remains in solution.

A further increase in *Eh* leads to the disappearance of magnetite in the sediments and formation only of $Fe(OH)_3$. The sediment-solution system ceases to be potential-determining for the sedimentary basin, and *Eh* varies as a function of fluctuations in the composition of the hydrosphere and atmosphere. In these conditions the migrational capacity of iron is reduced and precipitation of hydroxides becomes possible at lower and lower pH values. For instance, if *Eh* changes from +0.028 V ($Fe_3O_4 + Fe(OH)_3$ in equilibrium) to +0.160 V ($Fe(OH)_3$ in equilibrium) with a constant value of $pH = 5$, most of the iron is removed from solution (the equilibrium activity changes from 167 to < 1 g/l) although there is no free oxygen.

Thus in highly reducing conditions iron can migrate in a wide pH range (from 0 to 6) and precipitates as sediment in the form of oxides and hydroxides only in neutral environments. The acidity of the environment, as a natural geochemical barrier governing the precipitation of iron, is appreciably reduced. Variation in the redox potential as a result of the overall evolution of the atmosphere, hydrosphere, and biosphere plays a large role.

Silicate iron sediments are formed in the case of joint deposition of iron and silica. According to the data of thermodynamic calculations, chemical deposition of pure amorphous silica ($SiO_{2(a)}$) occurs only in the case of fairly high concentrations of dissolved SiO_2 (around $10^{-2.7}$ g·ion/l) in a wide pH range (from 0 to 10).

In the presence of active forms of silicic acid in oxidizing environments, an association of amorphous $Fe(OH)_3 + SiO_2$ is formed, and in reducing environments, finely dispersed crystalline silicates.

Preliminary estimation of the relative stability of crystalline iron silicates (fayalite, Fe_2SiO_4; ferrosilite, $FeSiO_3$; grunerite, $Fe_7Si_8O_{22}(OH)_2$; minnesotaite, $Fe_3Si_4O_{10}(OH)_2$; and greenalite, $Fe_3Si_2O_5(OH)_4$, respectively representing the main groups of natural silicates—pyroxenes, olivines, amphiboles, and chlorites) showed that in low-temperature conditions mainly greenalite, which in thermodynamic formulations can provisionally be considered the analog of the iron chlorites that occur extensively in sediments, is stable. Previously we considered minnesotaite (iron talc) to be the stable phase in silicate sediments. Analysis of experimental and petrological data (see Appendix) showed that minnesotaite forms, usually, in diagenesis or low-rank metamorphism. Klein and Bricker (1977) also take mainly pure iron greenalite as the stable phase in plotting diagrams, although in some cases the stability fields of minnesotaite, "stilpnomelane" with various ratios of $Fe^{2+}:Fe^{3+}$ in the molecule, and greenalite with an admixture of Mg^{2+} are also shown. The main problem here is the reliability of the thermodynamic constants used. A calculation from the constants given in the work cited (Klein and Bricker) for the reaction:

$$Fe_3Si_4O_{10}(OH)_2 + 2(H_2O)_l = Fe_3Si_2O_5(OH)_4 + 2\ SiO_{2(a)} \qquad (3.5)$$

gives a value of $\Delta G^0_{298} = 0$, which indicates that stability of these phases at 25°C is equally probable and does not permit establishment of the boundary of the transformation of minnesotaite into greenalite. The divergences between approximate estimates of ΔG^0_f of greenalite, in kcal/mol, recalculated to the formula $Fe_6Si_4O_{10}(OH)_8$), obtained by various authors are very large: -1440.0 (Tardy and Garrels, 1974); -1426.0 (Nriagu, 1975); -1417.3 (Eugster and Chou, 1973). At the same time, in most of the works cited the value of the ΔG^0_f of minnesotaite (-1069.9 kcal/mol) that we obtained earlier (Mel'nik, 1972b), which is consistent with the ΔG^0_f value of -1434.4 kcal/mol for $Fe_6Si_4O_{10}(OH)_8$, was used. The differences are quite obvious, and the use of inconsistent constants obtained by different methods and by different investigators leads to inadmissible errors. Still greater is the uncertainty as to the position of the phase boundaries of iron silicates of complex composition (stilpnomelane, ripidolite, "ferric" minnesotaite, "Al-greenalite") and different degree of dispersion. Out of these considerations we have limited ourselves to the plotting of "stability" diagrams of phases of simple composition.

From Fig. 39 it is seen that as a result of the formation of the primary iron silicates, the field of $Fe(OH)_2$ disappears from the diagram and the field of Fe^{2+} is somewhat reduced, which reflects a decrease in the migrational capacity of iron in the waters in a reducing environment. Deposition of the primary silicates is possible only in the absence of free oxygen. Replacement of the field of Fe_3O_4 by the field of $Fe_3Si_2O_5(OH)_4$ theoretically explains the limited occurrence of magnetite in recent silicate iron sediments and oolitic ores (Strakhov, 1960; Yurk et al., 1960; Stashchuk et al., 1964), in which iron silicates play an important role. The iron hydroxides hydrogoethite and goethite are widespread in these ores.

Carbonate iron sediments are formed when iron is precipitated in the presence of dissolved carbonic acid or as a result of interaction of the primary sediments with organic matter in the course of diagenesis. The determination of the stability of primary iron carbonates with respect to iron oxides and hydroxides was made on the basis of using the functional dependence of P_{CO_2} and a_{Fe} on pH, calculated by the method of Garrels and Christ (1968) for the system carbonate-water. According to this dependence, in conditions of $FeCO_3$ in equilibrium with water the value of partial pressure of carbon dioxide decreases in proportion to increasing pH.

If the activity of iron remains constant due to the great buffering capacity of the sedimentary basin or as a result of constant arrival of "fresh" batches of iron-bearing solutions, then at the beginning of deposition of siderite the pH depends only on P_{CO_2} according to the relationship:

$$Fe^{2+} + (CO_2)_g + (H_2O)_l = FeCO_3 + 2\,H^+ \tag{3.6}$$

$$pH = 4.57 - \tfrac{1}{2}\log a_{Fe^{2+}} - \tfrac{1}{2}\log P_{CO_2}$$

The greater the partial pressure of carbon dioxide, the more acid the environment at which deposition of carbonate begins. At $P_{CO_2} = 0.1$ bar (for amorphous greenalite-siderite sediment) and iron activity of 10^{-2} g · ion/l, deposition of siderite begins at $pH \approx 6$ (Fig. 39c). The field of deposition of carbonate is bounded by the lines of deposition of magnetite according to reaction (3.1) and of amorphous amakinite* according to the reaction:

$$Fe^{2+} + 2(H_2O)_l = Fe(OH)_2 + 2\,H^+ \tag{3.7}$$

$$pH = 7.01 - \tfrac{1}{2}\log a_{Fe^{2+}}$$

* Amakinite: $(Fe, Mg, Mn)(OH)_2$ — Kozlov and Levshov, Zapiski Vses. Obshch. 91, p. 72, 1962 (Translator's note).

The lower limit of partial pressure of CO_2 at which the formation of siderite in oxide-carbonate sediments is possible also is limited:

$$FeCO_3 + (H_2O)_l = Fe(OH)_2 + (CO_2)_g \tag{3.8}$$

$$\log P_{CO_2} = -4.88; \; P_{CO_2} = 1.3 \cdot 10^{-5} \text{ bar}$$

When $a_{Fe^{2+}} = 10^{-2}$ g · ion/l, deposition of siderite occurs in the pH = 6–8 range; if the activity of iron is reduced this range theoretically can be extended somewhat into the field of more alkaline environments, but the combination of circumstances necessary for this is hardly likely.

Taking into account the established relationship of the values of pH, P_{CO_2}, and a_{Fe} in the water, it must be presumed that in present conditions formation of siderite is practically possible only in a narrow range of nearly neutral environments. From Fig. 39 it follows that the redox potential of such environments would have negative values.

Formation of siderite in the system Fe–H_2O–CO_2 does not change the relationships between oxide and hydroxide sediments.

As has already been mentioned, the appearance of carbon dioxide in oxide systems leads to deposition of iron at lower pH. Thus if the hypothetical buffer mechanism existed in the sedimentary basins, maintaining the pH at a constant level, then an increase in the partial pressure of carbon dioxide would have led to deposition of iron in the form of carbonates. The actual picture is more complicated, inasmuch as carbonate and bicarbonate equilibria themselves act as buffers, an increase in P_{CO_2} leading to a decrease in pH and corresponding solution of carbonate and conversely, a drop in P_{CO_2} causing an increase in pH and deposition of carbonate.

Thus chemogenic deposition of siderite could occur only in oxygen-free environments at low *Eh* values in the early stages of development of the atmosphere and hydrosphere, in the Archean and beginning of the Proterozoic. The main cause of deposition was periodic variations in P_{CO_2} and pH.

With some increase in *Eh* the characteristic association of $FeCO_3$ and $Fe(OH)_3$ arises; this probably played a very important part in the composition of the Proterozoic cherty iron-formations. In this case deposition of certain mineral forms of iron was regulated by fluctuations in P_{CO_2} and *Eh* (or P_{O_2}). Siderite and iron hydroxide are formed at relatively high P_{CO_2} (> 0.1 bar), in slightly acid environments—pH < 6. At lower P_{CO_2} and correspondingly higher pH the associations siderite + magnetite or magnetite + hydroxide appear.

And finally, it should be mentioned that only siderite (at pH < 7) or magnetite (at pH > 7) can occur in equilibrium (metastable) with free amorphous carbon, but not iron hydroxide, which is in good agreement with the data on recent and ancient sediments.

Sulfide iron sediments can be formed both by direct chemogenic deposition from solution and in the course of diagenesis of the original sediments (Curtis and Spears, 1968; Skripchenko, 1969).

Primary iron sulfides are encountered relatively rarely in Precambrian rocks, and it is not ruled out that they too were formed as a result of diagenetic transformations, as has occurred at the time of diagenesis of recent sediments (Strakhov, 1960; Berner, 1964a, b; Stashchuk et al., 1964).

Deposition of primary iron sulfides depends on many natural factors—pH, *Eh*, concentration of iron and sulfur, presence of reactive forms of carbon dioxide and silicic acid, etc. To ascertain the conditions of chemogenic formation of iron sulfides, a special examination of the effect of all these factors is needed. We will limit ourselves to determining the conditions of formation of sulfide iron sediments from waters with a constant sulfur content, and then we will examine the effect of a change in its concentration.

Chemogenic deposition of iron sulfides occurs as a result of interaction of iron compounds with dissolved forms of sulfur, the stability field of which was calculated by the method of Garrels and Naeser (1958). On the basis of analysis of reactions between iron compounds and forms of dissolved sulfur corresponding to certain values of pH and *Eh*, diagrams of the relationships between iron compounds in the system Fe–H_2O–S were plotted for $a_S = 10^{-3}$ g·ion/l (see Fig. 39). From the diagram it is seen that with such a high sulfur concentration in the waters, or even if the sulfur activity is not so high (~ 30 mg/l), iron sulfides can be deposited at any pH values, mainly in reducing conditions. It is typical that mainly pyrite is deposited in the stability field of water; in these conditions pyrrhotite is deposited only in highly alkaline environments (pH > 11) with very low values of *Eh* (< -0.6 V).

In environments that are close to neutral, magnetite is replaced by pyrite to a substantial extent—on the diagram (see Fig. 39) only a narrow strip of magnetite field is seen. The formation of iron sulfides from high sulfur concentrations in the waters also leads to a sharp drop in the migrational capacity of iron in acid environments. In this case it is interesting to note that the Fe^{2+} field is intersected by the FeS_2 field, indicating that pyrite is dissolved in acid waters under both oxidizing and reducing conditions. The appearance of sulfur substantially limits the migrational capacity of Fe^{2+} at low values of *Eh*.

Oxide–carbonate–silicate–sulfide iron sediments can be formed under conditions of joint interaction of iron compounds with reactive forms of silicic

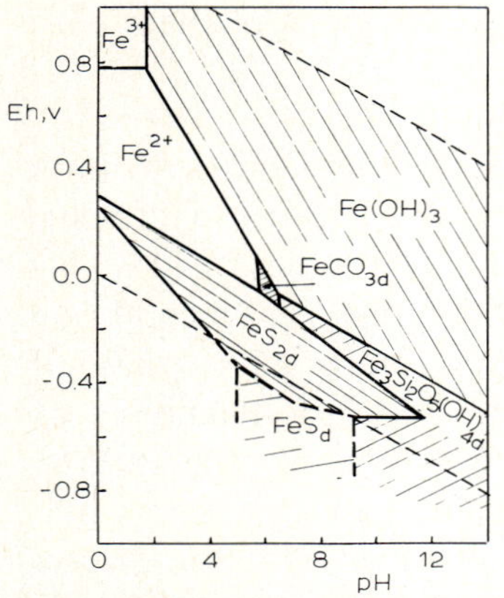

Fig. 40. Relationships between iron compounds in primary oxide-carbonate-silicate-sulfide sediment in the case of an excess of silica (graph of *Eh* versus pH for $a_{Fe} = 10^{-2}$ and $a_S = 10^{-3}$ g · ion/l).

acid, carbonic acid, and sulfur. The composition of such compounds is controlled mainly by the concentrations of these forms in the waters.

Analysis of the relationships between iron compounds in the system $Fe-SiO_2-CO_2-H_2O-S$ was made for aqueous environments and iron and sulfur contents of 10^{-2} and 10^{-3}, respectively. In such environments iron sediments of all the groups examined earlier are formed.

The formation of carbonates in the system is limited mainly by the presence of active forms of SiO_2 and by the concentration of S. On the composite diagram (Fig. 40) the siderite field is represented by a sector of small area in environments with a pH of 5.5 to 6.2 and *Eh* from +0.05 to −0.05 V, even at high partial pressure of carbon dioxide (1.0–0.01 bar); with other parameters siderite is replaced by $Fe(OH)_3$ and iron silicates or sulfides.

The relative stability of finely dispersed carbonate and silicate sediments is determined by the reaction:

$$Fe_3Si_2O_5(OH)_4 + 3\ CO_2 = 3\ FeCO_3 + 2\ SiO_2 + 2\ H_2O \tag{3.9}$$

for which the equilibrium value of P_{CO_2} at 25°C is 0.06 bar ($10^{-1.2}$), which corresponds to pH = 6.15 for an iron content of 10^{-2} g · ion/l under conditions of equilibrium of pure siderite with water. Consequently, the presence of active forms of SiO_2 leads to the formation of iron silicates and a decrease in the stability field of siderite, which is retained only in slightly acid environments.

The relative stability of carbonate and sulfide sediments is determined mainly by the sulfur content of the waters and the alkalinity of the environment. The quantitative relationships in the replacement of siderite by sulfides as a function of variation in the iron and sulfur contents in the waters are seen from the sections on the composite diagram for pH = 6 at $a_S = 10^{-3}$ (Fig. 41) and $a_{Fe} = 10^{-2}$ g·ion/l (Fig. 42). Analysis of these diagrams shows that siderite is replaced by iron sulfides even in the case of extremely low sulfur contents: in waters with pH = 6 siderite begins to be replaced by pyrrhotite at $a_S \geqslant 10^{-4.5}$, and by pyrite at $a_S \geqslant 10^{-8.5}$ g·ion/l. The formation of siderite is almost independent of the sulfur content of the waters only in a narrow field of *Eh* values close to zero, where sulfate ions are stable. The small sector of the siderite field on the composite diagram (see Fig. 40) corresponds precisely to this region.

The relationships between the silicate and sulfide sediments are also determined mainly by the content of sulfur in the water, and pH and *Eh*. For given sulfur activities the sulfide field is curtailed due to formation of silicates only in highly alkaline environments (pH > 11). Deposition of silicates in environments close to neutral occurs in the stability field of magnetite. Magnetite is not formed in primary sediments in the presence of active forms of SiO_2. As the sulfur content in the waters decreases, the boundaries between the iron sulfide and iron silicate fields shift toward neutral environments. In waters with pH = 8 pyrrhotite begins to be replaced by silicates at $a_S \leqslant 10^{-4.5}$, and pyrite at $a_S \leqslant 10^{-7}$ g·ion/l.

It is typical that $Fe(OH)_3$ occurs in association with siderite or greenalite but not with iron sulfides. The absence of a significant amount of sulfides in Precambrian iron-formations indicates not only low concentrations of sulfur, but also certain limits of *Eh* values, which range from −0.2 to +0.2 V.

Examining the sections of the diagrams in coordinates of log a_{Fe} − *Eh* for

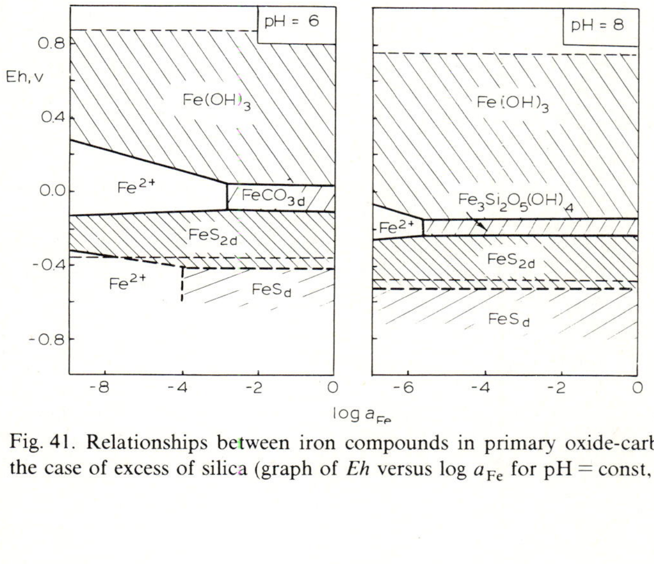

Fig. 41. Relationships between iron compounds in primary oxide-carbonate-silicate-sulfide sediment in the case of excess of silica (graph of *Eh* versus log a_{Fe} for pH = const, $a_S = 10^{-3}$).

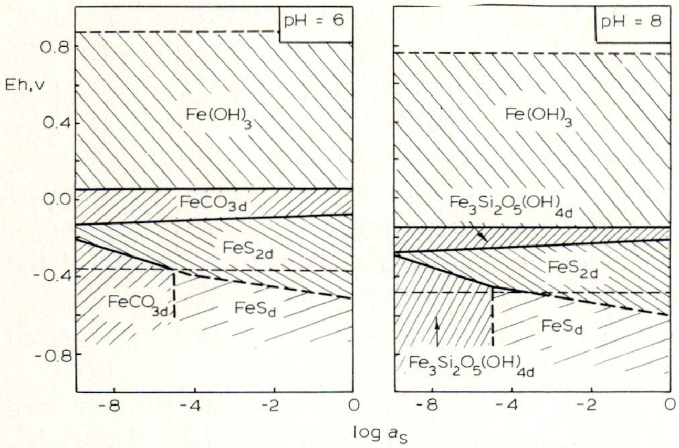

Fig. 42. Relationships between iron compounds in primary oxide-carbonate-silicate-sulfide sediment in the case of an excess of silica (graph of *Eh* versus log a_S for pH = const, $a_{Fe} = 10^{-2}$).

constant pH values, we see that transport of iron in ionic form is very limited under these conditions (see Fig. 41).

At pH = 6, which presumably occurred in the ancient Precambrian sedimentary basins, migration of iron could have been accomplished only in the form of Fe^{2+} at an *Eh* not lower than -0.10 and not higher than 0.0 to $+0.10$ V. A lowering of *Eh* would have led to massive deposition of pyrite, and an increase in deposition of Fe^{3+} hydroxides. The actual conditions of chemogenic formation of cherty iron sediments consisting of siderite, iron

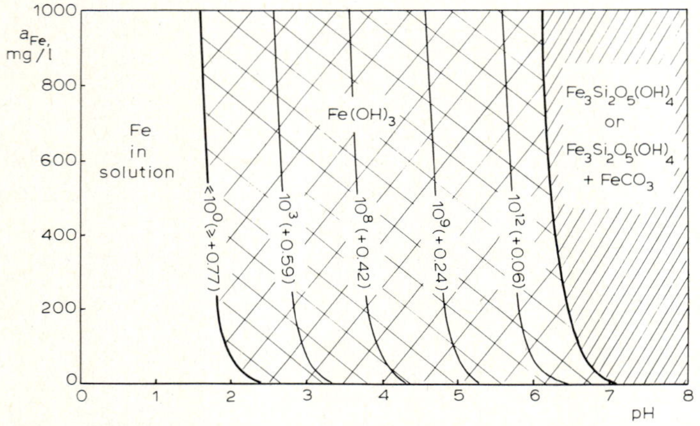

Fig. 43. Conditions of precipitation of iron as a function of pH, a_{Fe}, and *Eh*. Figures on curves of beginning of deposition of $Fe(OH)_3$ designate $Fe^{2+}:Fe^{3+}$ activity ratio and value of *Eh* (V) (in parentheses).

hydroxides, and minor disseminated sulfides were limited to extremely narrow limits of *Eh*.

The conditions of deposition of iron from ionic solutions are depicted more visually in Fig. 43, on which curves of the beginning of deposition of hydroxide and silicate-carbonate sediments are entered. Amorphous $Fe(OH)_3$ hydroxide can be deposited in a wide pH range from solutions with various ratios of ferrous and ferric iron activities. However, already at $pH \geqslant 3$, $a_{Fe^2} \gg a_{Fe^3}$ in the solution. Consequently, if free oxygen is introduced into the solution, precipitation of $Fe(OH)_3$ hydroxide, in which this oxygen is completely fixed, should theoretically occur. The actual course of this process is determined by the rate of the reaction:

$$4\,Fe^{2+} + O_2 + 10\,H_2O = 4\,Fe(OH)_3 + 8\,H^+ \tag{3.10}$$

or

$$4\,Fe^{2+} + O_2 + 4\,H^+ = 4\,Fe^{3+} + 2\,H_2O \tag{3.11}$$

and then:

$$Fe^{3+} + 3\,H_2O = Fe(OH)_3 + 3\,H^+ \tag{3.12}$$

After the end of deposition the solution in equilibrium with the hydroxide contains practically no Fe^{3+} or dissolved oxygen, and the *Eh* values remain low.

Formation of silicate or carbonate-silicate sediments begins at a ratio of activities of Fe^{2+}: $Fe^{3+} = 10^{13.5}$ to $10^{14.5}$ (ferric iron is absent) which corresponds to an *Eh* range of -0.025 to -0.085 V. In this same range, formation of complex hydroxide–carbonate–silicate iron sediments is possible. The relative amount of hydroxide in such sediments depends on the oxygen supply, and that of siderite on the carbonate "capacity" of the sedimentary basin.

If the pH changes from 6 to 7 (by one), nearly all the iron is deposited (change in activity of Fe from ~ 1200 to 12 mg/l). In more alkaline environments the migrational capacity of iron drops sharply and even with optimal values of *Eh*, determined by the stable existence of greenalite, the Fe content of the solution at $pH = 8$ will not exceed 0.1 mg/l.

The gradient of the redox potential is the only geochemical barrier causing deposition of iron when pH is constant.

Forms of transport and conditions of deposition of silica

In aqueous solutions silica occurs in one valency state—4^+. Detailed investigations of the chemistry of aqueous solutions of silica, summarized by

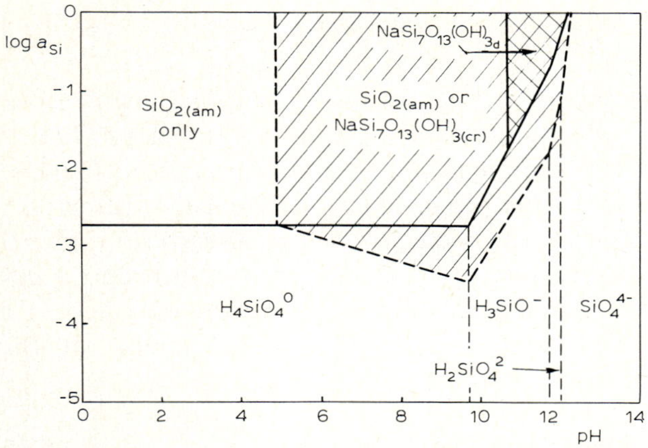

Fig. 44. Stability fields of ions and precipitates of amorphous silica. The field of possible formation of crystalline magadiite is hatched.

Iler (1959), showed that the stable existence of the monomeric form $Si(OH)_4^0$ or $H_4SiO_4^0$ is possible in a wide pH range (0–10). In alkaline solutions charged silicate ions are formed: $H_3SiO_4^-$, $H_2SiO_4^{2-}$, SiO_4^{4-} and others. The relationships between the various forms of silica in solution, calculated from the consistent thermodynamic constants, are shown in Fig. 44. From this figure it is seen that the solubility of amorphous silica at pH from 0 to 10 is constant and amounts to $\sim 2 \cdot 10^{-3}$ g · ion/l or ~ 120 mg/l. In more alkaline solutions the solubility increases sharply as the charged ions appear, and at pH = 12–13 it becomes very high.

Eugster and Chou (1973) suggested that silica was deposited from sea water in the form of sodium silicates of magadiite type ($NaSi_7O_{13}(OH)_3 \cdot 3\,H_2O$). This conclusion was based on experimental data and thermodynamic calculations (Bricker, 1969) from which it follows that for the usual sodium content and pH > 8, the solubility of crystalline magadiite is much lower than that of amorphous silica. However, freshly precipitated amorphous (dispersed) magadiite is formed only in highly alkaline environments (pH > 10.7), as is seen from the diagram we have plotted (Fig. 44) for sodium activity of ~ 1 g·ion/l, which corresponds approximately to the salt content of the present ocean. Crystalline magadiite actually is less soluble than amorphous silica at pH > 5, but considerably more soluble than crystalline quartz (for α-SiO_2 the equilibrium $a^0_{H_4SiO_4} = 10^{-4}$ g·ion/l). The actual composition of silica sediments depends to a substantial extent on their crystallinity, and not just on the sodium content of the solution.

At the same time the migrational capacity of silica in neutral and alkaline

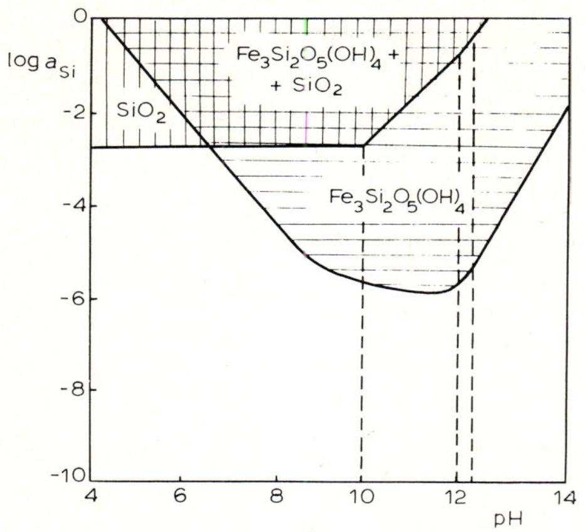

Fig. 45. Conditions of deposition of amorphous silica and finely divided greenalite (in the case of congruent solution and Fe : Si = 3 : 2 activity ratio).

environments (pH > 6.5) decreases much more intensively when amorphous (dispersed) silicates of greenalite type are formed (Fig. 45), which is more likely than formation of magadiite in the deposition of BIF, although these processes could go on in parallel, or alternate with each other cyclically. The field of greenalite is entered on the diagram (see Fig. 45) very schematically, inasmuch as the calculations were made for the process of congruent solution of silicate according to reactions of the type:

$$Fe_3Si_2O_5(OH)_4 + 6\,H^+ = 3\,Fe^{2+} + 2\,H_4SiO_4^0 + H_2O \qquad (3.13)$$

with a 3:2 ratio of iron and silica in solution. For other ratios of the dissolved components the stability field of greenalite will be shifted. An association of silicate plus amorphous silica in the primary cherty-iron sediment could have arisen only in the case of substantial concentration of $H_4SiO_4^0$ ($\geqslant 1.4 \cdot 10^3$ g·ion/l) in neutral or slightly alkaline environments (pH = 6–10).

Thus from the thermodynamic data it follows that if acid or neutral silica-bearing solutions arrived in the depositional basin, the action of the presumed geochemical barriers (gradients of pH, *Eh*, concentration) could not operate in isothermal conditions of chemogenic deposition of silica, to form cherty or cherty-iron sediments. Only a change in concentration due to evaporation of substantial volumes of water in closed basins could have led to deposition of silica. An easy calculation shows that to deposit chert bands 0.3–0.5 cm thick in that way, a 100-m water layer has to be evaporated. Thus the formation of thick piles of Precambrian iron formations would have required the evaporation of a fantastic amount of water from restricted basins with constant influx of silica-saturated solutions.

There are some experimental data on the solubility of silica at ordinary temperatures. Careful measurements by Alexander et al. (1954) showed that the solubility of very pure amorphous silica at 25°C, which is minimal in neutral solutions (~ 100–110 mg/l), increases sharply in alkaline (to 1120 mg/l at pH = 10.6) and increases somewhat in acid solutions (to 150 mg/l). It is suggested (Iler, 1969) that the solubility increases in acid solutions due to formation of the SiF_6^{2-} ion or complex anions with other acids, impurities of which are inevitably present in reagents or are introduced in the course of the experiments.

In investigations on the stability of colloidal solutions of silica of different concentration, we carried out several series of determinations of the content of ionic (monomeric) silica in "equilibrium" with colloidal silica in solution and in gel-like sediment. Figure 46 gives the data obtained at various pH values three months after the solutions were prepared. Good correspondence of the final content of ionic silica in the solutions with the various initial concentrations was observed. For pH = 5–10 the "solubility" of the gel was close to the solubility of amorphous solid silica measured by Alexander et al. (1954). In the more acid region the "solubility" was much higher in our experiments. This apparently is explained by the metastable state of ionic silica (supersaturated solution) in mobile "equilibrium" with colloidal silica in the absence of gel-like sediment. Therefore it cannot be concluded from the experimental data that the migrational capacity of silica in ionic form is substantially higher in acid environments, especially in the normal zone of supergenesis where the conditions for formation of concentrated ionic-colloidal systems and supersaturated solutions do not exist.

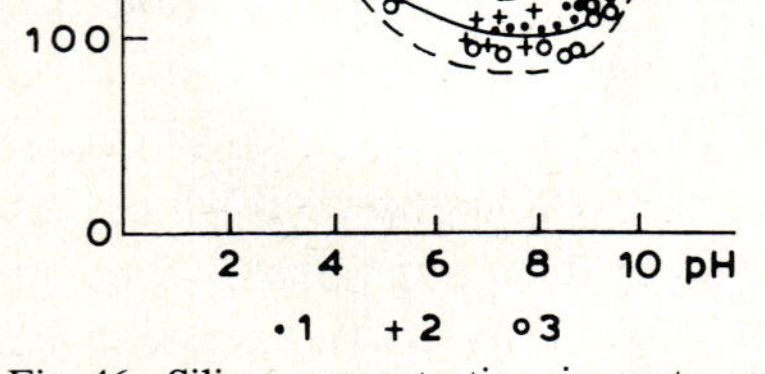

Fig. 46. Silica concentration in water as a function of pH: *I* = data of Alexander et al., *II* = our experimental data on content of ionic silica in ionic-colloidal systems with total initial SiO_2 concentrations (mg/l) of: *1* = 750; *2* = 1500; *3* = 3000.

Similar conditions can be produced when alkaline solutions saturated with silica are mixed with highly acid waters. Another genetic variant inevitably suggests acid thermal waters, in which the solubility of SiO_2 is a function of temperature (Khitarov, 1953), as the source of the silica. When thermal waters cool, mix with surface waters, and are partially neutralized, ionic-colloidal systems with anomalously high content of monomeric silica can arise. It is obvious that both these variants reflect specific conditions that are not typical of geochemical processes in the weathered layer.

Migration and deposition of iron and silica from colloidal solutions

The importance of the colloidal state of matter in the processes of formation and transformation of minerals, rocks, and ores of various genetic types has been examined in detail by Chukhrov (1955, 1965). Therefore general information on the structure, methods of obtaining, and stability of colloids, according to present theoretical ideas, will be discussed briefly. More attention is paid to an examination of the structure and stability of colloids of iron and silica, and in addition to a summary of published data, new experimental data obtained by the author together with Vorob'yeva in 1968–1978 are given. The stability of undialyzed colloidal solutions of iron and silica under the influence of various electrolytes was studied experimentally in a wide range of pH; results of experiments on the stabilization of iron colloids by silica colloids in acid and alkaline solutions are given; and the role of carbon dioxide in the transport of colloidal iron is ascertained. A special series of experiments was carried out to obtain banded iron-silica sediments from mixed colloidal solutions interacting with sea water.

Structure and stability of colloidal solutions

Colloidal solutions or sols occupy an intermediate position between true (ionic or molecular) solutions and suspensions. In colloidal solutions the particles of the dispersed phase range from 1 to 100 μm in size. Numerous investigations have shown that the individual particles of the dispersed phase are not molecules, but aggregates consisting of a great number of molecules. Therefore any colloidal solution is a heterogeneous system for which the conditions of formation should be insolubility of the dispersed phase in the dispersing environment (for instance, particles of iron hydroxide in water). In such heterogeneous systems there are physical interfaces between the dispersed phase and the environment, which determine certain specific properties of the colloidal solutions: scattering of light, low diffusion capacity, aggregative instability, ability to dialyze, etc.

According to thermodynamics, the instability of colloidal systems is caused by the presence of excess energy concentrated on the large phase interface. A tendency to reduce the total phase interface is inherent in the very nature of colloidal solutions (Kul'man, 1957), which leads to enlargement of particles and coagulation to form suspensions and sediments.

According to kinetics, the instability or stability of colloidal systems is determined by the balance of the forces of attraction and repulsion between the individual particles. The forces of attraction, causing the particles to stick together, are of the same nature as intermolecular forces and increase very rapidly as the particles approach each other. The forces of repulsion may be electrical, arising as a result of selective adsorption by the phase interface of one of the ions of an electrolyte present in the system. One of the factors keeping colloidal particles apart may be the formation on the interface of a solvate shell of molecules of the environment.

From what has been said it follows that in general the relative stability of a colloidal system is determined by whether the energy barrier is high enough to prevent the particles from coming together over short distances.

Structure of colloidal solutions. The concept of the structure of the particles of aqueous colloids is closely related to the study of adsorption and the double electrical layer.

A double electrical layer arises at the interface of two phases due to redistribution of the electrical charge when charged particles (ions, electrons) pass from one phase to another (Pisarenko et al., 1964). In colloidal solutions the particles of the dispersed phase enter into an adsorption reaction with electrolyte ions present in the solution. The electrolyte ions are selectively adsorbed on the surface of the particles and give it a certain charge. Thus the inner face of the double electrical layer is formed. Ions of opposite sign (counterions), which in part are concentrated on the surface of the particles and in part form a loose mobile shell some distance from the surface, constitute the outer face of the double electrical layer (Fig. 47).

The electrically neutral-structured colloidal unit thus built up, surrounded

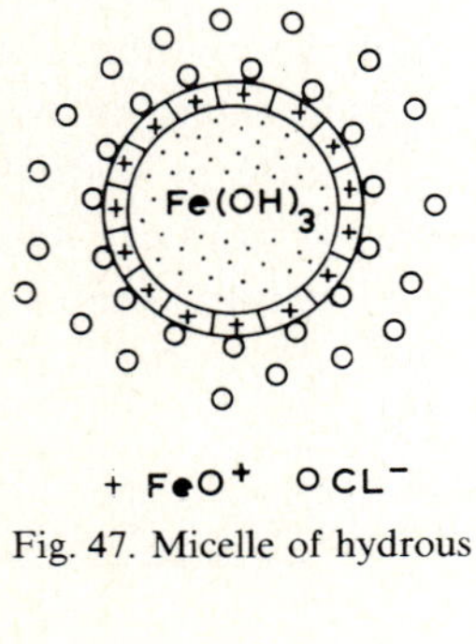

Fig. 47. Micelle of hydrous iron oxide (after Chukhrov).

by a double layer of ions, is called a micelle (Gameyeva, 1969). An insoluble nucleus forms the main mass of the micelle. The adsorbed ions of the inner face are called potential-determining. Together the potential-determining ions and the counterions constitute an adsorption layer. The mobile counterions form a diffusional layer. The counterions in the diffusional layer are called free, and those attached in the adsorptional layer on the walls of the particle, bound. The nucleus of the colloid together with the adsorptional layer, moving in an electrical field as a whole, is sometimes called a granule or particle.

The electrical charges in the micelle are distributed in such a way that the counterions bound in the adsorptional layer decrease the charge of the nucleus created by the potential-determining ions but do not completely neutralize it. Thus the sign of the charge of a granule is determined by the sign of the charge of the potential-determining ions. This excess charge of the granule in solution is neutralized by the counterions of the diffusional layer. The potential difference arising between the mobile (diffusional) and immobile (adsorptional) part of the double electrical layer is called the electrokinetic potential or zeta potential (ζ-potential). The state in which the zeta potential becomes zero is called isoelectrical.

The problem of the nature of the potential-determining ions is very complex and in many cases has not been solved definitively, therefore the chemical formulas of micelles often are tentative in character.

Stability of colloidal systems. Two types of stability are distinguished, kinetic and aggregative (Pisarenko et al., 1964).

Kinetically (or sedimentationally) stable systems are those in which the particles do not settle out appreciably under the force of gravity. Kinetic stability depends on the particle size and on the difference in density between the particles and the environment. All colloidal systems have very high kinetic stability.

Aggregative stability is the resistance to cohesion of particles, to coagulation, whereas kinetic stability is resistance to settling, related only to particle size. In colloid chemistry, stability usually is understood to mean the aggregative stability of a given system.

In one and the same colloidal system, two opposite tendencies are embodied—a tendency toward a decrease in the total phase interface and enlargement of particles, and a tendency toward self-hardening due to adsorption of stabilizers—usually charged ions. Thus the stabilization of colloidal solutions is caused by the presence of electrolytes or hydration of ions, in particular of the counterions of the diffusional layer bound to the granule. All the factors that will raise the zeta-potential and increase the hydration of the micelles will enhance the stability of the sol. And conversely, any external influence on colloidal solutions leading to a decrease in hydration will promote the process of breakdown of the colloidal system, i.e. coagulation (Voyutskiy, 1964).

The value of the zeta-potential, like the degree of hydration of the micelle, is determined by the thickness of the diffusional layer, which is very sensitive to the smallest changes in concentration of electrolytes in the dispersed phase. Dilution of the electrolytes in the sol can reduce the electrokinetic potential, and thus the charge, of a granule to zero. The discharged particles, describing Brownian motion and colliding with each other, agglutinate into larger aggregates. The clots that form settle rapidly, and the sol ceases to exist as a dispersed colloidal system.

Colloidal systems can also be destroyed under the influence of other factors: the action of light, changes in concentration, heating, freezing, etc. Raising the temperature, shaking, and increasing the concentration of the dispersed phase cause the colloidal particles to come closer together, which increases the extent of their aggregation. Thus, a certain drop in aggregative stability is observed upon heating. This is explained by the increase in intensity of Brownian motion and desorption of part of the stabilizer. Cooling hydrosols to temperatures above their freezing point usually has little effect on the stability and only freezing leads to coagulation (Kul'man, 1957). However, in most cases the effect of the enumerated factors is not great compared to the effect of electrolytes.

For appreciable coagulation it is necessary that the concentration of the electrolyte exceed some minimum value, called the coagulation threshold.

The coagulating part of the electrolyte is an ion which carries a charge opposite to the charge of the colloidal particle. For instance, for positive sols, for sols of certain metal hydroxides, the coagulants are anions. The effect of the ions of the coagulants is intensified many times as their valency increases (the Schulz-Hardy rule). Thus, the coagulating ability of divalent ions is about 50 times greater, and of trivalent ions 1000 times greater, than that of monovalent ions.

Coagulation of colloids is brought about not only by the effect of electrolytes but also by the interaction of oppositely charged colloids, which sets in when one colloid is adsorbed on the surface of another. The necessary condition for mutual coagulation is equality of opposite charges of the particles of the sol, as a result of which coagulation can occur only in a narrow zone of concentration ratios (Kul'skiy, 1960).

It has been established experimentally that whereas mutual coagulation takes place in certain quantitative relationships when various oppositely charged colloidal systems—hydrophile and hydrophobe—interact, quite a different process can take place in the case of other interrelationships: the addition of a small amount of a hydrophile colloid to a hydrophobe system increases its stability, in particular with respect to the action of electrolytes. This phenomenon is called a hydrophobe-shielded colloidal system, and the substances capable of producing the colloidal shield, shielding colloids. Adsorbed on the surface of the hydrophobe particles, they impart to it the properties of a hydrophile colloid.

Humic substances occurring in natural waters in the form of stable negatively charged sols exhibit shielding action with respect to colloidal solutions of SiO_2, $Fe(OH)_3$ and $Al(OH)_3$ (Kul'skiy, 1960). Colloids shielded by humus do not coagulate and, remaining in the state of sols, can be transported by the waters of rivers, seas, and oceans (Kuznetsov, 1964) for considerable distances, and the Schultz-Hardy rule becomes inapplicable to shielded colloids.

Other factors also affect the stability of colloidal solutions, in particular the pH of the environment. The surface molecules of the nucleus of the micelles of different colloidal systems can have acid, basic, or amphoteric properties. Colloids characterized by this feature are called, respectively, acidosid, basoids, and ampholitoids (Marchenko, 1965). It has been established that the coagulation threshold of acidoids (sols of weak acids) increases and that of basoids decreases when the pH is increased.

Obtaining colloidal solutions. Inasmuch as in particle size of the dispersed phase, sols occupy an intermediate position between true solutions and suspensions, they can be obtained either by uniting the individual molecules or ions of dissolved matter into aggregates, or by dispersing relatively larger particles.

Correspondingly, the methods of obtaining colloidal systems are divided into condensational and dispersional. The method of peptization stands apart; it consists of putting into solution sediments whose original particles already are of colloidal size. The main conditions for obtaining colloidal systems are insolubility or fairly low solubility of the dispersed phase in the dispersing medium and the presence in the system of substances capable of stabilizing the colloidal particles.

The condensation method makes it possible to obtain more highly dispersed systems than the dispersion method, and true lyophobic sols are always prepared by this method. Colloidal solutions are obtained by the condensation method as a result of chemical reactions of nearly all known types. But it should be noted that sols are by no means always formed, but only in the case of certain concentrations of the original substances, order of their mixing, temperatures of interaction, and a combination of several other conditions. The main method of preparing sols of heavy hydroxides is hydrolysis of solutions of salts, which takes place more completely and more rapidly at high temperatures and in dilute solutions.

Peptization is less important in the formation of natural colloidal solutions, inasmuch as conversion of the coagulant into sol again is characteristic of fresh highly dispersed sediments under the influence of special substances — peptizers. Peptization usually is observed when substances capable of forming sols act on sediments formed by coagulation of colloidal particles.

Dispersing is still less likely in the formation of colloids in the Earth's crust. It is difficult to obtain particles of the small size needed to obtain a stable sol by mechanical dispersing (grinding). Moreover, the composition of such particles will not be uniform.

Thus the condensation methods of formation of colloidal solutions, mainly from true ionic or molecular solutions, are most important in nature. The sols obtained nearly always are "contaminated" by various impurities, usually electrolytes, including components of the original true solutions.

Purification of colloidal solutions is based on the ability of contaminating ions and molecules to penetrate freely through special membranes which hold back colloidal particles (dialysis). Inasmuch as low-molecular impurities in sols usually are electrolytes, dialysis can be accelerated by imposing an electrical field on the liquid to be dialyzed (electrodialysis). Prolonged dialysis leads not only to the removal of impurities from the sol, but also to the removal of an electrolyte-stabilizer which could lead to coagulation.

In using the numerous experimental data on the stability of various colloids it should be kept in mind that the experiments often were conducted with pure dialyzed colloids, which hardly are encountered in nature. The properties of dialyzed and undialyzed colloids are rather different.

Structure and stability of colloidal solutions of iron

Colloidal solutions of iron are fairly stable, are easily obtained in laboratory conditions, and play a large part in the migration of this element in the supergene zone (Strakhov, 1960). Compounds of trivalent iron, mainly the hydroxide, usually form colloidal solutions; colloids are not typical of the more soluble compounds of divalent iron.

Structure of colloids of iron hydroxide. Colloidal solutions of iron hydroxide usually are obtained experimentally by hydrolysis of a ferric chloride solution. The particles that are formed consist of molecules of $Fe(OH)_3$, which constitute the insoluble nucleus — the main mass of the micelle. The number of molecules in the nucleus is not constant and may range from tens to 500.

Different opinions exist concerning the potential-determining ions of iron hydroxide colloids. According to Kroyt (1933) and Zigmondi (1933), the stabilizing electrolyte is ferric chloride and thus the potential-determining ions are Fe^{3+} cations. Another group of investigators (Rabinowitsh and

Kargin, 1928; Kargin and Kiseleva, 1938) believes that sols of $Fe(OH)_3$ always contain a certain amount of hydrochloric acid which acts as the stabilizing electrolyte, and the positive charge of the granule is produced by adsorption of H^+ ions. The opinion prevailing at present is that the stabilizer is the product of incomplete hydrolysis of perchlorate of iron (FeOCl) formed according to the equation:

$$Fe(OH)_3 + HCl = FeOCl + 2\ H_2O \tag{3.14}$$

The molecules of the electrolyte-stabilizer dissociate according to the equation:

$$FeOCl = FeO^+ + Cl^- \tag{3.15}$$

If the number of molecules of $Fe(OH)_3$ is designated by m, and the number of molecules of the ionic stabilizer entering into adsorptional interaction with the nucleus by n, then the micelle of iron hydroxide can be represented in the form of the formula:

Granule

adsorption layer

$$\{m[Fe(OH)_3] \cdot n\ FeO^+ \cdot (n-x)\ Cl^-\}^+\ x\ Cl^-$$

nucleus — potential-determining ions — counterions

micelle

The granule as a whole has a positive charge due to adsorption of FeO^+ cations by the surface of the nucleus. It is assumed that m is always many times larger than n.

Obtaining colloids of iron hydroxide. Stable colloidal solutions of iron are obtained only under certain conditions, a slight change in which leads to coagulation. Thus, when a ferric chloride solution is neutralized by an alkali according to the reaction:

$$FeCl_3 + 3\ NaOH = Fe(OH)_3 + 3\ NaCl \tag{3.16}$$

it is difficult to obtain an hydroxide sol inasmuch as the formation of

$Fe(OH)_3$ takes place very rapidly and aggregates of molecules, which are not able to accomplish adsorption of stabilizing ions, can go past the zone of colloidal dispersion very rapidly and precipitate as sediment.

Colloidal solutions are easily obtained if $Fe(OH)_3$ forms at a finite rate, for instance, in hydrolysis of a ferric chloride solution. The step-like character of this process rules out rapid formation of sediment:

$$FeCl_3 + H_2O = FeOHCl_2 + HCl$$

$$FeOHCl_2 + H_2O = Fe(OH)_2Cl + HCl$$

$$Fe(OH)_2Cl + H_2O = Fe(OH)_3 + HCl \tag{3.17}$$

In the chain of hydrolysis there is the compound $Fe(OH)_2Cl$, which is more soluble than $Fe(OH)_3$. This intermediate compound dissociates according to the equation:

$$Fe(OH)_2Cl = FeO^+ + Cl^- + H_2O \tag{3.18}$$

Thus at the moment when a large number of colloid-dispersing nuclei appear in the solution there is a certain concentration of FeO^+ and Cl^- ions, from which FeO^+ ions are adsorbed as potential-determining inasmuch as they have an affinity with the material of the nucleus and are able to build on to the crystal lattice of the latter.

In laboratory conditions hydrolysis usually is carried out as follows: a solution of $FeCl_3$ is added drop by drop to distilled water heated to boiling, stirring constantly. The whole liquid quickly takes on the red-brown color characteristic of hydrous ferric oxide, but remains transparent. If the solution is allowed to cool, the color fades somewhat, as some of the $Fe(OH)_3$ changes back into $FeCl_3$. To prevent the reverse reaction, the solution is boiled for a few minutes to remove HCl with water vapor, or it is removed by dialysis. In this way one obtains stable colloidal solutions of iron hydroxide containing from 5 to 5000 mg/l of iron.

At room temperature colloidal solutions of iron hydroxide can be obtained only by way of prolonged dialysis (Glazman et al., 1958). And finally, experiments are known in which sols were obtained by peptization, by treating freshly precipitated, washed $Fe(OH)_3$ sediment with ferric chloride while heating. A dilute solution with a certain amount of HCl acts on freshly precipitated $Fe(OH)_3$ as ferric chloride does. The sols of $Fe(OH)_3$ obtained by peptization are no different in structure from the sols obtained by hydrolysis.

Stability of colloids of iron hydroxide. The main cause of coagulation of sols is the action of electrolytes on them, as a result of which two equilibria in the

TABLE VIII
Action of electrolytes on positive sol of hydrous iron oxide of 0.823 g/l concentration (after Kablukov et al.)

Coagulating electrolyte	Coagulation threshold (precipitating capability, mmol/l)	Coagulating electrolyte	Coagulation threshold (precipitating capability, mmol/l)
KI	16.0	KCl	9.00
KBr	12.5	K_2SO_4	0.205
KNO_3	12.0	MgSO	0.095
NaCl	9.25	$K_2Cr_2O_7$	0.195

solution are shifted: the equilibrium between bound and free counterions and the equilibrium of hydration:

$$(n-x)\,Cl^- \rightleftarrows x\,Cl^- \tag{3.19}$$

An increase in the concentration of counterions in the medium leads to an increase in $(n-x)$, which produces a decrease in the charge of the particles. In the general case the conversion of micelles of iron hydroxide to the isoelectrical state is represented thus:

$$x \to 0;\ (n-x) \to n,$$

$$\{m\,[Fe(OH)_3]\cdot n\,FeO^+ n\,Cl^-\}^0$$

In that state the sol is least stable and easily decomposed, and the dispersed phase comes down as sediment.

Let us give some experimental data on the electrolyte concentrations that produce coagulation of iron colloids. From Table VIII it is seen that the maximum coagulating action is exerted by divalent anions and, primarily, by disseminated sulfate ion.

As has already been mentioned, the results of experiments on the stability of colloidal solutions are difficult to compare inasmuch as they were made with different concentrations of the sol, different pH, and with sols of different purity. Therefore we carried out several series of experiments with undialyzed colloidal solutions of Fe_2O_3 of various concentrations and strictly controlled pH values. The dependence of the stability of the colloidal solutions on time, the relationship of concentration of the sol and pH, the effect of electrolytes on sols of different concentration, and several other questions were studied. The results obtained are given in summary.

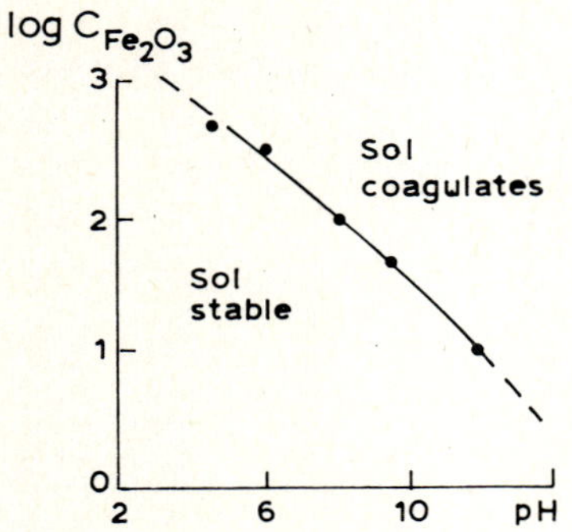

Fig. 48. Stability of colloidal solutions of iron hydroxide of various concentrations (mg/l) as a function of pH.

(1) Acid undialyzed sols of iron hydroxide obtained by hydrolysis of a solution of $FeCl_3$ are characterized, in the absence of coagulating electrolytes, by considerable stability even at high iron concentrations. The pH values of those solutions depend on the concentration of iron and range from 3.6 at 5 mg/l to 1.45 at 5000 mg/l. In highly acid solutions some of the iron occurs in ionic form. Observations in the course of two years revealed no visible changes— the solutions retained their color and transparency, the pH value did not change appreciably. Thus the migrational capacity of ferric iron in acid solutions is considerably increased when colloids are formed.

(2) An increase in pH of the solution leads to a decrease in the stability of the iron hydroxide colloids, and a certain relationship exists between the maximum concentrations of sol and the alkalinity of the medium (Fig. 48). Concentrated colloidal solutions ($C_{Fe_2O_3} > 500$ mg/l or $\log C_{Fe_2O_3} > 2.7$) are stable only in an acid environment ($pH < 4.5$), more dilute solutions are stable in a wide range of pH. Neutral waters with $pH = 6$–8 may contain up to 0.8–0.1 g/l of colloidal ferric oxide in the absence of coagulating electrolytes (except NaCl). The change in pH can serve as a geochemical barrier, causing deposition of considerable amounts of $Fe(OH)_3$, when acid colloidal solutions are neutralized in sedimentary basins.

(3) Sols of iron hydroxide are extremely unstable in the case of reaction with electrolytes, among which the main coagulant is sulfate ion (Fig. 49). With an overall $MgSO_4$ content of 10–30 mg/l ($\log C = 1.0$–1.5), colloidal iron is completely precipitated, both from concentrated solutions ($C_{Fe_2O_3} >$ 500 mg/l) and from dilute ($C_{Fe_2O_3} \approx 10$ mg/l). Salts obtained by evaporation of sea water have a similar effect. Inasmuch as the sulfate content in the present seas and oceans is tens or hundreds of times greater than the coagulation threshold, it is to be expected that all colloidal iron is already deposited from solutions of any concentration in near-shore zones or at the

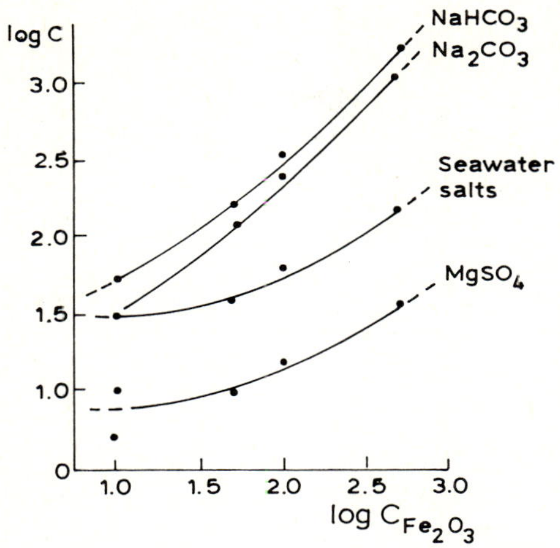

Fig. 49. Minimum concentrations of individual electrolytes causing precipitation of iron hydroxide (coagulation threshold, mg/l) as a function of concentration of sol.

mouths of rivers. Here the geochemical barrier is the gradient of sulfate ion content. Other components of sea water are less effective—sodium chloride has no obvious coagulating effect, and the effect of carbonates likewise appears only at higher concentrations, variation in pH operating additionally.

(4) Carbon dioxide has no essential effect on the stability of iron hydroxide colloids toward the action of electrolytes. The coagulation thresholds (concentrations of $MgSO_4$) prove to be the same for colloidal solutions of iron saturated with carbon dioxide ($P_{CO_2} = 1$ bar) as for solutions in the air.

Structure and stability of colloidal solutions of silica

The distinguishing feature of silica compared to iron hydroxide is its rather high solubility not only in acid but also in neutral solutions. As has already been mentioned, amorphous highly dispersed siliceous sediments such as gels, opal, and chaldedony are most soluble; for them true solubility reaches 80–200 mg/l (Krauskopf, 1956; Okamoto et al., 1957). The main form of dissolved silica is the $H_2SiO_4^0$ molecule, charged anions appearing only in highly alkaline environments. Molecular solutions are less stable systems than ionic due to the tendency of uncharged particles to aggregate, with subsequent polymerization and condensation. Therefore even in undersaturated solutions there are colloidal particles, whose concentration increases in the course of time. Colloidal silica usually is present in natural waters, although truly soluble forms predominate (Ginzburg and Kabanova, 1960).

Structure of silica colloids. Present ideas on the state of silica in aqueous solutions are examined in detail in the review by Sretenskaya (1970), where it is mentioned that one of the most typical features is the tendency to polymerize, i.e. to convert from a molecularly dispersed to a colloidally dispersed state. Polymerization is based on the process of condensation of silicol groups $Si(OH)_4$ with liberation of water and as a result formation of the siloxane bonds Si–O–Si.

Grinberg (1966) explains the mechanism of polymerization by the ability of silica to form aquo-hydroxy ions. Subsequent combination of such ions with the liberated water leads to formation of Si–O–Si linkages. Iler (1959) also reports that in polymerization discrete particles are formed, not linear molecules, neutral elements condensing until seed crystals of SiO_2 with coordination number 4 are formed, so that the polymer can be expressed in the form:

$$\left[Si(OH)_5 \cdot (SiO_2)_{n+1} \cdot Si(OH_5)\right]^{-2}$$

Sretenskaya draws the conclusion: explaining the mechanism of polymerization in different ways, all investigators agree that in aqueous solutions molecules of silicic acid are polycondensed, liberating water and forming molecules of polysilicic acids of various degrees of complexity. These complex molecules apparently are the nuclei of micelles of colloidal silica. They are very hydrophile, and hydration of the particles is a factor sufficient to make the colloidal system stable even in the absence of stabilizing electrolytes, i.e. when the particles have no electrical charge of their own.

Kul'skiy (1960) gives this form of the micelle:

Granule

adsorption layer

$$\{m[SiO_2] \cdot \; n\,SiO_3^{2-} \; \cdot 2(n-x)\,H^+\}^- \; 2x\,H^+$$

nucleus — potential-determining ions — counterions

micelle

The charge of a colloidal particle is explained by dissociation of the solid layer of the nucleus itself—the SiO_2 molecule—rather than by adsorption of ions from the solution. In a neutral environment a particle of colloidal silica is practically uncharged (Iler, 1959), as the hydrogen "counterions" are firmly bound and in fact form OH groups.

Obtaining colloids of silica. The main laboratory methods of obtaining colloidal solutions of silica are condensational and come down to hydrolysis of soluble salts. The difference between obtaining colloids of $Fe(OH)_3$ and SiO_2 is in the properties of the solutions of the original salts: an acid reaction for iron and alkaline for silica.

To obtain pure sols, dialysis or electrodialysis, ion exchange, and peptization at high temperatures are used (Iler, 1959).

In our experiments silica sols with a certain pH value were obtained by hydrolysis of sodium metasilicate at 100°C in the presence of hydrochloric acid, the concentration of which was calculated in advance.

Stability of colloids of silica. It depends both on pH and on the presence of electrolytes. From published data (Iler, 1959; Ginzburg and Kabanova, 1960) it is known that partial or complete coagulation of a silica sol can be expected in the range of $pH = 5-8$ and that slightly alkaline sols are more stable than slightly acid, in connection with the stabilizing effect of the negative charges of OH^-. An increase in pH to 11.5 causes the colloidal particles to dissolve, forming silicate ions.

The effect of electrolytes on the stability of colloidal solutions of silica remains insufficiently clear despite the substantial number of published works.

In some experiments and published reviews (Moore and Maynard, 1929; Krauskopf, 1956; Okamoto et al., 1957; Bien et al., 1958; Harder, 1965) the possibility of deposition of silica from undersaturated solutions is examined. However, many of these experiments proved to be irreproducible, and in some cases the existence of colloids was not strictly proven or equilibrium was not attained.

In other experiments more concentrated solutions were studied. Thus, Lovering (1923) showed that the addition of sea water to a solution of sodium silicate causes instantaneous coagulation at $C_{SiO_2} = 490$ mg/l, but deposition of silica was incomplete and 288 mg/l SiO_2 remained in solution. In the experiments of Kazanskiy et al. (1968) sea salt also was used as coagulant, and it was established that only $CaCl_2$ and $MgCl_2$ are effective precipitants of silica, the lower limit of concentration of the sol to be precipitated being 92–94 mg/l SiO_2. However, Allen and Matijevic (1969), who studied the effect of NaCl, KCl, $CaCl_2$, and other salts on the stability of silicate colloids at various pH, came to the conclusion that none of these salts is an active coagulant in acid environments. Even in the pH range where colloids are unstable by themselves, a much higher concentration of

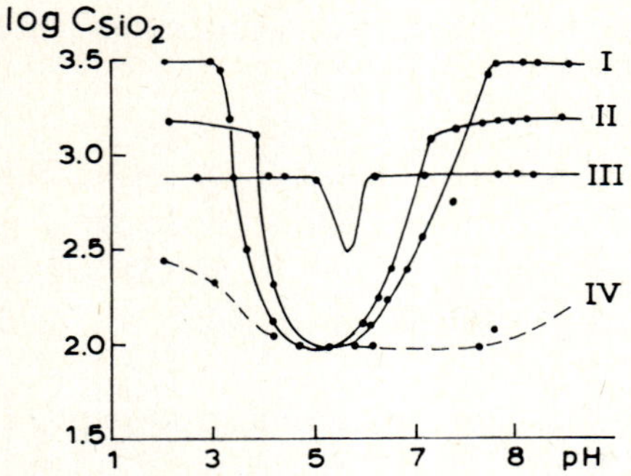

Fig. 50. Stability of colloidal silica as a function of pH. Total concentration of SiO_2 in initial solution, in mg/l: $I = 3000$; $II = 1500$; $III = 750$. IV = concentration of ionic silica.

electrolytes was needed to coagulate them than to coagulate typical lyophobic colloids.

To study the effect of pH on the stability of undialyzed silica sols, solutions of these initial concentrations were used: 300, 750, 1500, and 3000 mg/l SiO_2. The observations were made over several months, starting from the moment the sol was prepared.

The contents of total ($SiO_{2coll} + SiO_{2ion}$) and ionic silica in solution were determined after three months. The results obtained are presented in a graph (Fig. 50) and can be discussed briefly as follows:

(1) The stability of undialyzed colloidal solutions of silica in the absence of electrolytes (except NaCl) depends on the pH and concentration of the sol. Sols of all concentrations are quite stable in acid (pH = 2–3) and alkaline (pH = 8.5–10.0) environments.

(2) Other things being equal, a more concentrated sol is less stable.

(3) Maximum deposition of colloidal silica is observed in slightly acid solutions with pH = 4.5–6.0.

(4) A relationship exists between concentration of the original sol and the pH range in which this sol is least stable. For highly concentrated sols this range is wider (pH = 4–7 at $C_{SiO_2} = 1500–3000$ mg/l), and for less concentrated sols, narrower (pH = 5–6 at $C_{SiO_2} = 750$ mg/l).

(5) Low-concentration silica sols ($C_{SiO_2} \leqslant 300$ mg/l) are very stable and showed no signs of coagulation in the whole pH range studied (2–10) in the course of three months.

(6) The concentration of "residual" silica, analytically determined as the monomeric form of silicic acid, is about 100 mg/l in slightly acid and neutral solutions.

(7) The pH gradient can be an important geochemical barrier controlling the deposition of silica from colloidal solutions. Such deposition would take

place most intensively from initially concentrated acid solutions ($C_{SiO_2} \geqslant 1000$ mg/l) beginning at pH = 3–4, and from less concentrated solutions (C_{SiO_2} = 100–1000 mg/l) at pH ≈ 5.0–5.5. The rate of coagulation is inversely proportional to silica concentration, and for sols with $C_{SiO_2} \leqslant 300$ mg/l it probably is measured in years.

To study the effect of electrolytes, sols of several concentrations were also used (150, 300, 750, and 1500 mg/l), and the experiments were made at various controlled values of pH (2, 4, 6, 8, and 10). A special series of experiments in a neutral medium (pH = 6) was conducted separately to compare the coagulating effect of the individual ions of sea water. From the preliminary results of these experiments, the following conclusions can be drawn:

(1) Addition of electrolytes to silica sols extends the range of pH values in which these sols are unstable. For instance, "pure" sol with C_{SiO_2} = 750 mg/l is unstable at pH = 5–6, and in the presence of electrolytes coagulation is observed at pH = from 4 to 8 or 10.

(2) The resistance of silica sols to the action of electrolytes is reduced as SiO_2 concentration increases. In low-concentration sols ($C_{SiO_2} \leqslant 150$ mg/l) no coagulation is observed.

(3) The coagulating effect of electrolytes increases as the reaction of the medium changes from acid to slightly alkaline.

(4) The coagulating effect of electrolytes is produced by cations, the relative effectiveness of which increases in the order Na^+–Ca^{2+}–Mg^{2+}. Figure 51 shows as an example the dependence of completeness of coagulation of a silica sol in an almost neutral solution (pH = 6) on the concentration-

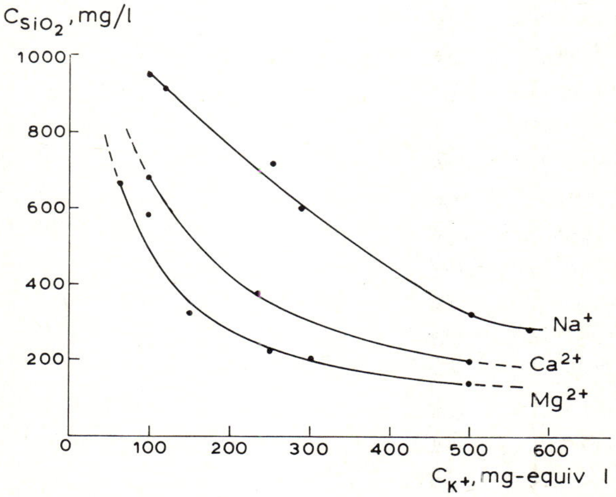

Fig. 51. Stability of colloidal silica at pH ~ 6 as a function of concentration of coagulant cations. Initial concentration of SiO_2 = 1000 mg/l.

tion of cation-coagulants (SiO_2 in the solutions was determined two months after adding the electrolytes). Compared to sols of iron hydroxide, silica sols are less resistant to the action of electrolytes. However, the concentration of salts in the present ocean is sufficient to cause coagulation. When the actual concentrations of salts in the present sea waters are taken into account, it turns out that Na^+ will have the greatest coagulating effect, Mg^{2+} is close to it, and Ca^{2+} has relatively little effect.

(5) Depending on various factors, the main ones of which are the SiO_2 concentration and pH, the product of coagulation may be a gel or a flocculent precipitate of amorphous silica. In the $pH = 5.5–6.5$ range, where the rate of coagulation is very high (the process of separation of the solid phase from solution takes a few days or weeks), a flocculent amorphous precipitate is formed. The formation of gel-like clots of silica usually occurs as a result of slow (a few months or even years) coagulation of sols, observed in slightly acid environments ($pH = 4$).

Stability of mixed iron-silica colloidal solutions

The mutual influence of colloids of iron hydroxide and silica occurring together in solutions is of dual character and in some cases leads to mutual coagulation of the sols, but in others to shielding of the iron hydroxide sol.

The idea, deeply rooted in the geologic literature, that oppositely charged colloids of iron hydroxide and silica mutually coagulate to form banded sediments due to the kinetic-sedimentational effect is the result of the well known experiments by Moore and Maynard (1929). However, these experiments were conducted with very dilute solutions undersaturated with SiO_2, to which the laws of colloid chemistry cannot be applied, as recent investigations have shown. The reason for partial deposition of silica from such solutions evidently should be sought in sorptional trapping of SiO_2 by iron hydroxide during coagulation.

In an experimental study of mixed iron-silica colloids we found the opposite phenomenon, stabilization of colloids of iron hydroxide by silica colloids, which is manifested very clearly in undialyzed iron hydroxide sols of average concentration with a ratio of $Fe_2O_3:SiO_2 = 1:3$. Colloidal iron in mixed solutions proved to be more resistant to the action of electrolytes than in isolated sols of iron hydroxide. Only colloidal silica shows any stabilizing effect on sols of iron hydroxide; dilute solutions undersaturated with $Si(OH)_4^0$ are not stabilizers. It is characteristic that colloidal silica is capable of stabilizing colloidal iron in the same pH ranges in which "pure" silica sols are stable: in acid ($pH < 4$) and alkaline ($pH > 8$) environments. In slightly acid environments ($pH = 5–6$) iron-silica sols are unstable and decompose to form mixed sediments, which sometimes are not uniform due to different rates of coagulation and deposition.

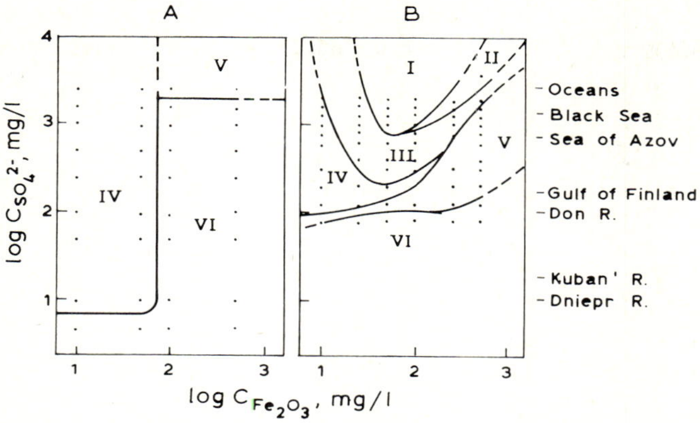

Fig. 52. Coagulation of mixed colloidal solutions of iron and silica. Environment: A—slightly acid; B—slightly alkaline. Concentration of recent waters indicated at right. *I* = total precipitation of iron and silica with formation of banded sediments; *II* = same, with formation of mixed sediments; *III* = total precipitation of iron and partial of silica; *IV* = total precipitation of iron, silica in solution; *V* = incomplete precipitation of iron, silica in solution; *VI* = no precipitation observed.

Experiments were made in slightly acid (pH ~ 4) and slightly alkaline (pH = 8–11) environments with sols of different concentration, prepared by mixing colloidal solutions of $Fe(OH)_3$ and SiO_2 with a given pH value. Solutions of $MgSO_4$ of appropriate concentration were used as the coagulating electrolyte. The results obtained are shown in diagrams (Fig. 52) and can be presented briefly as follows:

(1) In dilute solutions ($C_{SiO_2} < 200$ mg/l) there is no stabilization, iron hydroxide coagulates completely under the influence of very dilute solutions of $MgSO_4$ (< 10 mg/l in acid mediums and 100 mg/l in alkaline), and silica remains in solution.

(2) More concentrated solutions (100–500 mg/l Fe_2O_3 with 300–5100 mg/l SiO_2) are very stable both in themselves, and toward the action of electrolytes. After long storage (two years) of alkaline solutions, no sediment appears, there is no change in color or transparency, and complete coagulation can be produced only by solutions of electrolytes of substantial concentration ($C_{MgSO_4} > 1000$ mg/l). In acid environments the sols also are stable or coagulate very slowly and incompletely, forming a uniform iron-silica gel.

(3) Coagulation is different in acid and alkaline environments. In alkaline media products of incomplete coagulation, mixed sediments, sometimes banded, often appear.

(4) The optimal concentrations were determined at which iron and silica sediments separate rapidly from a mixed sol as it coagulates, forming banded jaspilite-like layers. In these experiments, differential deposition of iron

hydroxide and silica was very clearly observed only in alkaline solutions with an average concentration of the sol ($C_{Fe_2O_3} = 100$ mg/l) and high concentration of the electrolyte ($C_{MgSO_4} > 1000$ mg/l). It is not ruled out that complete differentiation of $Fe(OH)_3$ and SiO_2 also occurs in the case of coagulation of more dilute sols.

(5) The geochemical barrier which causes rapid deposition of colloidal iron from neutral or slightly acid dilute ($C_{Fe_2O_3} < 100$ mg/l) mixed sols stabilized by SiO_2 is the gradient of concentration of electrolytes, mainly $MgSO_4$. In this case colloidal silica practically does not coagulate and is deposited very slowly. As a result sols of SiO_2 will easily cross the concentration barrier which causes the separation of chemogenic deposition of iron and silica in space and time.

(6) Joint deposition of SiO_2 and Fe^{2+} can be presumed in the case of a high gradient of $MgSO_4$ concentration, for instance in the case of intensive mixing of fresh water with sea water of normal or slightly low salinity. According to the experimental data, in such conditions mixed iron-silica sediments without the typical banding are formed. Apparently their natural analogs are iron ores of Kerch type.

(7) Interaction of relatively concentrated alkaline mixed sols with sea water of normal salinity is necessary for the formation of banded sediments of jaspilite type.

(8) The data obtained do not preclude the possibility of mutual coagulation of colloids of iron and silica for different quantitative relationships in the solution or for another mechanism of formation of mixed sols. Obviously in that case colloidal silica proper would occur in the solutions and its concentration would be higher than the equilibrium concentration.

Geochemical data on the forms of transport and conditions of deposition of iron and silica in natural waters

In examining the particulars of the distribution, migration, and deposition of iron and silica in natural waters it is advisable to distinguish normal sedimentary processes and volcanic sedimentary processes. By normal sedimentary processes we mean the cycle: weathered layer–transport–deposition in a sedimentary basin. The main distinguishing feature of volcanogenic-sedimentary processes is the endogenetic source of the material and exogenetic method of its deposition.

Normal sedimentary process

In the works of Strakhov (1960) on the theory of lithogenesis, iron compounds are assigned to the third group, which is characterized by very low solubility of most natural compounds in water, negligible amount in a dissolved state in river and sea waters, and ability to yield colloidal solutions. An appreciable amount of ferrous iron is sometimes carried by ground water, but when it reaches the surface it is rapidly and almost completely oxidized and then precipitated in the form of hydroxides, which are transported mechanically.

Silica is more soluble than ferric iron, and together with carbonates is assigned to the second group of compounds, which migrate to a substantial extent in dissolved form. Natural waters usually are sharply undersaturated in SiO_2, which apparently is explained by kinetic factors, i.e. by the low rate of solution of quartz and silicates. Chemogenic deposition of silica from present sea waters is not very likely; biologic extraction predominates in them.

Transport and deposition of iron

Migration of iron with river waters. Recent river waters contain Fe in from barely perceptible traces (0.01 mg/l) to 7.82 mg/l; the usual content is 0.2–1.5 mg/l. Kontorovich (1968) mentions that the average Fe content in the rivers of the world is 0.73 mg/l, including 0.53 for the rivers of North America, 0.61 for Europe and India, and 1.95 for South America; in the rivers of the Central Siberian plateau the content is 0.32 and of the West Siberian lowland, 1.01 mg/l. In a work by Konovalov et al. (1968) it is emphasized that iron in dissolved form is a microcomponent of river water and its concentration reaches the order of tens, rarely hundreds of micrograms per liter. According to Strakhov's data the percentage of ionic iron is very small even in those cases where more soluble ferrous iron is carried in substantial amount by oxygen-free ground waters feeding rivers; the Fe^{2+} is rapidly and almost completely oxidized and precipitated in the form of $Fe_2O_3 \cdot H_2O$. Mokiyevskaya (1959) also mentions that under anaerobic conditions iron is dissolved relatively rapidly in water containing CO_2 and forms the bicarbonate, which partially dissociates in aqueous solution. With a decrease in P_{CO_2}, related to a rise in temperature or occurring as a result of photosynthetic activity of plants, $FeCO_3$ is precipitated from solution. The carbonate is unstable under anaerobic conditions; Fe^{2+} is easily oxidized to Fe^{3+} with precipitation of $Fe(OH)_3$. Other versions of deposition of Fe^{2+} are possible. Thus, Castano and Garrels (1950) concluded on the basis of laboratory experiments that a substantial amount of ferrous iron can be transported in aerated river waters with $pH = 7$ or less. However, when such waters arrive in a marine environment with solid calcium carbonate, the iron mainly is deposited wholly in the form of Fe^{3+} oxide, and deposition occurs both directly from solution and by way of direct replacement of $CaCO_3$. Calcium carbonate in equilibrium with sea water determines the pH of the solution (buffer system), but at fixed P_{O_2} and *Eh* of the solution. In the experiments cited, the equilibrium value is $Eh \sim 0.550$ at pH ~ 7.8. However, in most cases the conditions necessary for stability of Fe^{2+} are not fulfilled in river waters, and the importance of that ion in the transport of iron, like that of simple complexes of Fe^{2+} with inorganic anions, is not great.

From the geochemical data it follows that a substantial part of dissolved iron migrates not in ionic form, but as $Fe(OH)_3$ sol shielded by organic matter (also colloidal) or by a sol of SiO_2, and also as colloidal (or truly dissolved) complex ferro-organic compounds, chiefly soluble iron humates (Strakhov, 1960).

A distinct dependence of the optimum conditions for chemical transport of iron on the oxidation-reduction setting of the environment and mobilizing action of organic matter has been established. The content of dissolved Fe in rivers flowing over a swampy lowland reaches 14 mg/l, and in Strakhov's opinion this figure is not extreme and still higher iron concentrations can be reached in nature. According to the data of Glagoleva (1959), in rivers draining a swampy locality and carrying a substantial amount of organic acids (for instance, the Pripyat' in the Ukraine), the content of dissolved iron sometimes is very high, and sometimes even exceeds the content in suspension. The form of occurrence of the iron in solution apparently is colloidal and molecular–dispersed complex metallo-organic compounds. Kontorovich (1968) points out that the Siberian rivers with swampy catchment areas carry considerably more Fe (up to 0.9–1.35 mg/l) than rivers flowing on the southern part of the lowland and having slightly swampy or not at all swampy catchment areas (0.1–0.2 mg/l). The presence of organic matter stabilizes solutions of less mobile elements, including Fe. Thus, 73% Fe_2O_3 was found in the ash of a petroleum ether extract of an average sample of Ob' water. Waters with a low content of C_{org} always contain little Fe, and those with high C_{org} may contain either much or little of it. Undoubtedly an increase in Fe content in waters is governed to a decisive extent by an increase in the organic matter content.

Perel'man (1968) points out that in neutral waters true dissolved molecules of undissociated $Fe(OH)_3$ can also occur in predominant amount (compared to Fe^{3+} ion). Probably the molecules of $Fe(OH)_3$ are stabilized by organic acids. It is important that transport in the form of true ionic solutions is possible only for divalent iron, while trivalent iron migrates mainly in the form of colloids and organic compounds. However, Castano and Garrels (1950) believe that aerated waters provide conditions adverse to the preservation of organic matter itself. The available data suggest that organic matter decomposes at the same rate at which it arises.

And finally, in addition to dissolved ionic and colloidal forms, part of the iron is always transported in the composition of various particles mechanically, both in a suspended state and dragged along the bottom. The mineralogic forms of iron transported in suspension are diverse: hydrogoethite in different degrees of hydration, silicates, semi-weathered and weathered aluminosilicate minerals — hornblende, hydromicas, magnetite, titanomagnetite, etc. (Strakhov, 1960). According to the data of Glagoleva (1959), in most modern rivers Fe migrates mainly in suspended form and only partly in a dissolved state. The suspended matter consists partly of fragmental material, but mainly of colloidal hydroxides. Study of recent and ancient marine deposits confirms the predominant transport of Fe in mechanical ways. Thus, in the deposits of the Sea of Okhotsk most of the Fe arrives with clastic material and with the products of volcanic activity (Ostroumov, 1955).

Only in those rivers where there is very little suspended matter are the amounts of dissolved Fe and iron in suspension comparable to any extent (Pechora, West Dvina, Ob'), but in most rivers by far the greatest amount of Fe occurs in suspension. The suspended form is the main form of its migration (Konovalov et al., 1968).

In conclusion, if we limit ourselves to considering the transport of iron in the dissolved state (in the form both of true ionic and molecular solutions and of stabilized colloidal solutions), it is feasible to refer to the scheme of geochemical behavior of iron in the conditions of a humid climate according to Strakhov (1960). Dissolved iron is carried from regions with abundant vegetation, for instance from swamps, almost entirely in the form of ferrous bicarbonate, accompanied by a large amount of carbonic acid. When the ground waters arrive in well aerated river waters the CO_2 is volatilized and the oxygen of the air oxidizes Fe^{2+} to Fe^{3+}, ferric oxide is precipitated from solution chiefly in the form of a gel, and some residue continues to migrate with the water, shielded by organic colloids or as a ferro-organic compound. Clots of gel forming part of the suspended matter also migrate in large part.

The development of ore accumulations of iron in sediments due to migration in the dissolved state is possible, if high Fe concentrations (10–14 mg/l or more) are combined with low overall turbidity of the rivers (absence of contaminating non-ore impurities). It is usually believed that the solutions of iron compounds were colloidal before deposition, or necessarily went through a colloidal stage during deposition. When solutions seeping from swamps arrived in river or sea water with a sharply different physicochemical environment, an abundance of free oxygen, oxidation of Fe^{2+} to Fe^{3+} and rapid coagulation of $Fe(OH)_3$ began.

Iron in the waters of the present seas and oceans. Inasmuch as suspended iron occurs chiefly in the form of particles of subcolloidal size, it often passes beyond river mouths without being deposited. The Fe concentration in sea water is 20–50 times less than in river water, which can rightly be explained by simple mixing, dilution of the river water, without resorting to actual chemical precipitation (Strakhov, 1960).

Outside the zone of mixing the Fe content is negligible—10–20, more often 3–6 mg/m^3. Skopintsev and Popova (1960) point out that the Fe content in sea water is not more than 50–60 μg/l according to the data of various authors.

The forms of occurrence of iron in sea waters were investigated in detail by Mokiyevskaya (1962). The total amount of Fe is the sum of dissolved, colloidal, and suspended iron of organic and inorganic origin:

$$Fe_{total} = Fe_{dis} + Fe_{col} + Fe_{sus}$$
$$= (Fe_{inorg} + Fe_{org})_{dis} + (Fe_{inorg} + Fe_{org})_{col} + (Fe_{inorg} + F_{org})_{sus}$$

Iron of inorganic dissolved compounds (bicarbonates, sulfates, chlorides, fluosilicates, etc.) may enter into the dissolved form of iron of inorganic origin (Fe_{dis}^{inorg}), but their existence is governed by an acid environment with a pH not higher than 3. As a rule the pH in sea water is close to 8 ($\pm$0.5). Under these conditions iron compounds are easily hydrolyzed and converted into hydroxides, which form colloidal solutions in sea water. In appropriate conditions colloidal hydroxide condenses to clots of gel and converts to the suspended state. Therefore there are practically no ionic forms of iron (Fe_{sol}^{inorg} proper). As early as 1937 Cooper (1937) concluded, on the basis of the solubility product and activity of ferrous and ferric iron ions and $FeOH^{2+}$ ions, that until equilibrium is reached sea water may contain about 10^{-8} μg/l of iron ions in true solution at pH = 8.5; the amount of ionic Fe in ferric form is still less—10^{-12} μg/l, which corresponds to the extremely low solubility of $Fe(OH)_3$. That amount of Fe in the ionic state is not detectable by present chemical methods. Strakhov (1960) also points out that ferric iron can in no way occur in sea water in the form of an ionic solution, but must form "colloidal suspensions". However, direct determinations of Fe dissolved in sea waters yield values substantially higher than the calculated values. Thus, according to Skopintsev and Popova (1960) the content of dissolved iron in the Pacific Ocean is fairly constant and averages 3.4 μg/l (with fluctuations of 0.7–5.2). Contents close to the lower limit were established by the authors in the waters of the Black Sea. Simons et al. (1953) determined 2.6 μg/l of dissolved Fe in the surface waters of the Atlantic Ocean and Gulf of Mexico (fluctuations from 1.0–7.0). The discrepancies are explained by various authors (Brusilovskiy, 1963; Perel'man, 1968) by the presence in true solution of molecules of undissociated $Fe(OH)_3$, which has already been mentioned . However, in the opinion of most investigators

(Skopintsev and Popova, 1960; Strakhov, 1960) most of the dissolved form of iron is in Fe-organic compounds (Fe_{sol}^{org}), as in rivers. Sometimes colloidal formations in which the existing hydroxides or sulfur compounds are "under the protection" of organic compounds of the sea water also belong to this group. Strakhov believes that most of the iron occurs in the form of ferro-organic compounds and a very insignificant part in the form of the hydroxide $Fe(OH)_3$; for instance, in the Barents Sea only 1–3 mg/m^3 out of 10–20 mg/m^3 is $Fe(OH)_3$. In contrast to $Fe(OH)_3$, the organic compounds of Fe dissolved in the sea do not saturate the water and might be further accumulated, but hydrolysis prevents this. Fe_{dis}^{org} may be represented by complex salts of organic acids of the type of tartrates, ascorbates, oxalates, citrates, and humates, and also products of decomposition of hemoglobin.

Colloidal iron consists mainly of colloidal solutions of hydroxides and phosphates of Fe (Fe_{coll}^{inorg}) and colloidal solutions of the enumerated organic compounds (Fe_{coll}^{org}). Direct determination of colloidal iron in sea waters is difficult and usually dissolved iron means the sum of $Fe_{dis} + Fe_{coll}$.

In many cases a large part of the Fe in sea water is not fixed in Fe_{dis} (including Fe-organic compounds), but enters into the composition of suspended particles (Fe_{sus}) of variable diameter—from large (more than 100 μm) to colloidal-dispersed (0.1–0.2 μm). Inorganic iron (Fe_{sus}^{inorg}) included in the composition of clastic turbidity, clots of Fe hydroxide and phosphate gels, and $Fe(OH)_3$ adsorbed on the surface of mineral particles, and Fe_{sus}^{org} in organic detritus, fecal pellets and living organisms, are part of this form (Mokiyevskaya, 1962). The total content of Fe_{sus} usually exceeds that of Fe_{sol}. According to the data of Skopintsev and Popova, the average content in the open ocean is 4.5 μg/l for a total of $\sim$8 μg/l; the highest content of Fe_{sus} is 10–50 μg/l in the upper layers of the near-shore region (all figures for the Pacific Ocean). Interesting data on the Fe content in suspension in the Baltic Sea are given by Yemel'yanov (1968). Maximum Fe is confined to the sources of supply (clastic)—almost exclusively related to limonite, hematite, flocs and globules of iron hydroxide, micas, various clay aggregates, and rock fragments. At a distance of 20–30 km from the coast suspensions of biogenic type already predominate: organic detritus, globules of iron hydroxide of unclear origin. In straits and bays Fe-organic particles and clots constitute up to 80–90% of all ferruginous material. Everywhere the maximum amounts of Fe are confined to the surface layers of the water (0–25 m), where the concentration of suspended matter and its content of clastic material are maximal. The author concludes that in the Baltic Sea, as in rivers, most of the iron occurs in the form of suspended matter.

Conditions of deposition of iron from natural waters. Below we will limit ourselves to an examination of the conditions and forms of deposition of iron only from solution, without going into the conditions of formation of the sediments containing mechanically transported suspended particles of iron compounds. In the general case deposition of dissolved matter from sea water is accomplished chemically, biogenically, and by sorption.

According to Strakhov (1960), in the waters of the present sedimentary basins dissolved iron occurs chiefly in trivalent form, and the proportion of truly dissolved iron (ionic—Fe^{3+} etc.) is insignificant; other forms are unstable. Enlargement of the micelles of colloidal solutions when they collide leads to slow settling to the bottom. Gradual hydrolysis of ferro-organic compounds leads to formation of $Fe(OH)_3$, which later coagulates and settles out. Ultimately iron introduced into the basin in the form of ferro-organic compounds reaches the bottom in the form of hydroxide gels.

Chemical deposition of iron from ionic solutions is limited to specific regions with anaerobic conditions, which are not characteristic of the vast basins of marine sedimentation or of regions where anaerobic conditions are superseded by aerobic. The development of living matter usually is the controlling condition.

One can give as an example the chemical deposition of iron in the special environment of swamps. According to the data of Kovalev and Generalova (1969), in the peat bogs of Belorussia the equilibrium form of Fe is determined by the combined effect of the waters and microbiological activity. In the summer, in the period of the biological maximum, oxygen dissolved in the peaty waters is almost entirely consumed by microorganisms which in turn leads to an appreciable reduction in *Eh* and a shift of equilibrium toward prevalence of compounds of metals with low valency. In the cold half of the year biological activity of living substances and the process of decomposition of organic matter are minimal, the concentration of O_2 increases and that of CO_2 decreases. Under these conditions deposition of some Fe compounds that are not very abundant in sediments of the usual type, in particular ferrous hydroxide (ferrohydrite), is observed. In the presence of phosphate ion vivianite is formed; in peat-bog deposits there exists a very mobile equilibrium between ferrous hydroxide and siderite, apparently determined by the CO_2 regime.

Transport of iron in carbonate waters, mainly in the form of Fe^{2+} bicarbonate, is more common. The decrease in P_{CO_2} due to the overall reduction in pressure when ground waters come to the surface, when carbon dioxide is consumed as a result of photosynthetic activity of plants or even, as Mokiyevskaya (1959) mentions, when the temperature rises, leads to deposition of $FeCO_3$. In Strakhov's opinion such a process could lead to the formation of oolitic hydrogoethite-chamosite-siderite ores. The iron migrated in mobile form as Fe^{2+}, which accumulated in solution in a reducing environment. Formation of the ores was related to the draining of high-iron waters formed in swampy regions. The near-shore parts of the sea with complex bottom relief were the trap for the iron; when the physicochemical conditions changed abruptly, there began oxidation of Fe^{2+} to Fe^{3+}, formation of a colloidal solution, and its coagulation. Malakhovskiy (1956) does not rule out the possibility of direct deposition of siderite along with rhodochrosite, vivianite, and in part with chlorite and silica in similar conditions in the formation of the Kerch ores. However, colloidal and metacolloidal compounds of Fe^{3+} played the leading role in the transport of the ore masses, the final product of deposition being the hydroxide.

Biogenic accumulation of Fe in sediments (along with buried organisms) is very unimportant, although according to Boychenko et al. (1968) the concentration of Fe by plants from sea water is 10^4-10^5, and from river water 10^3-10^4, compared to the content in the surrounding medium. However, dry plant matter contains a total of $10^{-2}-10^{-1}$% iron.

Transport and deposition of silica

Migration of silica with river waters. In recent summaries of the silica content in thermal and cold ground waters of the U.S.S.R. and other countries (White et al., 1956; Ginzburg and Kabanova, 1960; Bogomolov et al., 1967) it is mentioned that the waters of platform structures are undersaturated in amorphous silica, the content of which ranges from units to 20–40 mg/l. This state of undersaturation is maintained in river waters also. Ginzburg and Kabanova (1960) report that the waters of arctic lands contain less SiO_2 than those of the subtropics: circumpolar rivers contain from 1.65 to 7.5 mg/l SiO_2, and the subtropical mountain streams of South America from 9.0 to 47.5 mg/l; the overall variations in the river waters of the dry land are still greater—from 1.8 to 137 mg/l.

It is typical that the SiO_2 content is subject to very wide fluctuations depending on the time of year, intensity of rainfall, and total mineralization of the waters. Ginzburg and Kabanova concluded that one and the same bulk content of SiO_2 can be produced by different causes, which are reflected both in the form in which the SiO_2 exists and in its amount. The SiO_2 content of mineral waters is highest, especially those of thermal origin, highly hydrochloric-acid waters containing a substantial amount of fluorine and boron. The absolute SiO_2 content is low in waters of meteoric origin (in the rainy period), and also in waters of glacial origin. In natural waters, usually undersaturated in SiO_2, a distinct predominance of "soluble" (ionic, molecular) silicic acid over colloidal is established. This undersaturation is explained by several factors, of which the main ones are (Ginzburg and Kabanova, 1960): "(1) climatic conditions (in the case of a large amount of precipitation in a short time and dilution of the solution); (2) rocks resistant to weathering in the conditions of high and middle latitudes or rocks that are only slightly soluble (quartz sands and sandstones); (3) slow establishment of equilibrium between the aqueous solution and the solid phase of rocks containing silica; and (4) activity of organisms which use silica to build their skeleton".

Silica in the waters of the present seas and oceans. Strakhov (1960, 1966) repeatedly mentioned that sea waters are still less saturated with silica than ground waters and river waters. The surface waters of the seas and oceans contain 0.5–0.6 mg/l SiO_2 in total, and its concentration often falls below analytical zero (0.2–0.3 mg/l). Everywhere the total concentration of silicic acid increases with depth, the maximum as a rule being observed at depths greater than 2000 m (Bogoyavlenskiy, 1966). Thus, very large concentrations are found in the northern part of the Pacific Ocean at depths of 2000–3000 m, where they are estimated to be of the order of 10–11 mg/l, but not more than 12 mg/l SiO_2. In the other ocean basins the maximum concentrations fall within 3 to 8 mg/l SiO_2. The average for all the waters of the three main oceans is about 5 mg/l (85 μg-at/l). In the same work Bogoyavlenskiy (1966) points out that the largest amount of dissolved silicic acid also is observed in the northern half of the Pacific Ocean, where its average content in the whole water column reaches 6 mg/l; in the other oceans the concentration is lower (2–5 mg/l SiO_2).

The main form of occurrence of silica apparently is dissolved $H_4SiO_4^0$. The existence of complex polymeric forms and sols is not considered to be very likely.

Another form is suspensions of amorphous silica, consisting of the remains of organisms (diatomaceous algae, radiolarians, silicoflagellates), and the content of amorphous silica decreases rapidly with depth, especially in the upper 100 m of the water layer (Lisitsyn et al., 1966). There usually is 5 to 50 times less suspended than dissolved SiO_2. Thus by far the greater part of silica occurs in true solution, even in the surface ocean waters where organic life is especially well developed.

And finally, in near-shore zones some of the silica occurs in the form of tiny fragments of quartz and silicates, the transport of which is governed by the laws of mechanical differentiation.

Conditions of deposition of silica from natural waters. At the present time it is considered to be firmly established that only biological extraction of silica from river and ocean waters sharply undersaturated in SiO_2 is possible (Strakhov, 1960, 1966). In the oceans this process takes place mainly in the uppermost layer, where photosynthesis occurs. Diatomaceous plankton sink down when they die and the silica gradually begins to dissolve (Bogoyavlenskiy, 1966). Despite strong solution of suspended silica during deposition, 0.01–0.1 of the silica reaches the surface layer of the bottom sediments.

Accumulation of silica sediments on the bottom can be expected only in certain zones of the ocean with intensive accumulation of biogenic silica in suspension and a unique hydrodynamic regime (Lisitsyn et al., 1966).

Volcanogenic-sedimentary process

For a correct idea of the physicochemical conditions of accumulation of iron-ore sediments, an analysis of the forms of transport and conditions of deposition of iron and silica in recent active volcanic regions is necessary. Such an analysis should include establishing possible sources of the ore material (vents of active volcanoes, fumaroles, hydrothermal volcanic waters), the character and intensity of the process of migration and forms of transport of the ore components, and the conditions of deposition of the ore components in the course of their migration to the sea basin and when the river waters mingle with sea waters.

Transport and deposition or iron

In regions of recent active volcanism, hydrothermal processes which bring about substantial removal and redeposition of many elements, and in particular of iron, from volcanic rocks are extensively developed. Study of the physicochemical conditions of migration and deposition of elements from such waters is complicated by the continual change in properties of the waters themselves (temperature, pH, *Eh*) as they leave the eruptive centers. In this case, naturally, there is practically no equilibrium in the system studied, except the state of equilibrium established between the congealing products of deposition and the pore solutions.

Recent investigations convincingly indicate that in recent volcanic regions the process of removal and redeposition of iron is going on on a vast scale. Study of this process, correlated with detailed geologic investigations of the cherty iron-formations, makes it possible to interpret their origin from the standpoint of the exhalative-sedimentary theory. According to this theory, the iron arrives in the water basin in the form of volcanic exhalations during terrestrial or submarine eruptions, and also when enormous masses of eruptive rocks in areas of volcanism are reworked by hydrothermal waters. Deposition of iron occurs as a result of a change in physicochemical conditions (Butuzova, 1968; Formozova, 1968).

The hydrothermal waters of recent volcanic districts are the product of mixing of juvenile subterranean solutions given off from the magma and surface (vadose) waters in which volcanic gases are condensed and dissolved (Naboko, 1963, 1968). The latter are formed from volatile substances with low melting temperatures and boiling points, dissolved in the magma (H_2S, HCl, HF, CO, CO_2, SO_2, HBO_3, metal chlorides and fluorides). When the magma cools (crystallizes), volcanic gases are given off due to the increase in internal pressure of volatiles during boiling and distillation.

More than 3/4 of the gaseous mixture given off from the magma in an eruption consists of water. We do not know the actual relationships between the other main components, but in decreasing order they can be arranged thus: gases of the carbon group, sulfur gases, nitrogen, chlorine, hydrogen (Markhinin, 1967b). Concerning the gases of the carbon group it should be mentioned that CO predominates in high-temperature fumaroles, and CO_2 in low-temperature.

When volcanic gases dissolve in ground and surface waters, highly heated, very acid (pH less than 2) hydrochloric and sulfuric acid solutions are formed; at the surface there are no bicarbonate ions in these solutions, the stronger acids—HCl and H_2SO_4—supplant them. At depth, where the solubility of CO_2 increases, the concentration of carbonate ions in the water increases in connection with the increase in pressure, as indicated by the abundant formation of carbonates when deep springs emerge at the surface.

The temperature regime of hydrothermal waters determines both the amount of juvenile solutions coming directly from the magma chamber when water vapor condenses or from deeper horizons of the Earth, and the amount of dissolved magmatic gases. The ionic and gaseous state of the thermal waters also is determined by exchange reactions with the surrounding rocks—a direct relationship between the composition of the waters and that of the rocks in contact with them is observed.

The acidity of thermal waters usually ranges within pH = 1–3, but more acid waters are often encountered (pH less than one and even sometimes less than zero) (Ivanov, 1956). The mineralization of such waters varies with the variation in acidity. In thermal waters which are the condensate of the high-temperature stage with pH less than one, the mineralization is more than 12 g/l. However, from recent manifestations in the Kamchatka volcanic arc only a few cases can be cited where the ionic force of the waters exceeded 0.5. Only the pore solutions are highly mineralized; the freely circulating waters have low ionic force (Sidorov, 1967). As lava cools the acidity of the waters decreases and thus so does their mineralization: whereas at pH = 2 the mineralization amounts to 0.6–1.2 g/l (chloride-sulfate composition), at pH = 5–6 the mineralization amounts to 0.2–0.3 g/l (slightly acid waters, increased role of CO_2) (Basharina, 1963).

Iron comes into volcanogenic waters both as a result of solution of volatile compounds of the magma, $FeCl_3$ and FeF_3, and as a result of leaching of enormous bodies of igneous rocks in volcanic areas. The presence of areas of bleached rocks of substantial size, from which the iron has been leached almost completely, is typical of regions of recent volcanism.

Migration of iron in waters is regulated chiefly by the amount of atmospheric precipitation diluting the hot springs (Zelenov, 1960). In turn, the amount of atmospheric precipitation and ground water depends on the climate. In high mountain regions, where the catchment area is not large, most of the gases go off into the atmosphere, and hot springs that are formed are not highly acid and cannot leach a large amount of iron from the eruptive rocks. Nor are conditions favorable for the transport of iron for great distances created in tropical regions, where a tremendous amount of rain falls, leading to great dilution of thermal waters. The optimal conditions for intensive removal and migration of iron are regimes of temperate latitudes with the active participation of ground waters.

Thus the transport of iron is regulated by the variation in acidity of hydrothermal waters. Inasmuch as the physicochemical properties of iron itself limit the possibility of its transport in dissolved form, a necessary condition for the migration of iron over considerable distances is that the volcanic waters maintain their acidity until they reach the sea basin. Thus, on Paramushir Island (Kurile Islands) the Yur'yeva River, formed by the confluence of hot springs of Ebeko Volcano (the temperature of the springs is about 100°C), maintains the acidity of its waters until they reach the Sea of Okhotsk, as a result of which 35–50 tons of iron are brought to the sea annually (Zelenov, 1958). Table IX presents the measurements of acidity and iron content of the waters of this river from source to mouth. The Gorshkov River, formed by the confluence of numerous acid streams (pH = 2.6 to 2.7 and temperature 20–30°C), contains 1–15 mg/l Fe^{2+} and 6–10 mg/l Fe^{3+}.

Study of the composition of the waters of Ebeko Volcano by Borisenkov (1967) showed that the Fe^{2+} content of various springs ranges from 7.9 to 243 mg/l, and Fe^{3+} from traces to 200 mg/l, with the acidity of the waters from 0 to 3.6.

Analysis of numerous data on the iron content of volcanogenic waters in the Kurile Islands (Markhinin, 1967a) permits drawing the following conclusion: within limits of acidity of the waters from 0 to 2.5, the Fe^{2+} content ranges from a few units to 250 mg/l, and Fe^{2+} from traces to 200 mg/l. However, waters are encountered which contain 500–700 mg/l Fe^{3+} (Upper

TABLE IX
Iron content in the waters of the Yur'yeva River, according to Zelenov

Component	Source	700–800 m from source	1700–1800 m from source	Mouth
pH	1.3	1.4	1.6	1.4
Fe^{2+} (mg/l)	150	20	10	5
Fe^{3+} (mg/l)	120	200	240	320

Yur'yeva springs on Vlodavets Volcano). In the waters of the "Machekha" crater of Teben'kov Volcano an Fe content of 1548.2 mg/l is even observed, but it can be considered anomalous. In very acid springs the iron content can reach very considerable values, as in Lake Idjen (Indonesia) where water with pH = 0.02 contains 2.3 g/l Fe_2O_3 (Zelenov, 1960).

As the pH increases, the amount of iron dissolved in the waters drops off sharply. Usually the iron content in slightly acid waters with pH = 3–5 is a few tens of milligrams per liter. In Kamchatka and parts of the Magadan district on the Sea of Okhotsk, rivers with pH = 4.5–5.3 contain from 1 to 7.8 mg/l of iron (probably due to its leaching from ashy material of recent eruptions, in which iron occurs in an adsorbed form) (Gushchenko, 1965).

The iron content in the waters also depends on several other factors, in particular on temperature and composition of the surrounding rocks, composition of the volcanogenic rocks, and the presence of gases of the carbon group in these waters. These same factors determine the form of transport of iron by volcanogenic waters.

In the opinion of most investigators, in connection with the prevalent chloride-sulfuric acid composition of volcanogenic waters iron is transported in them mainly in the form of hydrochloric and sulfuric acid complexes. Only in deep hot springs containing a large amount of CO_2 does iron presumably migrate in the form of bicarbonate complexes. Evidence for this is the formation of iron-bearing carbonate (ankerite) in zones below the limit of steam formation, while chiefly carbonates without iron (calcite) are formed in the near-surface zone.

Attempts were made to determine experimentally the nature of the influence of the anion composition of volcanogenic waters on the removal of iron from volcanic rocks. For this purpose samples of basalt, andesite, and dacite were leached with sulfuric, hydrochloric, and carbonic acid solutions under conditions of high temperature and pressure (Naboko and Sil'chenko, 1960).

From the data of the experiments it can be concluded that the main factor governing the removal of iron from volcanogenic rocks is just the acidity of the waters; the composition of the waters has some influence on the removal of iron only under deep conditions at high temperatures and pressures. From these same data it can be concluded that the composition of the rocks has little effect on the removal of iron.

In the conditions of submarine eruptions, which usually take place under the pressure of a water column of up to 10,000 m, the greater part of the gases and volatile components given off is dissolved in sea water, in contrast to terrestrial conditions where only a small portion of volcanic gases is taken up by atmospheric or ground waters. The rest is spontaneously given off to the atmosphere.

The chemical change produced in sea water by volcanic exhalations is mainly replacement of the hydrogen ions of acid gases dissolved in the water by ions of the alkaline-earth metals and alkali metals Na, Ca, and Mg (Zelenov, 1963). This process goes on more actively at high CO_2 pressure.

When acid volcanic gases dissolve in sea water the acidity of the latter is not significantly altered inasmuch as the vast masses of water very rapidly virtually neutralize the H^+ ions that are released.

In submarine eruptions hot springs carry mainly the products of differentiation of the magma; often they are rich in silica, manganese, iron, and aluminum. However, in connection with the mixing of hot springs with sea waters, compounds of these elements are rapidly hydrolyzed, which prevents their being transported for substantial distances.

In the conditions of submarine eruptions iron arrives in the sea basin mainly in the form of volatile components of the magma—$FeCl_3$ and FeF_3. These compounds are easily dissolved both in condensed volcanic steam and in the sea water itself. Moreover, iron is brought from depth by submarine exhalations either at the time of or after an eruption (Formozova, 1963).

The data accumulated so far are insufficient to establish the limits of the distances to which ore elements can be transported in the conditions of submarine eruptions. It only is known that these distances are determined by the hydrodynamics of the area of the eruptions. Observations on the behavior of currents stained by the products of solution of gases or made turbid as a result of hydrolysis of various elements suggest that these distances usually are not more than a few kilometers, in isolated cases reaching two or three tens of kilometers (Strakhov, 1963). The limited distribution of volcanogenic products and, in particular, of iron compounds is a typical feature of submarine volcanism.

Iron is deposited from volcanogenic hydrothermal waters either on their way to the sea basin or when they mix with sea waters upon entering the basin. The main cause of precipitation of iron as sediment is a decrease in acidity of the water, due to dilution with atmospheric or sea water or due to reaction with the wall rocks, to pH values exceeding those of the beginning of hydrolysis of iron.

Dilution of thermal waters by meteoric does not usually lead to any significant change in acidity inasmuch as a tenfold dilution is needed to change the pH by one unit. In the Kurile Islands and Japan, where from 1000 to 2000 mm of precipitation usually falls in a year, the pH of thermal waters is essentially changed, by up to 2 to 5 in all. Iron removed from the rocks at depth in ferrous form is rapidly oxidized at the surface to ferric, and with the change in acidity is deposited on the flanks of volcanoes and in depressions in the form of limonite.

When iron is deposited from thermal waters on the way to the sea basin, iron-rich sediments form on the bottom of the streams due to precipitation of iron from solution in hydrolysis and due to mechanical destruction of the rocks. The amount of iron in such sediments usually increases with distance from the eruptive centers.

When acid volcanic waters are diluted by enormous amounts of sea water the acidity decreases; the abrupt change in physicochemical conditions leads to coagulation and deposition of iron. According to the data of Zelenov (1958), when the Yur'yeva River enters the Sea of Okhotsk the iron content, which reaches 240 mg/l at the mouth, drops sharply to a few tens of milligrams per liter at the site of the mixing of the waters of the river and of the sea, and at that place the pH of the waters increases to 4–4.5; when the Gorshkov River enters the Sea of Okhotsk the iron content is reduced to 4 mg/l with pH = 3.2.

Deposition of iron in the form of suspensions when hot springs mix with sea waters first occurs in the upper layers of the sea, in which case almost pure iron hydroxide is formed. Gradual dilution leads to expansion of the region where suspended iron sediment is deposited in the water. With depth, the iron hydroxides are replaced by clastic silica.

The formation of limonite sediments by mixing of river and sea waters is observed even in the stream channels. In the caldera of the volcano Bogdan Khmel'nitskiy (Iturup Island in the Kuriles) the Limonite Cascade deposit has been formed, the reserves of which increase by 1 ton per day. The limonite thus formed contains practically no impurities, inasmuch as associated elements, in particular Al, are deposited at higher pH values. Only in places of substantial dilution is limonite deposited along with other hydroxides.

The hydrochemical regime of the sea is of tremendous importance in the accumulation of sediments. Thus, when the hot springs of the Kurile Islands and Japan enter the Sea of Okhotsk the suspended iron hydroxides are disseminated, and do not form any appreciable accumulations (Zelenov, 1960). Basins with a quiet hydrodynamic regime are favorable for the accumulation of iron. In a bay of the Mediterranean Sea with such a regime iron-ore sediment has gradually accumulated as a result of constant supply of iron by low-temperature hot springs of carbonic acid composition, formed in fumarolic activity of the volcano Santorin (Butuzova, 1966). The formation of iron sediments occurs as a result of mixing of the volcanogenic and sea waters. Iron entering the basin in the form of carbonates and bicarbonates is oxidized by oxygen present in the sea water to Fe^{3+}, coagulates, and is deposited in the form of clots of hydroxide gel, forming a sediment with a very small mechanical admixture of pyroclastic material.

Under the conditions of submarine eruptions acid exhalations cannot greatly increase the acidity of the huge masses of sea water with which they mix; at best the acidification of the waters goes as far as pH values close to neutral. Under such conditions iron compounds are easily hydrolyzed, which explains the low mobility of iron in submarine volcanism; iron forms ore concentrations only in effusive-tuff accumulations. In this case the hydrodynamic regime of the basin also may play a considerable part, promoting the accumulation or dissemination of sediments.

In the opinion of Oftedahl (1958), the dominant role in the process of formation of sediments in submarine eruptions belongs to iron. Metalliferous gases arriving in the basin are dissolved in sea water and form various compounds in the immediate vicinity of the gas vents. In conditions close to neutral, iron compounds have an extremely poor migrational capacity and form sediments in the immediate vicinity of the volcanoes.

In studying the exhalations of underwater fumarole fields, as illustrated by the volcano Banua-Wuhu in Indonesia (Zelenov, 1965), it was established that the submarine hot springs of this volcano are similar in composition to the ocean water but contain a large amount of dissolved carbon dioxide and silica (the silica content in the hot springs reaches 8 as against 0.8 mg/l in the sea water). Deposition of iron along with manganese already begins at a distance of 1 m from the bottom: the suspended matter which is formed colors the jet from the spring yellow, then is deposited at the foot of the jet or carried away by the current. In this way ferrous iron minerals are deposited (magnetite, siderite, silicates, sulfides, chlorites), i.e. the minerals which in the usual sedimentary process are only formed diagenetically (Strakhov, 1960).

Such great diversity of stable mineral types in the sedimentary process leading to the formation of sedimentary iron deposits (Oftedahl, 1958) often makes it impossible to establish the direct relationship of these deposits to a volcanic center.

Transport and deposition of silica

Cherts very typically are restricted to volcanic formations. Volcanic removal of SiO_2 affects not only the chemogenic deposition of silica, but also its biogenic extraction. In regions of intensive ash deposition diatomaceous algae are prolific, putting large amounts of silica into the sediment (Khvorova, 1968). In geosynclinal regions the accumulation of thick cherty sediments—jasper and phthanite— everywhere accompanies submarine effusions, consisting of basic volcanic rocks. Sometimes large accumulations of cherty sediments do not accompany basalts, but tend to accompany spilite formations.

The possible ways in which silica can be supplied to volcanic regions have been examined in detail by Petrova (1958), who distinguishes: (1) introduction of SiO_2 with steam and gases at the time of a volcanic eruption (magmatic silica); (2) introduction of silica in the composition of fumaroles and hot springs in the intervals between eruptions and after them; and (3) formation of silica in submarine eruptions as a result of interaction of incandescent lava and pyroclastics with sea water (reaction silica).

Apparently the second method of introduction is most important, especially in the case of areal development of fumaroles and prolonged post-eruptive activity. As has already been mentioned, chert deposits far from regions of active volcanism were distinguished by Shatskiy (1954) as a special group of remote-volcanic chert formations, present in many geosynclinal systems.The mechanism of transport of silica is largely unclear. In the case of eruptive exhalations some of the silica is carried in the form of volatile compounds of $SiCl_4$ and SiF_4 type, which form silicic acid when dissolved in thermal waters. However, most of the silica is carried in dissolved form, first in supercritical steam at high pressures, and later in hydrothermal fluid solutions. A substantial part of the silica comes from the wall rocks, when they interact with the thermal solutions.

The content of silicic acid (undissociated molecules) in the waters of hot springs ranges from 0.14 to 0.17 g/l (Naboko and Piyp, 1961). In an appendix to his monograph, Markhinin (1967a) gives a summary of chemical analyses of the waters of solfataric springs of the Kurile island arc. Here the H_2SiO_3 content is 0.236 g/l (average of 106 analyses), with variations from 0.004 to 0.910 g/l. In hot waters ($T = 80–100°C$) the silicic acid content is appreciably higher—0.284 g/l (average of 43 analyses) with variations from 0.077 to 0.638 g/l. According to Naboko's data (1968) the recent hot springs of Kamchatka, formed at a temperature of $\sim 200°C$ at depth, contain from 0.204 to 0.339 g/l H_2SiO_3 at a pH = 7.6–8.2, and those of Wairakei (New Zealand) contain 0.502 g/l at $T \sim 280°C$ and pH = 8.6. These data indicate that solutions supersaturated with silica at low temperatures can be supplied to sedimentary basins from volcanic sources.

In a compendium (Petrova, 1958) it is mentioned that various views have been expressed concerning the mechanism of deposition of silica in marine conditions, and the dispute comes down to a discussion of the possibility of chemogenic precipitation of silica and its role in the formation of cherts. Some investigators deny such a possibility completely, others attach paramount importance to inorganic deposition (Twenhofel, 1926). There also are intermediate opinions; in particular, many believe that in areas of submarine volcanoes chemogenic deposition from hydrothermal solutions can take place in the post-volcanic stage (Strakhov, 1954, 1966; Krauskopf, 1963). Recent investigations in the Red Sea depression (Miller et al., 1966) confirm the possibility of submarine introduction of silica "from below" even where there are no outpourings of lava (Petrova, 1958).

Dzotsenidze's conclusion (1965) that silica of volcanic origin plays the leading role in the chemical deposition of authigenic SiO_2 is interesting. This conclusion was drawn on the basis of a careful analysis of the investigations by Bruyevich (1953) and Bezrukov (1955) on the variation in silica content in the waters of the Atlantic, Pacific, and Indian Oceans. According to the data of his analysis, Dzotsenidze suggested that the increase in silica content from the waters of the Atlantic Ocean to the Indian and then to the Pacific can be explained by proximity to regions of recent submarine volcanism on the Kurile–Kamchatka ridge.

As in the case of iron, the hydrodynamic regime of the water basin has a considerable effect on the processes of accumulation of silica. In the case of simultaneous deposition of silica and iron, the sols of their hydroxides are usually converted to gels having the composition of ferruginous chlorites. In the case of coagulation of silica sol around submarine active springs with constant discharge, only small lenticular intercalations of spilite are formed.

Analysis of the energy of processes of diagenesis of cherty iron sediments

According to the works of Strakhov on the theory of lithogenesis, freshly formed sediments represent a physicochemical system that is out of equilibrium. The processes of alteration of the sediments leading to equilibration under the thermodynamic conditions of the Earth's crust before conversion into rock are called diagenesis. However, after diagenetic processes proper have ended the equilibrium between coexisting solid phases might be metastable in character. Highly dispersed and amorphous compounds, minerals with defects in the crystal lattice may be left in the rock. Consequently equilibration phenomena also occurred in the stage of katagenesis, protometamorphism, and low-temperature regional metamorphism.

The source of the energy which ensures that diagenetic transformations, proper, of the original sediments will go on is the "internal energy of the system itself, stored in the stage of sedimentogenesis and liberated in the processes of equilibration of the system itself" (Strakhov, 1960). Therefore the field of diagenetic phenomena should be limited to spontaneously occurring processes which in individual cases may be accelerated if T and P are increased. The essential difference from metamorphism is the absence in diagenesis of mineral transformations caused only by a change in temperature and pressure.

The mineralogical and geochemical particulars of diagenetic processes taking place in recent sediments have been studied in fairly great detail. Besides the well known summarizing monograph by Strakhov (1960), the compendium by Larsen and Chilingar (1967) and the works of Berner (1964a, 1967, 1969, 1970), Stashchuk et al. (1964), Ostroumov (1955), Volkov and Sevast'yanov (1968), and Presley and Kaplan (1968) should be mentioned.

Physicochemical study of diagenetic processes, determination of the variation in thermodynamic properties as a function of the character and duration of transformation of the sediment, and analysis of the evolution of the stability of individual minerals and mineral associations constitute a very complicated task. Until recently there were hardly any reliable thermodynamic constants for the amorphous and highly dispersed phases, experimental data were unreproducible and considerably divergent. As a result, the data of mineralogical investigations and theoretical diagrams of mineral equilibria in sediments were in poor agreement with each other, which gave rise to criticism of the very methods of thermodynamic analysis (Stashchuk, 1968). The development of a system of consistent thermodynamic constants of sediments (Mel'nik, 1972b) made it possible to some extent to eliminate several defects in theoretical formulations, and for the first time diagrams were obtained which better illustrated the actual processes, and the low abundance of such minerals as asmakinite or magnetite in recent sediments found an explanation. However, it should be kept in mind that only the first steps have been taken in the direction of quantitative thermodynamic analysis of diagenetic processes (Mel'nik et al., 1973). A closer association of theoretical and experimental investigations is needed in the future, with study of specific sediments, their composition, character of the solutions, and dynamics of ion exchange.

Mineral transformations of iron sediments in diagenesis

Let us examine the particulars of diagenesis of natural iron sediments.

Inasmuch as the number of natural bodies where primary or secondary iron hydroxide minerals are observed is very great, we will limit ourselves to recent sediments of water basins and to the Kerch iron ores as examples.

Recent sediments of water basins. In recent basins iron sediments consist mainly of the iron hydroxides $Fe(OH)_3$ or $Fe_2O_3 \cdot nH_2O$, but in very rare cases silicates and carbonates of Fe^{2+}, pyrite, and hydrotroilite enter into the composition of the sediment; all together they constitute reactive (mobile) iron, which actively takes part in the diagenetic processes. A mixture of clastic minerals, which decompose negligibly and take practically no part in the processes of diagenesis, constitute another group.

In most open water bodies iron is deposited in the form of hydroxide and the surface layer of the deposits contains only Fe^{3+}. Let us give some examples.

In the deep-sea sediments of the Atlantic Ocean (Regnell, 1961) off the west coast of North Africa iron is deposited mainly as colloidal hydrous oxide, flocculent or adsorbed by organic matter.

In the deposits of the oxygenated zone of the Black Sea (Volkov and Tikhomirova, 1966) the surface layer is oxidizing in character and contains reactive Fe exclusively in trivalent form.

In the bottom deposits of the Pacific Ocean (Ostroumov and Volkov, 1967) an oxidizing layer from 1 cm to a few meters thick is always recorded.

In addition to iron hydroxide the sediment and pore solution contain other oxidants, in particular dissolved free O_2, SO_4^{2-}, hydroxides of Mn, and others. At the same time recent sediments always contain a certain amount of dead organic matter and living organisms, chiefly bacteria, which need oxygen (Strakhov, 1960).

Equilibration may proceed differentially in iron sediments, depending on the amount and reactivity of organic matter.

In the case of a negligible content of organic matter, its poor reactivity, or slight bacterial activity, the environment in the sediment remains mainly oxidizing at all stages of diagenesis, and iron remains in trivalent form. Diagenesis comes down to crystallization of amorphous unstructured hydroxides into crystalline mineral forms. This process is accompanied by an increase in particle size and a decrease in solubility, specific surface and isobaric-isothermal potential. Such a course of diagenesis is possible in sediments with a large content of organic matter if porosity is high and there is much mixing, etc., i.e. in the case of fairly rapid removal of reducing agents and a supply of free oxygen. The final product of diagenesis in an oxidizing environment is iron hydroxides, hydrogoethite, and less often completely crystalline goethite; hematite does not occur as an essential mineral in most recent sediments.

Depending on the content of organic matter, bacterial activity, intensity of aeration, and such, in a reducing setting a number of conjugate oxidation-reduction processes go on, caused by reactions in which living or dead organic matter necessarily takes part. These processes also proceed toward the attainment of equilibrium in the conventionally closed system which the sediment, separated from the layer of aerated waters of the sedimentary basin by an impermeable, or more correctly semipermeable "membrane", represents. As a result, first glauconite is formed in the sediments from iron hydroxides, then ferric-ferrous iron silicates begin to form (in different investigations they are called "chamosite", "ferrichlorite", "hydroferrichlorite", and sometimes all the silicates are lumped together by the general term "leptochlorites"), next siderite, and finally pyrite. Each newly formed mineral either replaces those formed earlier, or is developed independently. In other words, in the course of diagenesis a gradual reduction of oxidized forms of iron to ferrous compounds takes place.

Deposition of the Kerch ores. The iron sediments originally accumulated in lagoons of a Kimmerian basin under the conditions of a hot and humid climate. Constant shifting of the shoreline and deepening of the basin as a result of tectonic movements led to a change in the physicochemical setting and formation of iron ores of various origins. Among the ores there are distinguished: "tobacco" ores, formed in the course of diagenesis of iron sediments in leptochlorite-facies conditions; carbonate ores, formed during diagenesis under conditions of an excess of organic matter; and oolitic ores, formed when the ore layer was reworked as a result of oscillatory movements.

Shnyukov (1965) mentions that the original material consisted of aggregates of complex chemical composition formed from colloidal flocs of $Fe_2O_3 \cdot H_2O$, $SiO_2 \cdot Al_2O_3 \cdot H_2O$, $MnO_2 \cdot H_2O$. From these aggregates the iron chlorites, hydrogoethite, and iron and manganese carbonates which constitute the main mass of the ore material arose in the course of sedimentary-diagenetic changes. Three stages of mineral transformation are distinguished: (1) oxidizing (progressive formation of hydroferrichlorite, beginning as early as the first moments of diagenetic transformations of the colloidal sediments); (2) reducing (formation of authigenic iron minerals—silicates, carbonates, sulfides—in the deeper horizons of the sediments, continuation of formation of chlorite); and (3) stage of redistribution of authigenic minerals (metasomatic replacement of hydroferrichlorite by iron and manganese carbonates in conditions of a reducing environment and excess of CO_2 created as a result of oxidation of organic matter, subsequent replacement of the carbonates by phosphates and sulphides).
occur in parallel with the processes of continuous reduction of Fe^{3+} to Fe^{2+} and formation of ferrous iron minerals. Interruption of these processes and migration of components in the reworking of ore layers naturally complicates this scheme.

Yurk and Shnyukov (1961), who analyzed the mineral parageneses of the Kerch ores, concluded that the environment of mineralization was slightly reducing and close to neutral, which was the reason for the universal formation of leptochlorites with a high content of ferric iron.

Precambrian cherty iron sediments. Data on diagenetic processes that went on in ancient rocks which have not retained their original form can be obtained only indirectly, on the basis of a number of assumptions. In any version of the chemogenic hypothesis, the original cherty iron sediment must have been amorphous-dispersed, and the processes of diagenesis were an important step in the formation of the crystalline rock. However, it is quite unclear whether essential mineral transformations occurred, or whether diagenesis was limited to recrystallization of the original sedimentary minerals. The extreme points of view on this question have been most clearly expressed by James (1954) and Plaksenko (1966).

James considered all the typical parageneses of the BIF to be primary sedimentary and distinguished oxide, silicate, carbonate, and sulfide facies, deposited at different depths of the basin, where different oxidation-reduction conditions existed.

Plaksenko suggested, in accordance with the theoretical ideas of Strakhov, that the primary sediments consisted of iron hydroxide which then was reduced to a ferrous form, producing the whole gamut of observed mineral associations in diagenesis under the influence of organic matter.

If we turn to the data on slightly metamorphosed BIF, it turns out that the actual mineral interrelationships were more complex. Thus, in the well known Biwabik BIF in the Mesabi and Gunflint districts, chert layers alternate with layers containing magnetite, siderite, ankerite, and iron silicates in variable amounts (Bayley and James, 1973). The original granules were of complex mineralogical composition and contained iron silicates, siliceous material, and magnetite in highly variable proportions; sometimes hematite also was present in rims. Beukes (1973) mentions that in the Penge BIF of Southern Africa the magnetite and other iron minerals crystallized from complex cryptocrystalline or colloidal sediments. French (1973) believes that the original iron-bearing sediment which precipitated from solution probably consisted of ferric hydroxide, silica gels containing iron, and colloidal particles of silicate and carbonate iron minerals. This unconsolidated primary sediment was converted to solid rock due to diagenetic processes, and these processes went on in the immediate vicinity of the sediment–water interface.

A few manifestations of slightly metamorphosed BIF in India, Brazil, and other areas have a predominantly hematitic composition. In such rocks the silica is often uncrystallized and consists of opal, "jasper", or other finely dispersed forms. However, the phenomena of diagenetic reduction are atypical of hematitic jaspers, likewise a facies relationship to iron cherts of various compositions is rarely observed.

Study of the distribution of residual carbon (graphitite) in chemogenic iron cherts of different mineralogical composition and in individual layers of those rocks has yielded no convincing evidence to support the hypothesis of reduction of homogeneous hydroxide sediments by organic matter. This same conclusion follows from data on the variations in isotopic composition of the oxygen in iron oxides and carbonates. These problems are considered in more detail in connection with the problem of the origin of the magnetite. At the same time many aspects of James' hypothesis appear unconvincing, especially with respect to the validity of the potential-determining systems capable of regulating the *Eh* in certain parts of the sedimentary basin.

In the analysis of the deposition of iron sediments it has already been mentioned that quite likely both iron silicates and carbonates and amorphous iron hydroxide were formed, which could convert to other forms both during the formation of the sediment and in subsequent diagenesis. Reduction of hydroxide could have been controlled by external (atmospheric) or internal (organic matter, free carbon in the sediment) oxidation-reduction buffer systems. All these variants need additional consideration in the thermodynamic analysis of diagenetic processes.

Experimental data on the properties and transformations of cherty iron-formations

Special experimental investigations of the properties, particulars of structure, and variations as a function of time, temperature and pressure have not been made so far for the iron cherts. However, the main components of such sediments—iron hydroxide and silica—have been rather well studied, and some data which are of interest to understanding the diagenetic processes are considered in this section. For the other components—magnetite, siderite, and sulfides—the very limited experimental data were examined in our previous work (Mel'nik, 1972b).

Iron hydroxide sediments

The most important characteristics which make it possible to compare sediments of the same composition but different degrees of dispersion and crystallinity are their structure, thermochemical properties, and solubility.

Structure of sediments. The properties of iron hydroxide sediments—color, density and degree of dispersion, texture, character of heating curves, diffractograms, and paramagnetic resonance spectra—depend on many factors, the most important of which are not only the aging time, but also the pH of the medium, character of the original solutions (ionic or colloidal), and presence of electrolytes.

A large number of published works deal with the iron hydroxides (Levina and Yermolenko, 1955; Meleshko, 1959; Feitknecht and Michaelis, 1962; Krause and Lezuchowska, 1962; Feitknecht and Schindler, 1963; Lezuchowska, 1963; Chalyy and Lukachina, 1970), but previously it was not clear which of the crystalline phases—goethite, lepidocrocite, or hematite—is thermodynamically stable in aqueous solutions at low temperatures. Therefore additional experimental and theoretical investigations were needed to determine the equilibrium parameters more precisely (Yepatko and Mel'nik, 1965; Mel'nik, 1972b; Vorob'yeva and Mel'nik, 1977), from which it was concluded that α-FeOOH is in equilibrium with pure water below 100°C.

Recently Chukhrov et al. summarized data on the synthesis of iron oxides in conditions comparable to natural, in the collective monograph "Supergene iron oxides in geological processes" (Petrovskaya, 1975). The results of experimental investigations, in conjunction with geologic data, not only made it possible to draw conclusions as to the conditions of formation and transformation of the individual natural oxides, but also to identify new compounds which are metastable phases. A very important fact is that the nature of the original hydroxide sediments depends on the mechanism and conditions of their formation.

Chukhrov et al. assign a special role to ferrihydrite (2.5 $Fe_2O_3 \cdot 4.5\ H_2O$), which is believed to be the typical product of rapid oxidation of Fe^{2+} in slightly acid, neutral, and slightly alkaline solutions with the participation of iron bacteria. The oxidation process also is accelerated by the catalytic action of silica. In the course of time ferrihydrite spontaneously converts to hematite, but in solutions with Fe^{2+} ions ferrihydrite is transformed into stable goethite in the absence of significant amounts of oxygen.

Goethite also is easily formed by alteration of silicates, sulfides, and other minerals in an oxidizing environment, where hardly any iron goes into solution in the Fe^{2+} form or where the solution is too acid.

The formation of lepidocrocite is possible only in the case of slow oxidation, in particular in solutions with a high carbon dioxide content, in the absence of iron bacteria and significant amounts of free silica. Such conditions are very common in the weathering of siderite.

Magnetite is formed in alkaline solutions in the case of slow oxidation, sometimes together with lepidocrocite.

On the basis of the data obtained Chukhrov et al. (1977) assume that in all epochs of the geologic history of the Earth, including the Proterozoic, iron was supplied to the sedimentary basins chiefly in the form of ferrihydrite, which either was converted to hematite (in the absence of organic matter) or was reduced by organic matter during diagenesis to siderite and iron silicates, as was assumed in the works by Strakhov (1960) and Plaksenko (1966).

Our experiments also established that the process of "aging" of freshly precipitated iron hydroxides does not proceed in the same way in sediments obtained from ionic and colloidal solutions of Fe^{3+}. Whereas "colloidal" sediments remain X-ray-amorphous for a long time, sediments from ionic solutions relatively rapidly acquire a crystal structure which is most clearly manifested in alkaline, is less ordered in acid, and almost X-ray-amorphous in neutral environments.

In the experiments which we carried out with Vorob'yeva, the change in iron hydroxide sediments in the parent solution (concentration of NaCl $\sim$ 17 –20 g/l) was studied in the course of 8.5 years.

All the original sediments remained X-ray-amorphous for several months.

Sediments obtained at room temperature from ionic solutions of $FeCl_3$, "aged" for 7 months in a neutral medium (pH = 6–8) remained X-ray-amorphous; in an acid medium (pH = 2–3) slightly crystallized goethite appeared, and in an alkaline medium (pH = 9–10) goethite also was formed, but better crystallized. After 3.5 years the goethite in acid and alkaline media was better crystallized, but in the neutral medium slightly crystallized hematite appeared. These results can be explained by a different mechanism of formation of the crystalline phases. The high solubility of iron in acid and alkaline media, with the formation of ionic forms (Fe^{3+} or FeO^+ and $Fe(OH)_4^-$), leads to the crystallization of thermodynamically stable goethite from solution. In a neutral medium no ionic forms are produced in the solution, the solubility of $Fe(OH)_3$ is extremely low, and during "aging" the particles enlarge, giving off water and forming dispersed hematite. We found no ferrihydrite in these sediments, although its formation as a protohematite phase is not ruled out.

The sediments obtained by coagulation of colloidal solutions behaved completely differently. Regardless of the pH of the parent solution, these sediments did not change externally and remained X-ray-amorphous for several years. Ferrihydrite was detected, by the method of electron microdiffraction, in sediments kept for 8.5 years in neutral (pH = 6–8) and alkaline (pH = 9–11) environments. In an acid environment (pH = 2–3) the colloidal coagulant (ferrihydrite) was converted to akaganeite—β-FeOOH—in the same length of time. Study of the fresh sediments showed that the X-ray-amorphous colloidal iron hydroxides have a fairly well ordered structure—they are paracrystalline substances very resistant to further transformations at low temperatures, despite the high degree of dispersion. It is not ruled out that this ordered (ferrihydrite) structure is intrinsic to the iron hydroxide colloid itself (micelle), which in the absence of coagulants is capable of existing in solution indefinitely.

Differences in the iron hydroxide sediments obtained from ionic and colloidal solutions and "aged" at different pH are also recorded by the methods of proton magnetic resonance (PMR). Typical PMR spectra are determined by protons of the OH groups. No lines were found that might be related to other types of water. A relationship of the second moments of the PMR lines to the increase in pH was established, especially in precipitates obtained from ionic solutions (Fig. 53a). Interpreting the relationship found, it can be presumed that as pH increases, the magnetic structure of the hydroxides is enhanced. It is typical that fresh precipitates obtained from colloidal solutions have a more ordered structure than fresh precipitates from ionic solutions.

Kinetic "aging" of iron hydroxides precipitated from true solution is depicted in Fig. 53b, which illustrates the dependence of the second moment

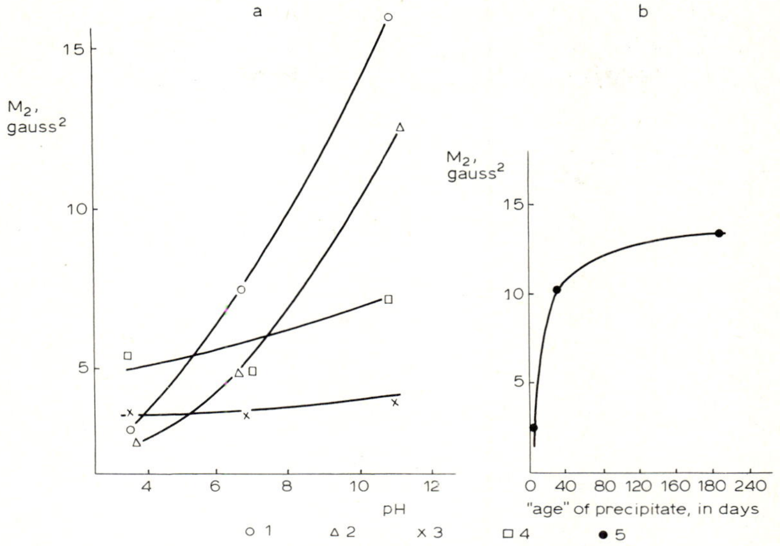

Fig. 53. Dependence of second moments of PMR lines on pH of the solution (a) and on the "age" (b) of iron hydroxides. *1* = precipitates from true solutions of $FeCl_3$, "age" 7 months; *2* = precipitates from true solutions of $FeCl_3$, "age" 1 month; *3* = precipitates from true solutions of $FeCl_3$, freshly prepared; *4* = precipitates from colloidal solutions of $Fe(OH)_3$, freshly prepared; *5* = precipitates from true solutions of $FeCl_3$, pH = 10–11 (Mel'nik et al., 1973).

of the PMR line on the "age" of the precipitate in an alkaline medium. A sharp increase in the second moment, apparently related to the ordering of the crystalline and magnetic structures of the hydroxide, is observed during the first 1 or 2 months; in time the rate of the process decreases. In acid and particularly in neutral media the picture is less clear, which is in good agreement with the X-ray data.

From these data it follows that when iron is precipitated in acid and neutral environments the first products should be X-ray-amorphous highly dispersed iron hydroxides, which in the course of time acquire the crystal structure of goethite or hematite. The mechanism of this process depends on kinetic factors (rate of oxidation of Fe^{2+}), form of migration of the iron (ionic or colloidal), and acidity of the parent solution. In neutral environments ferrihydrite possibly is formed as an intermediate metastable phase, especially if the iron migrates in colloidal form or in the form of the Fe^{2+} ion. The products of diagenesis of such a sediment may be both goethite (in the case of low *Eh* values typical of the Precambrian iron-ore process) and dispersed hematite (in the case of deposition of the oxide facies of BIF).

Thermodynamic properties and solubility of the sediments. Ferrier's (1965, 1968) experimental data showed that in thermodynamic investigations of the dehydration of goethite it was impossible to obtain unambiguous and comparable results if the degree of dispersion of the solid phases was not taken into account. From the heats of solution he established that for the reaction:

$$2\ \alpha\text{-FeOOH} = Fe_2O_3 + (H_2O)_l \qquad (3.20)$$

$$\Delta H_{344} = 3.06 + 0.00018\ \Sigma_2 - 0.00028\ \Sigma_1\ (\text{kcal/mol})$$

where 3.06 kcal/mol is the enthalpy of dehydration of macrocrystals, Σ_1 is the specific surface of enthalpy of goethite ($\Delta H = 0.28$ cal/m^2 for 2 α-FeOOH), and Σ_2 is the specific surface of enthalpy of hematite ($\Delta H = 0.18$ cal/m^2 for α-Fe_2O_3).

For very finely dispersed goethite and hematite (particles < 100 Å) the process of dehydration goes on with liberation of heat, for crystals of ~ 100 Å size the ΔH of dehydration $= 0$, and for larger crystals $\Delta H > 0$, approaching the 3.06 kcal/mol for macrocrystals as the particle size increases.

From these data one can estimate the change in enthalpy when finely dispersed crystalline goethite converts to macrocrystals. If it is assumed that the original particles (seed crystals) were of the order of 100 Å in size, then crystallization of such sediment to form crystals 10–50 μm in size would be accompanied by liberation of about 3 kcal/mol of heat by FeOOH.

When amorphous forms convert to crystalline, and finely dispersed materials to larger particles, the solubility should decrease and the isobaric-isothermal potential of mineral formation should increase (by an absolute value).

Analysis of determinations of the solubility and particulars of the structure of iron hydroxides makes it possible to divide the sediments into two groups.

The first group includes X-ray-amorphous compounds united under the tentative formula $Fe(OH)_3$, among which three varieties can be distinguished on the basis of solubility, water content, and protons: $Fe(OH)_3$-I—freshly precipitated oxide of Fe^{3+}, with maximum solubility and containing a large amount of water; $Fe(OH)_3$-II—active Fe^{3+} hydroxide with an aging time of 30 minutes to 200 hours, with lower solubility than the freshly precipitated hydroxide (Levina and Yermolenko, 1955; Leingweiler et al., 1961); $Fe(OH)_3$ -III—inactive Fe^{3+} hydroxide, with minimum solubility, formed in the course of long-term stabilization of iron hydroxide.

The second group includes crystalline compounds with a goethite structure. Among them there are distinguished: $(\alpha\text{-FeOOH})_n$—hydrogoethite, formed as a result of natural aging of amorphous Fe^{3+} hydroxides due both to processes of enlargement and polymerization of amorphous particles, and to condensation of OH^- groups; α-FeOOH—goethite, formed as a result of long preservation of hydrogoethite due to a decrease in activity of OH^- groups when they are converted to water particles, coordinated to the lattice (Lezuchowska, 1963).

Data on solubility obtained by radiochemical and potentiometric methods (Leingweiler et al., 1961; Schindler et al., 1963) were used for comparison of the properties of these varieties of hydroxides. Mathematical treatment of these data, including calculation and comparison of isobaric-isothermic potentials (ΔG_{298}), made it possible to obtain a functional relationship between variations in energy ($\Delta\Delta G_{298}$) and aging time of the hydroxides, and also to determine the energetic characteristics of amorphous sediments and goethites of different extent of crystallinity (Fig. 54). The decrease in solubility of iron hydroxides is reflected in the variation of $\Delta\Delta G_{298}$ and ΔG^0_{298}: points 2, 3, 4 correspond to the solubility of the products of aging formed after 2, 24, and 96 hours, which corresponds to $\Delta\Delta G_{298}$ values of −0.25, −0.66, and −0.93 kcal/mol. Extrapolation of the presumed linear dependence of $\Delta\Delta G$ on $\log \tau$ up to an aging time of 200 hours makes it possible to compare the values of solubility of the hydroxides obtained by different investigators (Leingweiler et al., 1961; Schindler et al., 1963).

From the graph of the dependence of variation in isobaric-isothermic potential on "aging" time, the duration of the individual stages of diagenetic

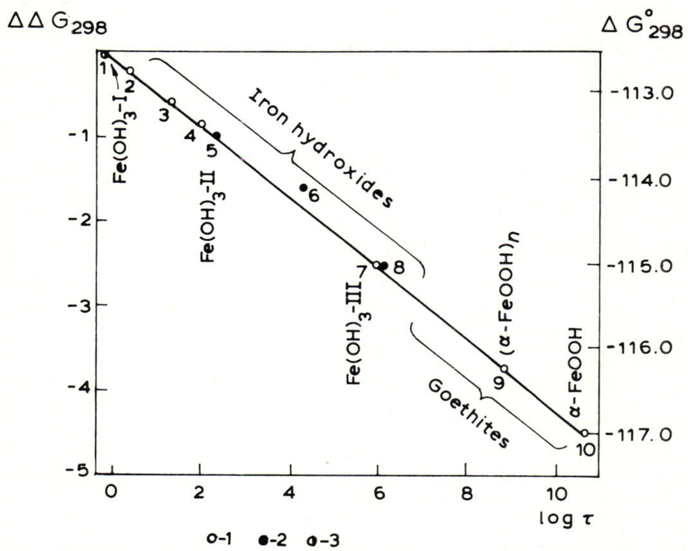

Fig. 54. Dependence of ΔG^0_{298} and $\Delta\Delta G_{298}$ (in kcal/mol) of iron hydroxides and goethites on duration of aging (recalculated to 1 mole of FeOOH). Original data: *1* = Leingweiler et al. (1961); *2* = Schindler et al. (1963); *3* = extrapolation using base constant $\Delta Z^0_{298} = -116.76$ kcal/mol for α-FeOOH (Mel'nik, 1972b) and data of Leingweiler et al., 1961; Schindler et al., 1963.

transformations can also be roughly estimated. The total duration thus found for transformation of the original iron hydroxide sediments into crystalline goethite amounts to some millions of years—a quite realistic value for geological processes.

Silica sediments

Structure of the sediments. In acid and neutral environments silica occurs in solution in the form of the monomer $Si(OH)_4^0$. In supersaturated solutions polymerization occurs, the polymeric particles interact with each other due to formation of hydrogen bonds and gradually enlarge, yielding a sol which then converts to a gel when a critical particle size is reached. The hydrogel consists of spherical primary particles 20–150 Å in size, bound to each other both by hydrogen bonds and by bonds of Si–O–Si type, the importance of which gradually increases as the hydrogel "ages".

Freshly precipitated hydrogel is characterized by small particle size and very large specific surface (of the order of 600–800 m^2/g). As the gel "ages" irreversible changes take place, chiefly enlargement of particles and decrease in specific surface. These processes are accelerated by an increase in temperature and pressure (Okkerse and de Boer, 1961; Akshinskaya et al., 1964; Chertov et al., 1966). An increase in temperature to 350°C leads to a tenfold decrease in specific surface in a few hours, and the particle diameter correspondingly increases. An analogous phenomenon is observed at low temperatures, but the process of enlargement of the particles is very slow, particularly in neutral environments. Thus, at a pH of the gel of 7.2, "aging" for 30 days led to a decrease in specific surface from 670 to 390 m^2/g, in which case the particle size increased from 41 to 70 Å (Sheynfayn et al., 1964).

The formation of crystalline quartz from a gel at low temperatures has not been established experimentally. Direct crystallization of quartz from sea water has been observed (Mackenzie and Gees, 1971). In the opinion of Harder and Fleming (1970), quartz is formed only by adsorption of SiO_2 by hydroxides of Fe, Al, Mg etc. from undersaturated solutions, while amorphous silica arises in supersaturated solutions. The mechanism of crystallization of gels to quartz was studied in detail in our laboratory by Mitsyuk (1974). It was established that in amorphous silica the process of quartz formation goes on at an appreciable rate in hydrothermal conditions —$T = 150$–250°C and $P = 0.5$–4.0 kbar—and is characterized by the following features:

(1) Crystallization goes through several successive intermediate stages. At first compaction of the gel substance occurs, ending with the formation of opaline silica. Then crystallization of the amorphous silica is observed, with the formation of unstable crystalline phases designated SiO_2-X, SiO_2-X_1,

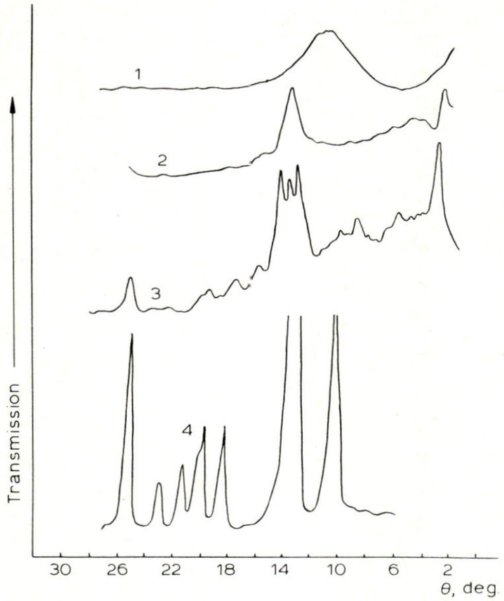

Fig. 55. Diffractograms of opaline silica (*1*), SiO_2-X_2 (*2*), SiO_2-Y (*3*) and quartz (*4*), after Mitsyuk (1974).

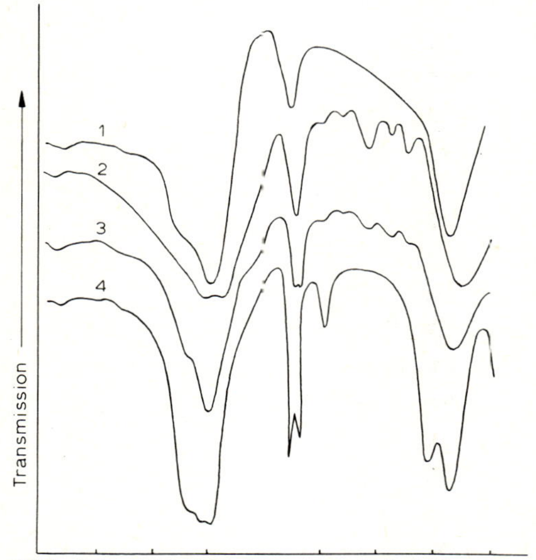

Fig. 56. Infrared spectra of opaline silica (*1*), SiO_2-X_1 (*2*), SiO_2-X_1 + quartz (*3*), and quartz (*4*), taken in the region of fluctuation of the Si-O bond (Mitsyuk et al., 1976).

and SiO_2-Y. With further hydrothermal treatment these phases are transformed into low-temperature cristobalite or, in the case of high rates of reaction, directly into quartz.

(2) Synthetic opaline silica is X-ray-amorphous (Fig. 55) and similar in its properties to natural opal, which consists of densely packed spherical particles (globules) about 1500–3500 Å in diameter.

(3) The SiO_2-X, SiO_2-X_1, and SiO_2-Y phases are hydroxyl-containing crystalline varieties of silica. The SiO_2-X phase was obtained for the first time and is a well crystallized variety of SiO_2-X. The SiO_2-Y phase, also obtained for the first time, is unlike any of the known silicate structures (Fig. 55). All these phases have a globular structure and particle size of the order of a few microns. According to infrared absorption spectra (Fig. 56), the structure of the intermediate phases is fairly disordered, with variable lattice parameters.

(4) The order of conversion of gel into quartz can be expressed by the following scheme:

$$\text{gel} \rightarrow \text{opaline silica} \begin{matrix} \nearrow SiO_2\text{-X}-SiO_2\text{-X}_1 \searrow \\ \searrow SiO_2\text{-Y} \nearrow \end{matrix} \text{cristobalite} \rightarrow \text{quartz}$$

(5) The kinetics of crystallization of silica are determined by the alkali concentration, temperature, and pressure. The rate of the process increases as these parameters are increased. Pressure exerts an especially strong effect. In some cases its effect is greater than that of temperature. The presence of Al and Fe ions slows the crystallization of quartz.

From the data given it follows that crystallization of silica hydrogel to form crystalline quartz goes on exceedingly slowly (Harder and Fleming, 1970) and even with the great duration of geological processes, amorphous or semicrystallized silica bands can be retained in unmetamorphosed BIF. The formation of the typical chert or mosaic structures probably can be explained by crystallization of amorphous silica during metamorphism, when seed crystals evenly distributed in the hydrogel mass appear simultaneously. The radial-fibrous spherulites observed in cherts also are formed by crystallization of amorphous silica without the participation of biologic processes (Oehler, 1976).

Thermodynamic properties and solubility of the sediments. From a comparison of the tabulated thermodynamic constants (Naumov et al., 1971) it follows that the transformation of glass into α-SiO_2 is accompanied by a 2.2 kcal/mol change in enthalpy. Values of the isobaric-isothermic potential of formation of amorphous silicas were calculated on the basis of a comparison of their solubility with the solubility of crystalline quartz. Such calculations were made by Kopeykin and Mikhaylov (1970) and by us (Mel'nik, 1972b) after analysis of the variation in solubility of various forms of silica as a function of pH (Alexander et al., 1954; Greenberg, 1957; Siever, 1962; Okamoto et al., 1963). In a new summarizing work Walther and Helgeson (1977) give a value of $\Delta G^0_{298} = -202.89$ kcal/mol for solid amorphous silica, and ΔG_{298} of the transition of amorphous silica → α-quartz = -1.76 kcal/mol. We note that the ΔG value of crystallization of amorphous silica varies little with temperature. Thus, Greenberg (1957) obtained $G = -1.22$ kcal/mol at 200°C. The G^0_{298} value of "quartz" glass and amorphous silica of opal type differ little from each other (~0.1–0.2 kcal/mol).

Thus the energetic effects in the "aging" of amorphous silica turn out to be relatively small, which explains the long existence of metastable forms in spite of diagenesis.

Thermodynamic analysis of the diagenesis of cherty iron sediments

On the basis of thermodynamic constants obtained for hydroxide compounds of iron with different "aging" time and also of experimental data, the physicochemical character of the diagenetic transformations of iron sediments of various compositions (oxide, silicate, carbonate, sulfide) can be traced. The results obtained are represented graphically in the form of stability diagrams of iron compounds as a function of variations in the main parameters governing the physicochemical character of the environment of diagenesis—pH, *Eh*, activity of iron and dissolved forms of sulfur and carbon dioxide.

Diagenesis of iron oxide sediments in an oxidizing environment results in crystallization of amorphous and dispersed hydroxides with the formation of stable goethite or metastable dispersed hematite. The diagram of the relationships between iron hydroxides of different degree of crystallinity (Fig. 57), reflecting their stabilization in the course of diagenesis, can be regarded as a model of the diagenetic transformations of the original sediments, which in the absence of reducing agents (free organics, hydrocarbons, sulfate-reducing bacteria, etc.) result in the formation of stable minerals. In recent conditions that is what happens in the case of highly porous sediments and intensive aeration providing a constant supply of oxygen. As a result of diagenesis under oxidizing conditions the stability fields of the solid phases are expanded, and the solubility of the iron hydroxides is reduced several thousand times (bv 3.5 orders of magnitude).

Analysis of the stability of oxide sediments when there is a change in the oxidation-reduction environment during diagenesis (Fig. 58) also indicates

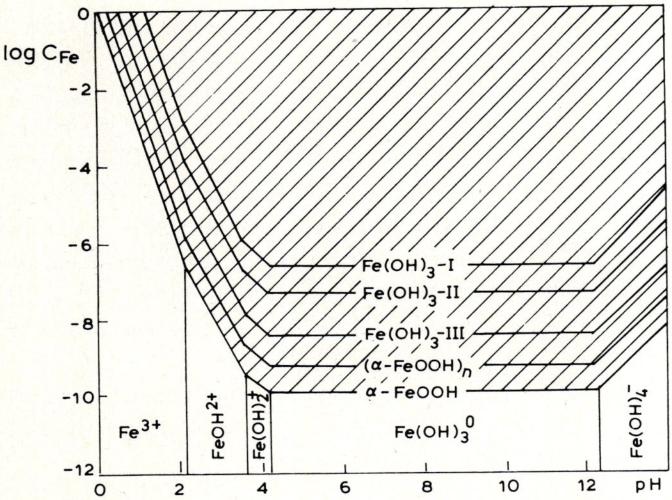

Fig. 57. Relative stability of hydroxides of trivalent iron in the case of diagenesis in oxidizing conditions.

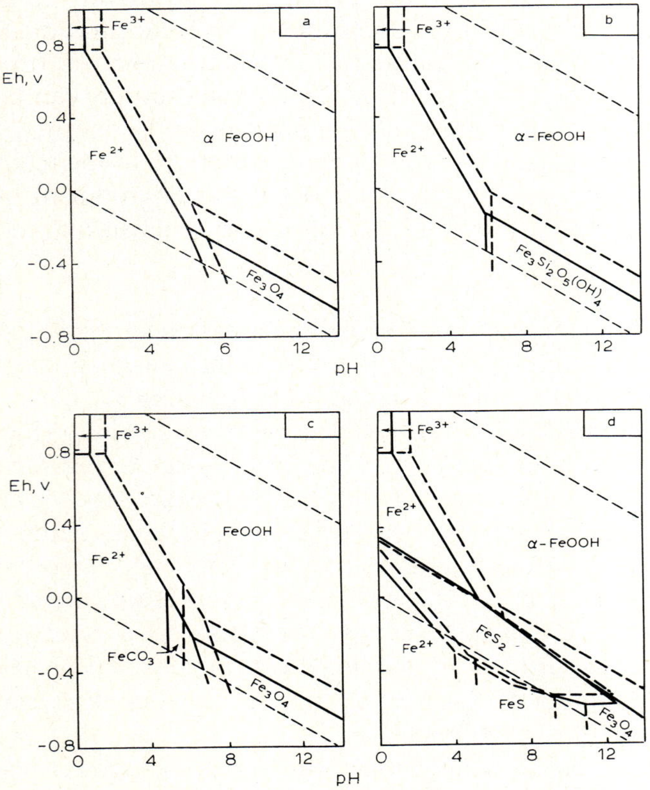

Fig. 58. Diagram of relationships between primary metastable sediments (dashed lines) and stable diagenetic minerals (solid lines) in an open system with $a_{Fe} = 10^{-2}$ g · ion/l: a—acid; b—silicate; c—carbonate (beginning of precipitation of $FeCO_3$ is shown for $P_{CO_2} = 1$ bar; d—sulfide ($a_S = 10^{-3}$).

an expansion of the field of goethite relative to the field of primary amorphous iron hydroxides. The expansion of the goethite field takes place due to a decrease in the fields of magnetite and the Fe^{2+} ion.

Crystallization of primary amorphous Fe^{3+} hydroxides as goethite is accompanied by liberation of energy equal to 4.55 kcal/mol or ~ 50 cal/g.

Crystallization of highly dispersed hydromagnetite:

$$Fe_2O_3 \cdot n\,(H_2O)_d \rightarrow (Fe_3O_4)_{cr} + n \cdot (H_2O)_l \quad (3.21)$$

is also accompanied by a decrease in solubility and liberation of energy: $\Delta G_{298} = -6.9$ kcal/mol or 30 cal/g.

In the course of diagenesis amorphous hydroxides of divalent iron are completely converted to crystalline magnetite. Thus if there are no complex-forming substances, diagenesis of iron oxide sediments in a reducing setting leads to the formation of magnetite or a goethite-magnetite association.

Diagenesis of iron silicate sediments in the absence of active forms of sulfur and carbon dioxide leads to some decrease in the silicate field due to expansion of the goethite field towards the region of reducing conditions (see Fig. 58). As in amorphous silicate sediments, the magnetite field in crystalline rocks is completely replaced by the greenalite field.

The decrease in solubility of greenalite itself as a result of crystallization is insignificant. The energy liberated in the course of diagenetic transformation of freshly precipitated highly dispersed greenalite into the dense crystalline modification is 5.7 kcal/mol or about 15 cal/g.

From a comparison of the equilibrium diagrams (see Fig. 58) it follows that the formation of greenalite (and in the general case, of minerals of the chlorite group) must be a common phenomenon in the diagenesis of hydroxide sediments in reducing conditions.

Diagenesis of iron carbonate sediments is controlled by the content of dissolved carbon dioxide and P_{CO_2} and by the presence of reactive organic matter.

In the course of diagenetic crystallization the stability field of siderite is enlarged (see Fig. 58) and for a given concentration of iron the $FeCO_3$ field occupies the region of neutral and acid environments. However, such high concentrations (10^{-2} g·ion/l) of iron hardly occur in the crystallization of amorphous carbonate. A lowering of a_{Fe} shifts the boundary of the stability field of $FeCO_3$ into the region of higher pH values and actually in diagenesis the formation of stable crystalline siderite would occur at pH = 6–7.

The energy liberated in the conversion of freshly precipitated siderite to crystalline material is 2 kcal/mol or ~ 17 cal/g.

Diagenesis of iron sulfide sediments. In the case of high contents of active forms of sulfur, primary sulfides are formed in a wide range of pH in reducing environments. In the case of diagenetic crystallization of sulfide sediments in the absence of reactive silicic acid and carbonic acid, a goethite-sulfide association is formed (see Fig. 58). In this case magnetite is unstable and is replaced by goethite if *Eh* is constant or by pyrite if *Eh* decreases.

Enlargement of the sulfide field in the course of diagenetic crystallization is insignificant and occurs chiefly due to a decrease in the field of the Fe^{2+} ion. Thus the main result of diagenesis of iron sulfide sediments is some decrease in the migrational capacity of iron in acid waters.

The energy given off in the course of diagenetic transformation of the original finely dispersed pyrite into the crystalline modification is 3.8 kcal/mol or ~ 30 cal/g.

Diagenesis of oxide–silicate–carbonate–sulfide sediments. If reactive sulfur, carbonic acid, and silicic acid are present at the same time in original iron sediments of any composition, the mineral associations formed are determined by the concentrations of these active forms. Figure 59a gives a diagram of the relationships between the crystalline iron compounds in the system $Fe{-}SiO_2{-}CO_2{-}S{-}H_2O$ for $a_S = 10^{-3}$ and $a_{Fe} = 10^{-2}$ g · ion/l.

Analysis of this diagram suggests that under the given conditions, diagenesis of the original sediments leads to the formation of a goethite-siderite-sulfide association. The fields of siderite and greenalite are represented by a very small sector: under the given conditions siderite is formed in slightly acid environments with *Eh* values close to zero; greenalite is formed only in alkaline environments with very low *Eh* values.

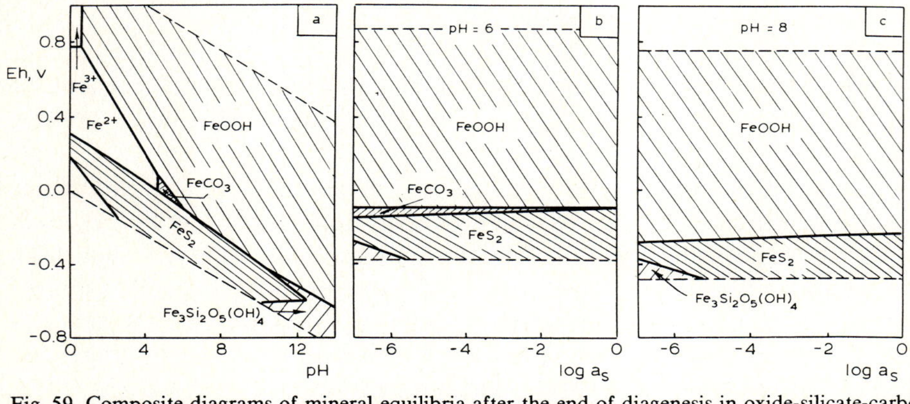

Fig. 59. Composite diagrams of mineral equilibria after the end of diagenesis in oxide-silicate-carbonate-sulfide rocks in an open system ($a_{Fe} = 10^{-2}$ g·ion/l): a—in coordinates of *Eh*-pH at $a_S = 10^{-3}$; b—in coordinates of *Eh*-a_S at pH = 6; c—in coordinates of *Eh*-a_S at pH = 8.

The relationships between iron carbonate and sulfide minerals formed in the course of diagenesis are determined mainly by the concentrations of active forms of sulfur in the waters. The existence of a very narrow wedge of siderite field between the fields of pyrite and goethite in the region of sulfate-ion stability indicates that siderite can form in association with pyrite and goethite in environments with *Eh* values close to zero.

The formation of silicates in the process of diagenesis is also controlled mainly by sulfur compounds in the waters. Thus, in environments with pH = 8 greenalite can be formed only when $a_S \leqslant 10^{-5.5}$ g·ion/l (see Fig. 59). In more alkaline environments the minimum values of a_S at which greenalite is formed are somewhat higher: at pH = 9 and 10, the $a_{S(min)}$ values are 10^{-5} and 10^{-4} g·ion/l, respectively.

In the absence of active forms of sulfur or when their content is less than $a_{S(min)}$, siderite or greenalite are formed.

For crystalline $FeCO_3$ and $Fe_3Si_2O_5(OH)_4$ the equilibrium value of P_{CO_2} is $10^{-2.13}$ bar, which corresponds to pH = 5.7 under conditions of siderite in equilibrium with pure water. Thus in the presence of active forms of silicic acid, iron carbonates can be formed in the course of diagenesis only in slightly acid mediums. The wide distribution of the goethite-siderite association in slightly acid environments is noteworthy. Probably these minerals predominated in Precambrian iron sediments deposited when the partial pressure of carbon dioxide was high and the *Eh* fluctuated due to periodic pulsations of the activity of photosynthetic organisms.

The question of the possibility of formation of diagenetic magnetite should be particularly considered. As is seen, there is no magnetite field on the diagrams characterizing the conditions of formation of complex iron sediments. Judging from calculations using the new consistent constants,

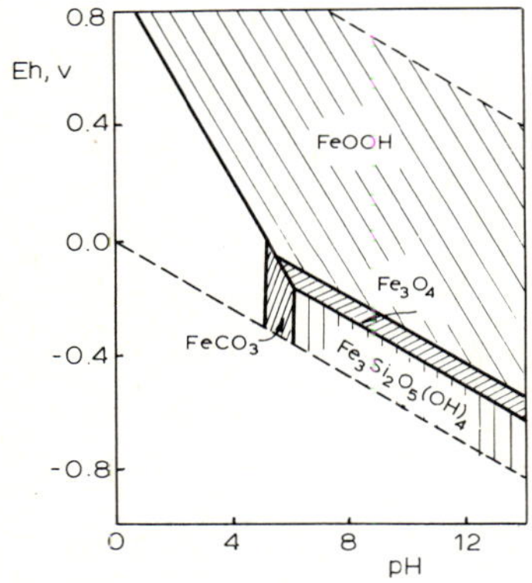

Fig. 60. Conditions of formation of magnetite in diagenesis of cherty iron deposits of complex composition ($a_{Fe} = 10^{-2}$ g·ion/l, a_{CO_2} = 1.0–0.02 bar, no active sulfur present).

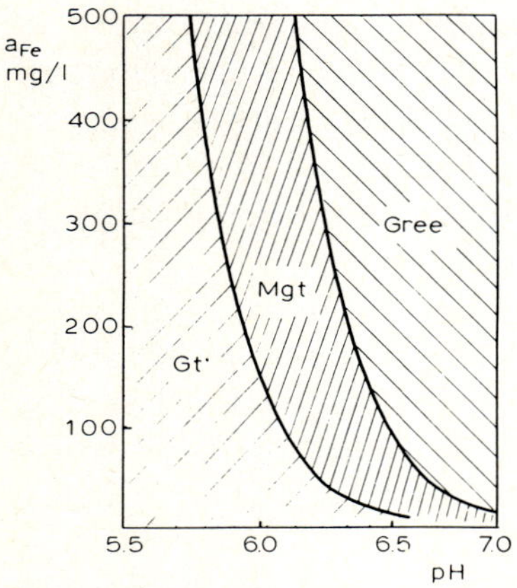

Fig. 61. Variation in activity of iron as a function of pH pore solution in diagenesis of iron-rich sediments of complex composition. *Gt* = (hydro)goethite; *Mgt* = magnetite; *Gree* = greenalite.

thermodynamically stable crystalline magnetite is formed in oxide-silicate rocks at temperatures above 150–180°C, that is, under conditions of epigenesis and low-temperature metamorphism.

However, the formation of magnetite under conditions of low-temperature diagenesis is not ruled out if kinetic factors furthering more rapid crystallization of Fe_3O_4 compared to the slower crystallization of iron silicates and amorphous silica are taken into account. Deposition of iron from ionic solutions of Fe^{2+} also should promote the formation of magnetite or siderite, inasmuch as iron hydroxides and silica precipitating from colloidal solutions are characterized by low reactivity and remain in the sediments in an amorphous state for a long time. Figure 60 gives a diagram characterizing the interaction between crystalline magnetite and diagenetic hydrogoethite, siderite, and greenalite; silica in the form of chalcedony is taken as the saturating phase. In the calculations for siderite and greenalite, ΔG_f^0 values intermediate between those of amorphous sediments and crystalline minerals were used (Table XIII).

As is seen, a magnetite field in this case separates the hydrogoethite and greenalite fields. The region of crystallization of magnetite falls in neutral environments (pH = 6–7) for medium and low concentrations of iron (Fig. 61), and the *Eh* values are negative everywhere (< -0.06 V). At corresponding values of P_{CO_2} (0.02–0.2 bar) a widely distributed magnetite + siderite association is formed.

The data obtained correspond in general features to geologic ideas concerning the directionality and character of diagenetic transformations in iron sediments in an open system, where the activity of iron and sulfur, the

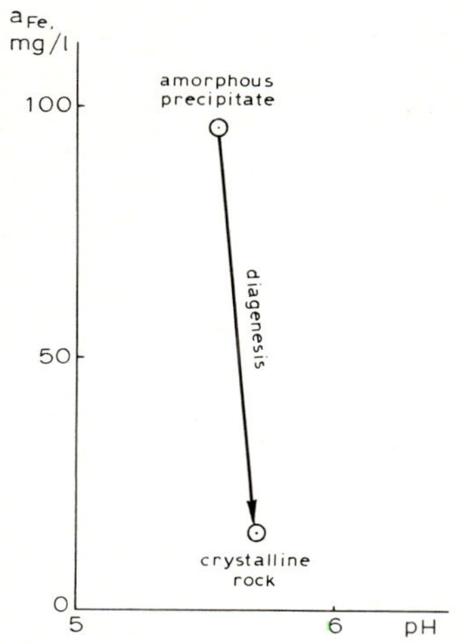

Fig. 62. Variation in activity of iron and pH of pore solution in diagenesis of a siderite-greenalite sediment (closed system).

carbonate content of the solutions, and the corresponding P_{CO_2} values are determined by external factors (the solution above the sediment is the potential-determining or buffer system).

The insignificant occurrence of sulfides in BIF indicates relatively low concentrations of sulfur in the waters of the sediments or little diffusional exchange with the bottom waters during diagenesis. The latter explanation is more plausible inasmuch as in recent iron sediments it is chiefly silicates that are formed during diagenesis under reducing conditions. In this case the composition of the equilibrium pore solution is controlled directly by the minerals (the rock is the potential-determining buffer system). The evolutionary trend of the composition of the pore solution during diagenesis of carbonate-silicate iron sediments in a closed system is shown in Fig. 62. In the transformation from amorphous sediment to crystalline rock, the activity of iron and the total carbonate content in the pore solution decrease (from ~ 100 to ~ 15 mg/l) and from $\sim 5 \cdot 10^{-3}$ to $\sim 8 \cdot 10^{-4}$ g·ion/l, respectively) and the pH increases somewhat.

On the basis of the experimental and theoretical thermodynamic data that have been considered, the following general conclusions can be drawn:

(1) Diagenesis of iron cherts does not lead to any changes in principle in the relationships between minerals and mineral associations. As a result of diagenesis, no new compounds arise which could not have been formed by chemical precipitation. From a comparison of the "stability" diagrams of the sediments and minerals it is seen that only the numerical values of the

equilibrium parameters (pH, *Eh*, activities) and the size of the stability fields of the individual solid phases are changed.

(2) Diagenesis in oxidizing conditions results mainly in dehydration, condensation, compaction, and crystallization of amorphous iron hydroxide and silica-gel sediments. The stable mineral forms—goethite and quartz—are formed in several stages via intermediate metastable phases.

(3) Reduction of oxide sediments is possible if the content of organic matter or free carbon is sufficient. The production of certain mineral forms depends on the activity of sulfur and pH. The formation of diagenetic magnetite is presumed only in neutral environments with low activity of sulfur and a slow rate of crystallization of the silicates. A rise in temperature (to 150–180°C) leads to the appearance of a magnetite stability field in silicate sediments.

(4) Iron retains a fairly high migrational capacity in the pore solutions of carbonate-silicate sediments after their burial and stabilization.

(5) The phase transformations during diagenesis are irreversible, they take place spontaneously and are accompanied by substantial liberation of energy (up to 50 cal/g) and a decrease in solubility (by 1 to 3 orders of magnitude).

(6) The rate of diagenetic transformations of iron cherts is determined not only by T and P, but also by the particulars of deposition, form of migration, dispersion of the particles, and pH of the environment. The time it takes to transform amorphous sediments into crystalline minerals is very great. For iron hydroxides it is several million years, and for silica it may be still more.

Further transformations of iron-rich sediments under the conditions of epigenesis and burial metamorphism, caused by an increase in T and P, will be considered in the analysis of the physicochemical conditions of metamorphism.

Physicochemical model of the formation of cherty iron sediments in the Precambrian

Comparison of the physicochemical and geochemical data on the forms of transport and conditions of deposition of iron and silica with modern ideas concerning the particulars of sedimentation in the Precambrian makes it possible to evaluate critically and to some extent place constraints on the numerous versions of existing genetic hypotheses. After such reexamination, the scheme of formation for the Precambrian iron-formations that is best substantiated from the physicochemical standpoint can be presented.

As a result of study of the distribution, temporal and spatial relationships, and mineralogic and lithologic particulars of the main BIF of the world, the following points are generally known and adequately substantiated:

(1) Typical BIF are characteristic only of the Precambrian, and the regularities of their formation should be considered in connection with the geochemical evolution of the conditions of sedimentation, atmosphere, ocean, and life in the early stages of development of the Earth's crust.

(2) The BIF mainly are metamorphosed chemogenic sediments which were normal deposits in ancient geosynclinal zones.

(3) Iron cherts are deeper-water formations than the clastic (chiefly shaly) strata with which they are often associated.

(4) A direct or indirect paragenetic relationship between the BIF and volcanic rocks of different composition and origin is often observed.

(5) Deposition of the iron cherts was controlled by the laws of chemical differentiation of matter, and that of the clastic sediments by the laws of mechanical differentiation.

To some extent these points can be considered to be geologic postulates, inasmuch as they do not stem directly from the general thermodynamic, experimental, or geochemical data, but on the contrary, place constraints on physicochemical formulations. However, it should be noted that specific geochemical investigations of the distribution of the essential and accessory elements and particularly of the stable isotopes (oxygen, carbon, and sulfur) in the rocks and minerals confirm the metasedimentary nature of the BIF.

Transport and deposition of iron

Ferric iron

The migration of trivalent iron in dissolved form is extremely limited.

Ionic form of transport. This is possible only in very acid solutions ($pH < 3$), the long existence of which is not likely at the Earth's surface due to interaction with rocks, due to dilution by surface waters, and due to the buffer effect of carbonate and silicate equilibria controlling the pH in the ocean.

The appearance of acid solutions in the weathered layer cannot be substantiated physicochemically even if a high content of carbon dioxide in the atmosphere is assumed: with $P_{CO_2} = 1$ bar the pH of the saturated aqueous solution (pure water with no impurities of other anions and cations) is only 3.9—a value clearly insufficient for intensive removal of ferric iron in the form of stable Fe^{3+} $FeOH^{2+}$, or $Fe(OH)_2^+$. Moreover, when "pure" (rain) carbonic acid waters react with rocks the pH value increases as the saturation with dissolved cations (mainly Mg^{2+}, Na^+, K^+, and Ca^{2+}) increases. The constant presence of other volatiles of the "acid fume" group in the Precambrian atmosphere is not very likely, due to their high solubility in water and subsequent rapid neutralization.

Acid volcanic waters occur everywhere in recent volcanic areas, and surely they were just as common in the Precambrian. However, these waters contain mainly ferrous iron, not ferric. Mixing with oxygen-bearing surface waters leads to dilution, neutralization, oxidation of Fe^{2+}, and subsequent deposition of Fe^{3+} in the form of hydroxide.

Colloidal form of transport of ferric iron in the Precambrian, apparently, was less important than in later geologic epochs. On the basis of an examination of the known methods of formation of colloidal solutions, it can be assumed that condensation phenomena predominate in the geochemical processes, and dispersion and peptization play a secondary role. Consequently one of the conditions for obtaining colloidal solutions is the presence of ionic solutions as an obligatory intermediate stage in the cycle: weathering → transport → deposition.

Ionic solutions of Fe^{3+}, which are stable at very low pH values, could not have formed in the weathered layer. Thus no colloidal solutions of ferric iron could have appeared.

Another possible variant of supergene migration of iron, observed in recent humid zones, calls for the formation of ionic solutions of ferrous iron in the weathered zone in the presence of organic matter and carbonic acid. When it reacts with aerated waters, iron is oxidized and yields colloids which migrate under the protection of organic stabilizers and in the form of metallo-organic compounds. Formal transfer of this model to the Precambrian raises serious objections, inasmuch as organic life did not appear on land until the Silurian–Devonian, when ultraviolet radiation dropped to an acceptable level. Thus in the Precambrian weathered layers there were no zones differing in oxidation-reduction conditions due to differences in content of organic matter and intensity of bacterial activity. Consequently stabilization of ferric iron colloids by organic compounds could not yet have been especially important. If unstable colloidal solutions of iron were produced in small amounts in the ancient weathered zone, they were unstable and easily coagulated under the influence of electrolytes.

When thermal volcanic waters react with aerated surface waters, the appearance of ferric iron colloids is quite permissible. However, there are no organic compounds of fulvic acid type in volcanic waters and only colloidal silica could act as a stabilizer. At the same time there often is sulfur in thermal solutions, the stable form of which in the presence of free oxygen is the SO_4^{2-} ion—the main coagulant of colloidal iron. For this reason the possibilities of colloidal transport of iron from volcanic sources to sedimentary basins are limited. A high CO_2 content in the hydrosphere and atmosphere does not exert a stabilizing effect on $Fe(OH)_3$ colloids.

Thus regardless of their origin, colloidal solutions of iron could hardly have accomplished the transport of a substantial amount of ore material. Coagulation of ferric iron colloids, even those stabilized by colloidal silica, already had occurred to a substantial extent in near-shore zones or even in rivers, under the influence of a relatively small increase in concentration of electrolytes.

The main coagulant was sulfate ion, which completely precipitates iron from dilute colloidal solutions ($\leqslant$ 100 mg/l Fe_2O_3) at very low concentrations (of the order of 5–15 mg/l SO_4^{2-}). In acid and neutral environments colloidal silica has practically no stabilizing effect. The idea that there were no electrolytes in the Precambrian sedimentary basins is not consistent with modern ideas on the evolution of the salt composition of the ocean and cannot explain such particulars of ancient sediments as the formation of the sulfide facies of the BIF and of sulfates in evaporites. Isotopic investigations also suggest the appearance of sulfates in ancient water bodies at the Archean–Proterozoic boundary. Moreover, the values of the concentration gradient of the electrolytes of sea water are different for the coagulation of colloids of iron and silica, which would have led to deposition of iron along with clastic sediments and the migration of colloidal silica to parts of the basins far from the shore. If mixed iron-silica colloids migrate together it is inevitable that there will be chemical differentiation, with the formation of spatially separated concentrations. Joint deposition of iron and silica to form banded sediments of iron hydroxide and amorphous silica is possible only if concentrated mixed colloidal solutions enter directly into sea basins with normal salinity. In this case the region of deposition of chemogenic sediments should be in the immediate vicinity of the source of the solutions and a considerable distance from the region of deposition of clastic sediments. Such a combination of different factors is theoretically possible only in the case of submarine fumarolic activity, the scale and cyclicity of which has not occurred in later geologic epochs. Let us also note that in this model the presence of substantial concentrations of ferric iron (Fe^{3+}) in the original solutions is presumed.

Ferrous iron

The main form of migration of ferrous iron in acid and neutral solutions is the simple ion, Fe^{2+}, the equilibrium concentrations of which can be very high in reducing environments devoid of free oxygen and are regulated mainly by fluctuations in pH and *Eh*. In the thermodynamic analysis of the migrational capacity of Fe^{2+} it has already been mentioned that near the lower limit of decomposition of water the equilibrium concentration of iron theoretically reaches several tens of grams per liter even in neutral solutions ($pH = 6–7$). Thus, in the conditions of the oxygen-free atmosphere of the early stages of evolution of the Earth's crust, when the redox potential of the ocean waters was regulated by conjugate reactions in which CH_4, CO_2, CO, H_2, and carbon participated, ferrous iron could have been carried in substantial amounts both from the weathering rocks on land and from volcanic sources, and could have migrated to the sedimentary basins without hindrance. The fact that no geochemical barriers capable of causing deposition of Fe^{2+} in any form could have existed in acutely reducing conditions or in the basin itself is very important. Only changes in external conditions, for instance fluctuations in pH or P_{CO_2}, the values of which apparently were directly or indirectly interrelated, could have caused deposition of excess ferrous iron in the form of carbonate. Wholesale precipitation of ferric iron was the result of an abrupt change in *Eh* due to the appearance of free oxygen.

Transport and deposition of silica

Examination of the physicochemical and geochemical data that have been given leads to the conclusion that the well known genetic scheme, in which silica is carried in ionic or colloidal form from subaerially weathered rocks and then deposited chemically in the sea basins, is not very realistic.

The silica content in waters of the supergene zone is determined by the solubility of the rock-forming silicates and quartz. A large part of the silicates of the original magmatic and metamorphic rocks dissolves incongruently, with the formation of hydrous minerals stable at low temperatures (hydromicas, kaolinite, etc.). Quartz apparently dissolves congruently. Regardless of the actual mechanism of solution, the maximum concentration of ionic silica in equilibrium with crystalline silicates and quartz is not more than a few milligrams per liter and remains virtually constant over a wide range of pH values—from 3 to 10. Such solutions, which are saturated with respect to the thermodynamically stable crystalline phases, for instance α-quartz, are at the same time acutely undersaturated in SiO_2 with respect to the actual chemogenic sediments—amorphous silica and finely dispersed hydrosilicates (the "equilibrium" concentration is of the order of 80–120 mg/l SiO_2). Inasmuch as the content of silica in the form of $H_4SiO_4^0$ in solution hardly depends on the pH and other factors (*Eh*, concentration of electrolytes), it is difficult to imagine the geochemical barrier that would cause deposition of substantial masses of amorphous silica from undersaturated solutions. The transformation of such ionic (molecular) solutions into colloidal ones seems unrealistic, inasmuch as it contradicts all the theoretical and experimental data on the obtaining of and stability of colloidal systems.

To sum up, the main mode of extraction of dissolved silica coming into the seas and oceans from the weathered layer is biogenic precipitation—a process widespread both in recent and in ancient sedimentary basins. Biogenic precipitation of silica from undersaturated solutions could also have been of definite importance in the formation of the Precambrian iron cherts, but no direct evidence has been obtained as yet.

If a volcanogenic source is assumed, one can avoid many contradictions and present a fairly sound model of chemogenic sedimentation, including migration of silica in thermal and then in cold solutions and subsequent deposition in sea basins.

The thermal waters of recent volcanic regions that have an acid reaction contain an average of about 200–300 mg/l SiO_2, in isolated cases up to 600–900 mg/l SiO_2, which is several times higher than the "equilibrium" concentration needed to precipitate amorphous silica. When the thermal waters cool part of the silica remains in ionic (molecular) form and part of it forms colloidal particles. Judging from experimental data, such ionic-colloidal solutions are quite stable in acid and slightly acid environments ($pH \leqslant 4-5$) in the case of average concentrations (100–1000 mg/l SiO_2), and could effectively transport silica for considerable distances. The nature of the geochemical barriers causing deposition of colloidal silica stands out quite clearly—it is the pH gradients which inevitably rise when acid volcanic and slightly acid or neutral river and sea waters mix, and the electrolyte concentration gradients, mainly of Na^+, Mg^{2+}, and possibly Fe^{2+}. It is typical that the combined action of these two factors, namely an increase in pH and increase in cation concentration, intensifies coagulation. As pH increases a certain amount of ionic silica, which occurs in acid solutions in the form of fluorides and other complicated complexes or in a metastable state, also precipitates. In contrast to the deposition of colloidal iron, coagulation of silica colloids and subsequent deposition of amorphous particles are very slow processes which take years even under laboratory conditions. The ionic silica remaining in solution after coagulation of the colloidal component can be precipitated only biogenically in slightly acid environments ($pH \leqslant 6.5$), or be fixed in part by sedimentary iron silicates in neutral environments ($pH = 6.5-8.0$). In the latter case the "equilibrium" concentration of silica in solution may be reduced to a few tens of milligrams per liter.

Formation of cherty iron sediments

Now it remains to put together the partial conclusions obtained and attempt to present a generalized physicochemical model of the formation of banded cherty iron sediments. It should be kept in mind that the transport and deposition of substantial amounts of iron and silica from low-temperature aqueous solutions evidently are governed by various factors that are not directly interrelated. Only a favorable combination of many factors in limited parts of the Earth's surface and in certain periods of geologic evolution led to chemical deposition of cherty iron sediments uncontaminated by clastic material.

Such a combination is most probable if one accepts the accumulational-biogeochemical version of the volcanogenic-sedimentary hypothesis (Mel'nik, 1973; Belevtsev and Mel'nik, 1976). It is suggested that during the long period of oxygen-free evolution of the hydrosphere and atmosphere all the iron that arrived, chiefly in the Fe^{2+} form, from the weathered layer and volcanic sources accumulated in the slightly acid solutions of the ancient water bodies. At the same time ionic (molecular) silica accumulated up to "equilibrium" concentrations.

When high iron and silica concentrations were reached, chemical deposition of iron and silica became possible, on the arrival of supersaturated volcanic solutions. The oldest (Archean) BIF, closely associated with volcanic rocks (formations of Algoma type) were deposited from supersaturated solutions formed when acid thermal waters mixed with ocean waters saturated with carbonic acid but devoid of free oxygen.

In the original thermal waters (as in the hot waters of recent fumaroles) both iron and silica occurred in true dissolved (ionic or molecular) form, producing rather high concentrations (of the order of $n \cdot 10^{-1}$–$n \cdot 10^{0}$ g/l); the environment was acid (pH < 2–3) and reducing ($Eh < 0.0$–$+0.2$ V; $P_{O_2}:P_{H_2} \ll 2$; Fe^{2+} ion predominated).

In the case of terrestrial or submarine effusions the physicochemical parameters of the thermal solutions (temperature, pressure, pH) changed abruptly, which led to partial condensation of $H_4SiO_4^0$ molecules to form complex polymeric forms and colloidal particles. In equilibrium with the Precambrian atmosphere, devoid of free oxygen and containing a high amount of carbon dioxide (P_{CO_2} between $n \cdot 10^{-3}$ and $n \cdot 10^{-1}$ bar), the natural waters remained slightly acid and reducing regardless of their origin. The acid (pH < 4) ionic-colloidal solutions were very stable and could carry silica for considerable distances. Iron also remained in the form of the stable Fe^{2+} ion and migrated along with the silica.

In the case of terrestrial effusions, when acid volcanic waters mixed with surface river waters and especially with sea waters in the near-shore zones, the pH value gradually increased due to dilution and neutralization and approached the equilibrium value, controlled by carbonate-silicate buffer reactions (pH within 6–8 depending on the type of buffer). As a result, colloidal silica already had begun to coagulate in the near-shore zones, and clots were deposited along with clastic components or were carried on by currents. However, maximum chemogenic deposition began some distance from the shore and did not coincide spatially with the maximum accumulation of the clay sediments which gave rise to thick shaly sequences. This offset is explained by the spatial separation of the main geochemical barriers causing deposition of colloidal silica, namely gradients of pH and of electrolyte concentration. Probably the first batches of silica were deposited due to neutralization of acid waters, and then, already in more distant sectors, deposition was intensified under the influence of electrolytes.

As a result, in the near-shore zones silica gels were intermixed with clastic components and did not produce separate accumulations. At a distance from the shore separate chert layers appeared among clastic sediments of argillaceous composition, increasing in number and thickness as the region where purely chemogenic sediments were accumulating was approached. The banding of the cherty-shaly sediments, so typical of the barren horizons or layers of the BIF, is explained by periodic deposition of clastic material rather than by cyclicity of silica deposition. Kinetic factors might also have played some part in the formation of clearly defined layers—very slow and continuous deposition of silica gel and relatively rapid deposition of dense clay particles arriving all at once at times of flood. Thus the microbanding (microstratification) arose; the larger rhythms probably are related to periodic shifting of the shore line and corresponding changes in depth under the influence of tectonic movements. If there had been no chemogenic deposition of iron, the ideal facies profile within the iron-ore sequence would have had this form: clay shale $\rightarrow$ clay shale with intercalations of silica $\rightarrow$ chert with intercalations of shale $\rightarrow$ chert.

After passing near-shore geochemical barriers the waters contained hardly any colloidal silica and in the deep parts of the basins the intensity of accumulation of chemogenic cherty sediments would have fallen off sharply.

There is still another consideration stemming indirectly from the physicochemical and geochemical data on the properties of silica in aqueous solutions. Coagulation of colloids and formation of relatively stable sediments are possible mainly from solutions "supersaturated" with respect to amorphous silica. Otherwise the opposite process will go on at the same time as coagulation—solution of the amorphous sediment or gel in the water layer to form an ionic (molecular) solution. Consequently, in contrast to the present ocean, the Precambrian ocean contained much more dissolved silica. If this silica was not primary, juvenile, there is reason to suppose that the period of chemogenic deposition was preceded by a long period of accumulation and concentration of silica which was never again repeated in geologic history after the Precambrian, when the ocean became sharply undersaturated in SiO_2 due to intensive biological extraction. It is not ruled out that biogenic deposition of silica had already begun in the Precambrian, in parallel with chemical deposition.

In the early stages of evolution of the ocean, all the iron arriving from volcanic sources and from intensively weathering rocks on land also accumulated continuously. In the reducing environment the iron migrated freely, inasmuch as in the Precambrian there were no real geochemical barriers (gradients of *Eh* or pH) capable of causing chemical precipitation before the appearance of free oxygen in the hydrosphere. Even in very slightly acid waters (pH $\leqslant$ 6) the concentration of Fe^{2+} in equilibrium with siderite or silicates of greenalite type was greater than the concentration of $H_4SiO_4^0$ in equilibrium with amorphous silica.

Thus before the beginning of photosynthetic activity of the first organisms iron could have been precipitated only at the end of a period of accumulation in individual zones where acutely supersaturated solutions, most likely of volcanic origin, were introduced. The geochemical barrier might have been a fairly steep gradient of the pH value, arising at the interface between acid thermal waters and neutral or slightly acid carbonate sea waters. Depending on fluctuations in partial pressure of carbon dioxide, silicates or carbonates of $FeCO_3$ or $(Fe,Mg)CO_3$ type were deposited. The oldest Archean iron-formations, usually closely associated with volcanic rocks, are the products of submarine fumarolic activity, deposited from supersaturated solutions at relatively short distances from volcanic centers. However, the mechanism of formation of the banding and the causes of differentiation remain unclear, which makes it difficult to substantiate this scheme physicochemically. One explanation for the formation of the banding in the deposition of volcanogenic BIF may be sinusoidal variation of pH. From analytical chemistry it is known that precipitation of poorly soluble sediments from aqueous solutions begins spontaneously at pH values higher than the equilibrium values. Wholesale deposition of iron according to the reaction:

$$3\,Fe^{2+} + 2\,H_4SiO_4^0 + H_2O = Fe_3Si_2O_5(OH)_4 + 6\,H^+ \qquad (3.22)$$

or

$$3\,Fe^{2+} + 3\,H_2O + \tfrac{1}{2}O_2 = Fe_3O_4 + 6\,H^+ \qquad (3.23)$$

is accompanied by liberation of protons and corresponding acidification of the solution. When pH drops below the threshold value, deposition of iron ceases. Some of the H^+ ions are actively sorbed by the fresh amorphous and dispersed iron sediments. In our experimental study of the "aging" of hydroxide sediments we detected acidification of the solution due to desorption of protons (in the course of several months the pH changed from 6–8 to 2.5–3.0). Thus fresh iron sediments might acidify the solution for a long time, preventing precipitation of the next batch of iron. As neutralization occurred the threshold pH value was again reached, spontaneous deposition occurred, and the cycle was repeated. It is not ruled out that in the deposition of silicate BIF, silica was also in part included in these cycles.

In the examination of the geologic data numerous claims have already been made regarding the relationship of deposition of the BIF to volcanic processes. Critics of the genetic interpretation of this relationship sometimes rely on quantitative calculations. Thus, Holland (1973) believes that to form the Hamersley BIF in Australia, about 1000 volcanoes 1.6 km apart from each other would have been needed, with a basin of 1600 km perimeter. Therefore the hypothesis of the volcanic origin of the iron seems unsatisfactory. This conclusion is based on two premises—the 6-million-year duration of deposition of the Hamersley BIF and the annual production of $\sim 35-50$ tons of iron by Ebeko volcano in the Kurile Islands. Both these premises are difficult to relate directly to deposition of BIF, inasmuch as it is obvious that volcanic processes were more active in the Precambrian and that we do not know the actual duration of the sedimentary iron ore process. The volcanogenic-sedimentary ore process has not lost its importance in our time—Shiikawa (1970) shows that most of the workable limonite deposits of Japan were formed, and in places are being formed today, by deposition of iron from acid iron-bearing waters of hot springs of volcanic origin.

The appearance and development of life led to a gradual change in the oxidation-reduction environment in the sedimentary basins. Oxidation of Fe^{2+} to Fe^{3+} apparently became possible only after oxidation of a substantial part of the methane in the atmosphere and of free carbon on land to carbon dioxide, and of sulfur and hydrogen sulfide in the hydrosphere to sulfate ion. It is not ruled out that before the appearance of typical photosynthetic plants there were very simple organisms that reacted directly with Fe^{2+}, causing oxidation and deposition of iron. Free oxygen had not yet arisen then, ferrous iron predominated in solution and *Eh* values remained low. The main components of oxide sediments were siderite in carbonate sediments and greenalite in silicate, associated with periodically deposited hydromagnetite. The formation of magnetite in sediments of complex composition depended on kinetic factors—rapid crystallization of the oxide and slow crystallization of silicates, arising chiefly in diagenesis.

The next important biogeochemical stage was the appearance of phytoplankton, mainly bluegreen algae—the main producers of free oxygen. Oxidation of most of the iron in the hydrosphere and deposition of the greater part of the iron cherts are related to this stage. There is every reason to suppose that a substantial amount of phytoplankton developed in narrowly localized parts of the ancient water bodies at a certain depth, controlled by the thickness of the shielding water layer, and at a certain distance

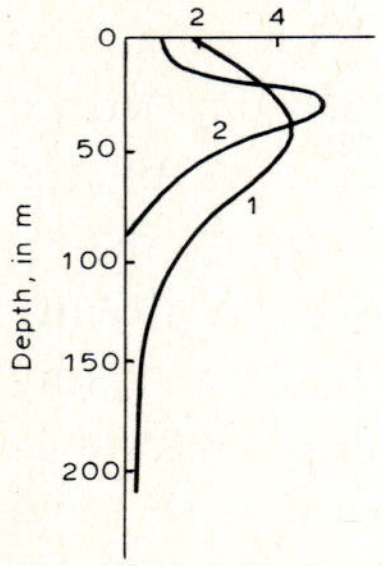

Fig. 63. Distribution of bacteria (*1*) and phytoplankton (*2*) by depth off California, according to Voytkevich and Zakrutkin (1970).

from the shore. An indirect relationship of "oases" of the first life to regions of active volcanic activity, from which the needed components, in particular silica, were supplied, is not ruled out.

The growth of phytoplankton is characterized by a clearly manifested cyclicity (Bruyevich et al., 1960). According to the data of Smetanin (1959) for the Kurile–Kamchatka basin, along with the thawing of ice and beginning of warming of the surface layer in the spring there occurs vigorous "blooming" of phytoplankton. The rate of formation of oxygen substantially exceeds the rate of its release into the atmosphere (saturation up to 175%, and according to the data of Zhizhchenko (1959), up to 250–400%). Maximum development of organisms is observed in near-shore parts where the intensity of warming is high, and at a certain depth from the surface of the water (Fig. 63).

Such periodic phenomena of intensive "blooming" in the conditions of the Precambrian did not lead to the release of oxygen to the atmosphere, but to oxidation of Fe^{2+} to Fe^{3+} directly in the water layer. As a result, vertical zoning of the redox potential could be produced in the ancient water bodies, causing a unique biogeochemical cycle of iron to arise (Fig. 64). In the deep parts of the basin a reducing zone with high Fe^{2+} content is distinguished. An oxidizing zone of photosynthesis is situated in the near-surface parts, where free oxygen used to oxidize Fe^{2+} is generated. Between these zones the variation in *Eh* occurs in jumps. If even the slightest traces of free oxygen appear (P_{O_2} 10^{-20} bar) the migrational capacity of iron drops sharply and in moderately acid solutions amorphous hydroxide—$Fe(OH)_3$—is deposited. From the beginning, as the hydroxide is slowly deposited as flocs, it is stablized, and its solubility correspondingly reduced. However, in the deeper parts of the water bodies below the zone of photosynthesis there is an intermediate zone where metabolism is difficult and organisms die. In recent water bodies it has been established (Skopintsev, 1949; Smetanin, 1959) that

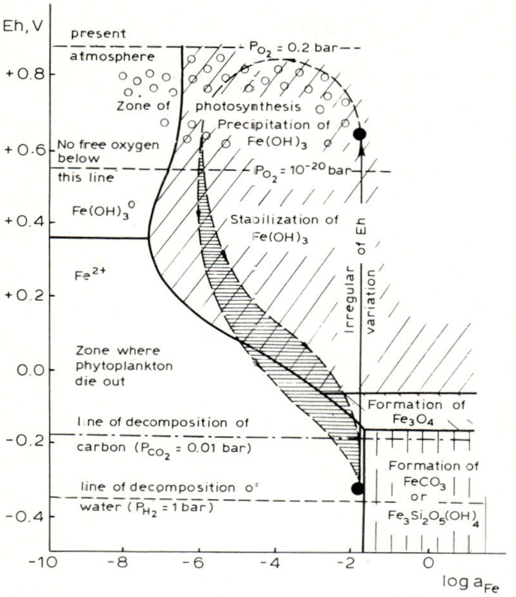

Fig. 64. Scheme of vertical zoning and biogeochemical cycle of iron in the formation of iron-rich sediments (pH = 6; siderite, greenalite, and hydrogoethite are diagenetic, magnetite crystalline).

most dead phytoplankton decomposes while still in the surface layer (0–100 m). In the zone of decomposition the *Eh* value drops and $Fe(OH)_3$ may either be dissolved (left-hand branch of the cycle) or be reduced to magnetite (right-hand branch), depending on the iron concentration.

The mineralogic composition of the sediment is determined by the depth of the basin, relationship of the zones, concentration of iron, and rate of deposition and burial (conservation).

Let us consider the effect of these factors on the schemes given (Figs. 64 and 65).

The initial stages of the ore process are characterized by high iron content, slightly acid solutions ($pH \approx 6$), and low *Eh* values.

Liberation of oxygen in the process of photosynthesis was accompanied by an increase in the biomass and corresponding absorption of carbon dioxide, which in turn led to a local drop in P_{CO_2} and increase in pH. As a result, chemical deposition of dispersed $FeCO_3$ occurred, along with biogeochemical precipitation of $Fe(OH)_3$. Parallel to this, in the zone where phytoplankton died out some of the hydroxide was reduced to $FeCO_3$, in slightly acid environments, and to Fe_3O_4, in neutral or slightly alkaline environments.

Inasmuch as there were a substantial number of other cations in the

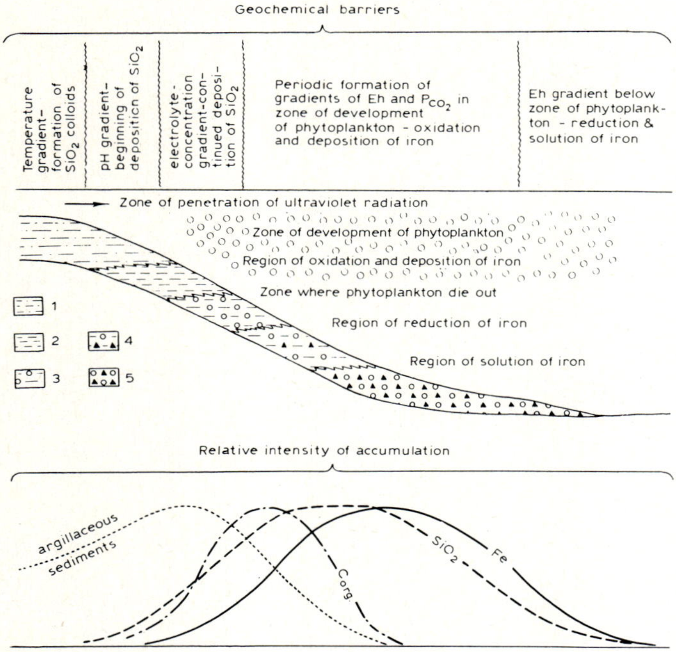

Fig. 65. Scheme of relationships of geochemical barriers, biogeochemical zones, and regions and relative intensity of sedimentation in the formation of the Precambrian BIF. Symbols of types of sediments: *1* = clastic sandy-shaly without chemogenic iron; *2* = argillaceous (schists) with admixed chemogenic Fe and SiO_2; *3* = argillaceous—iron-rich (iron-rich schists); *4* = argillaceous-cherty iron; *5* = cherty iron, including jaspilites (Mel'nik, 1973; Mel'nik and Belevtsev, 1976).

slightly acid and neutral waters of the Precambrian oceans, it was not pure siderite that was deposited chiefly, but magnesian-iron carbonates with a variable Mg content and some admixed Ca and Mn. Deposition of pure dolomite and calcite became possible only after a substantial part of the iron was deposited and the pH increased, as is shown on the scheme of the evolution of the ocean in the Precambrian (see Fig. 24).

The formation of primary dispersed magnetite in zones of "blooming" is not very likely, inasmuch as when free oxygen (biological potential-determining system) was generated, the *Eh* and P_{O_2} values changed in jumps, the partial pressure of carbon dioxide was fairly high, and the pH values did not exceed 6–7.

From the data given it also follows that the formation of iron sediments was limited to shallow depths where $Fe(OH)_3$ and $FeCO_3$ could not be dissolved when they passed through reducing zones. Preservation of sediments already deposited was aided by the conservation of amorphous silica, which was gradually being deposited in these same parts of the water bodies.

In regions far from shore accumulation of iron sediments ceased, in connection with the decrease in intensity of photosynthesis, solution of the precipitated hydroxides at depth (left-hand branch of the cycle in Fig. 64), and curtailment of deposition of amorphous silica (Fig. 65).

In the final stages of the ore process hydroxides also were deposited from solutions with lower iron contents, but reduction was incomplete and limited to partial formation of magnetite.

Thus in addition to zoning in the distribution of the sediments caused by the relationships of geochemical barriers in space (Fig. 65), an authigenic-mineralogical zoning also arose due to cyclic development of the ore process in time. In this scheme a gradual change in primary minerals is indicated: siderite → magnetite → hematite, complicated by the diagenetic development of greenalite after siderite and magnetite in neutral and slightly alkaline environments. Sulfides, formed locally in acutely reducing conditions, belong to the diagenetic minerals.

Thus secondary differentiation among the purely chemogenic iron sediments was possible, as a result of which the maximum of hydroxide accumulation was shifted to the deeper-water part of the basin and the maximum accumulation of Fe-Mg carbonates was shifted to the shallow-water parts, where finely dispersed clay sediments were being deposited. The maximum accumulation of organic matter was shifted still more into the region of deposition of clastic sediments.

Phytoplankton should be regarded as the main supplier of organic carbon in ancient marine sediments. However, it should be kept in mind that the distribution of C_{free} observed in the facies profile does not coincide spatially with the distribution of phytoplankton and production of oxygen, as most dead plankton decomposes in the surface layer, as a result of which the accumulation of organic matter would be considerably greater in the near-shore zone than in the deep sea.

Cyclic deposition of iron hydroxides and carbonates in relatively shallow basins against a background of continuous deposition of amorphous silica explains the nature of the banding and all the diversity of textures and structures of the iron-formations and shales. Kinetic factors—rapid deposition of dense particles of hydroxide and slow deposition of amorphous flocs of silica—produced distinct separation of individual bands. Cyclic sinusoidal variation in the pH, which was considered a possible cause of the formation of the banding in volcanogenic-sedimentary BIF, also furthered this process. Carbonates were not deposited as rapidly as hydroxides, which led to the formation of the typical siderite-chert bands in iron-ore and shaly sequences. Let us note that biological-chemical deposition of iron apparently occurred not only in deep-sea regions but also in near-shore sectors, but chemogenic deposition here was overwhelmed by clastic, and the iron compounds, like

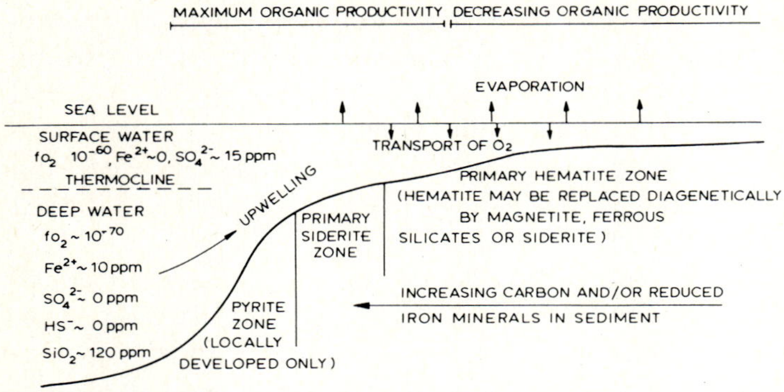

Fig. 66. Idealized depositional environment of a banded iron formation. The term "hematite" includes all ferric oxides and hydroxides (Drever, 1974).

amorphous silica as well, did not form independent bands. Indirect evidence of this is the high iron content in nearly all shales associated with iron-formations.

At the same time and independently, similar ideas concerning individual aspects of the origin of the BIF were published in the U.S.A. in a special issue of "Economic Geology" (Alexandrov, 1973; Garrels et al., 1973; Holland, 1973). In particular Holland, who regards the oceans as the possible source of the iron of the Hamersley BIF in Australia, believed that when saturated solutions welled up from the depths iron was oxidized and deposited in the form of $Fe(OH)_3$. Deposition of silica in the same zones is explained by the fact that as deep sea waters constantly rose toward the surface they became supersaturated with respect to SiO_2. Quantitative calculations show that even in neutral waters (pH = 7) with a low iron content (~ 3 ppm) this process can be accomplished in a geologically realistic time.

Holland's ideas were developed in a work by Drever (1974), who presented an idealized scheme of deposition of the BIF (Fig. 66). The banding is explained by variations in the intensity of upwelling, related to seasonal effects and individual storms. In the proposed model the newly arriving waters were oxidized relatively rapidly, forming iron layers, and prolonged evaporation led to gradual deposition of silica. The interrelationships of chemogenic and clastic deposits and especially the distribution of organic carbon in the sediments remain unclear in this model. The increase in carbon content in the deep zones of the scheme does not agree with the geologic data (the shallower-water clastic—shaly—sediments are enriched in carbon) and with modern observations on the development and death of phytoplankton. At the same time the ideas of Holland and Drever concerning the mechanism of deposition of iron and silica are in complete agreement with our model.

Chapter 4

Physicochemical conditions of metamorphism of the rocks of the Precambrian cherty iron-formations

Introduction

The rocks of the BIF are characterized by great diversity of composition, origin, structures, and textures. In addition to typical chemogenic iron cherts and carbonate rocks, clastic sediments, chiefly pelitic, and various volcanic rocks are also encountered in the make-up of the formations in question. Typical of the iron cherts proper are banded structures produced by alternation of bands of different composition: cherty, iron oxide, silicate, and carbonate. These rocks show frequent changes of mineral associations having different degrees of oxidation and different oxidation-reduction capacities, different original contents of water and carbon dioxide, and uneven distribution of corresponding minerals (carbonates and silicates). Both macro- and micro-anisotropy of composition and properties, the extent of which decreases as metamorphic reworking is intensified, are typical of the rocks of the BIF as a whole.

In analyzing the physicochemical conditions of metamorphism of complex heterogeneous stratified sequences it is necessary to take into account the possible existence of metastable but kinetically stable mineral associations, wide variations in the composition of the fluids in individual parts of the pile being metamorphosed, equilibria of mosaic character, and the presence of systems closed to water or with an insufficiency of it. These particulars require preliminary consideration of several controversial problems of the theory of metamorphism, especially reactions in the case of complex fluids and in the case of different pressures on the solid phases and fluid; analysis of the processes in which H_2O and CO_2 take part at high pressures, when the properties of these very important volatile components become different, also seems to be very important.

Main factors in metamorphism

A vast number of publications have been devoted to various aspects of the theory of metamorphism, even a cursory review of which goes far beyond the

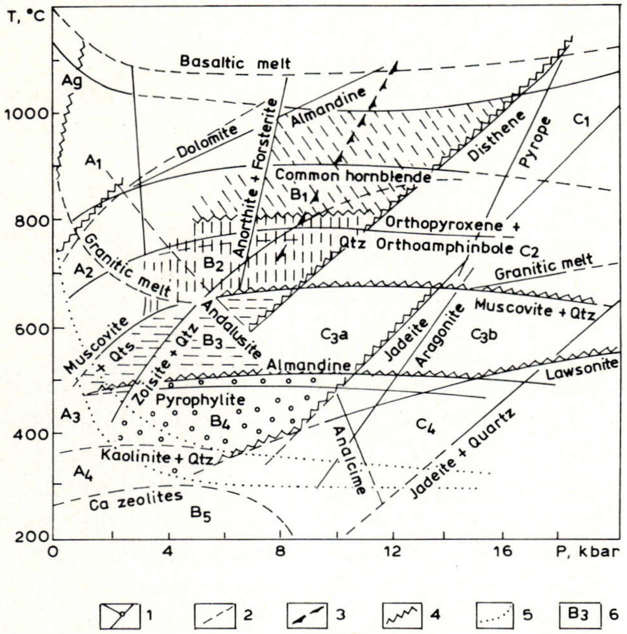

Fig. 67. Scheme of facies of contact- and regional metamorphism (after Sobolev et al.): *1* = lines of mineral equilibria bounding stability fields of most important minerals and associations (their names are written on the side of the line where they are stable), drawn taking account of recent experimental data for different values of P_{H_2O}—from 0.3 P_{total} in the high-temperature part (basalt melt, dolomite) to 0.9 in the low-temperature (lawsonite); *2* = same, for equilibria for which there are insufficient experimental data; for melting of basalt, dashed line refers to more basic basaltoids, for melting of granite, solid line is for $P_{H_2O} = 0.6\ P_{total}$, dashed line for $P_{H_2O} = P_{total}$; *3* = beginning of eclogitization of most basaltoid rocks; *4* = boundaries of facies (and subfacies C_{3a}-C_{3b}); *5* = probable boundaries of field of metamorphism; two lower boundaries—two possible positions of low-temperature kinetic threshold of metamorphism; *6* = letters and dots—fields of individual facies: *A* = low-pressure facies (contact metamorphism); *B* = medium-pressure facies ("ordinary" regional metamorphism); *C* = high-pressure facies.

scope of our specialized work. Moreover, the evolution of ideas on the main factors of metamorphism and the development of the facies principle are discussed completely enough in the summary monograph "Facies of metamorphism" edited by Sobolov (1970) and in other works (Yeliseyev, 1959; Winkler, 1969). The scheme of metamorphic facies proposed by Sobolev (1970) is generally in agreement with modern thermodynamic, experimental, and petrological data and can be accepted with certain corrections (Fig. 67). Most interesting in it for us is the group of medium-pressure facies (*B*), corresponding to "ordinary" regional metamorphism. Here overall pressures range from 3–5 to 10–15 kbar, and temperatures from 300–400 to 900–1000°C. These facies are distinguished within the group:

B_1—two-pyroxene gneisses (granulite facies). This facies corresponds approximately to the granulite facies of other authors: *T* is between 750–800°

and 900–1000°C, P from 4–5 to 12–13 kbar. The upper limit (T and P) is bounded by the lines of basalt melting and of the stability of almandine and dolomite, and also by the compositions of coexisting pyroxenes and the maximum alumina content of the orthopyroxenes.

B_2—*sillimanite-biotite gneisses (amphibolite facies).* The upper temperature limit is bounded by the equilibria of orthopyroxene + clinopyroxene + K-feldspar + quartz-garnet + biotite + amphibole, the disappearance of orthorhombic amphiboles and the biotite-sillimanite association. It corresponds to temperatures of 650–800°C.

B_3—*andalusite(sillimanite)-muscovite schists (epidote-amphibolite facies).* The upper temperature limit is set by the stability of muscovite plus quartz, corresponding to $T = 600$–650°C. The lower temperature limit is fixed by replacement of almandine by chlorite plus quartz, the disappearance of hornblende (it is replaced by actinolite plus epidote and albite) and staurolite (it is replaced by chloritoid), which corresponds to $T = 500$°C. The upper limit of pressure is 7.5–10 kbar.

B_4—*greenschists.* The temperature range is from 350–400° to 500–550°C. The upper temperature limit is set by the stability of chlorite plus quartz; pyrophyllite, pumpellyite, and oligoclase also are stable in the greater part of the facies. The lower limit of the facies can be set at several different temperatures depending on the pressure and local kinetic conditions and is determined by the absence of kaolinite, diaspore, and zeolites. Maximum pressures are up to 7–10 kbar.

B_5—*"zeolite facies" and regional epigenesis.* Temperatures from 100 to 300–350°C, pressures not more than 3–5 kbar.

In correspondence with this scheme and classification of facies, we adopted these parameters in the thermodynamic analysis and in plotting the diagrams of mineral equilibria, but some isothermal sections of the diagrams were plotted for temperatures expressed in degrees Kelvin (500, 700, 900, 1100°K), which was due to the constraints of the computer calculations.

It should be noted that in this facies scheme of Sobolev (1970) it is postulated that P_{H_2O} falls off systematically from the low-temperature facies B_4 and B_5 to the high temperature B_1 and B_2, where it drops to 0.2–0.3 P_{total}, and P_{CO_2} conversely increases in corresponding limits. It also is claimed that all the metamorphic reactions take place with the participation of fluids and through fluids, as a result of which only P_f is important. The authors of the work cited (Dobretsov and Khlestov) believe that in the case of non-confining pressures in the solid phase the pressure value varies at each point depending on the direction, and the pressure in the solid phase in a certain direction is indistinguishable from the pressure in the fluid. Therefore one can only speak of hydrostatic pressure of the system as a whole, equal to the fluid pressure, and of directed pressures in the solid phases. From these considerations it follows that in metamorphism:

$$P = P_f \approx P_{H_2O} + P_{CO_2} + P_i \tag{4.1}$$

A reduction in partial pressure of water can be caused only by additional components in the fluid, and a "deficit" of water in the system is regarded as a reduction in P_{H_2O} compared to the total pressure of the fluid phase.

It is obvious that with this approach, lithostatic pressure (nonhydrostatic pressure or loading pressure) is not a factor that has much effect on mineral equilibrium.

In addition to the ideas examined, the viewpoint that unequal pressures in general and lithostatic pressure in particular play an important role in metamorphism is widespread. In theoretical works (Barth, 1956; Sobolev, 1961; Fyfe et al., 1962; Semenenko, 1966) the principle of the shifting of equilibria under the influence of excess loading pressure on the solid phases, with the liberation of H_2O or CO_2 in metamorphic reactions, has been examined repeatedly.

In substantiating this principle one proceeds from the fact that lithostatic pressure is transmitted only to the solid phases (minerals), and the pressure of the volatile component is independent and is taken either as constant, or as systematically varying with depth. When $P_s \neq P_f$, equilibrium is determined by the well known relationship:

$$\left(\frac{\partial T}{\partial P_s}\right)_{P_f} = \frac{\Delta V_s}{\Delta S} \tag{4.2}$$

where ΔS is the change in entropy and ΔV_s the change in volume of the solid phases. Inasmuch as the value ΔV_s usually is negative in dehydration and decarbonation reactions, the value of P_s (when $P_f = \text{const}$) shifts the equilibrium into the lower temperature region.

Specific thermodynamic calculations support this relationship. In the well known work by Marakushev (1968) a system of mineral equilibria is created, based on the assumption that metamorphism of rocks is accomplished under constant moderate pressure of water (~ 1000 bar) and wide variations of pressures on the solid phases (up to 12 kbar), which are determined by depth. However, in this case the relationship between P_s and total fluid pressure is not specified clearly enough. If it is considered that $P_s = P_f$ and P_{H_2O} is a certain fraction of the fluid pressure, then a special case of the general relationship adopted by Sobolev (1970) is obtained. But if at any lithostatic pressures water predominates in the fluid and $P_s > P_f \approx P_{H_2O}$, then it is necessary to interpret the physical meaning of lithostatic pressure. This question has been examined in detail by Ostapenko (1977), on the basis of thermodynamic analysis and some experimental data. In particular, it was shown that when relationship (4.2) is used in petrology it implies an idealized model of the state of the rock, in which the loading pressure (P_s) is evenly distributed over the whole surface of the mineral grains. In such a model the places where grains are in contact with each other and the parts of grains which are the walls of the pores containing the volatile component are not differentiated. Excess pressure acts only in the places where grains touch and is not extended to those parts of the grains which are in contact with the pore solution. The conclusion reached, that in thermodynamic respects lithostatic pressure is virtually indifferent to dehydration and decarbonation reactions, seems quite sound, although it has provoked criticism (Marakushev and Fed'kin, 1969).

Let us also note that in reactions in which H_2O or CO_2 take part, when pressures are unequal ($P_s > P_f \approx P_i$), it likewise is obviously impossible to consider the value ΔV_s to be constant at any values of P_s.

Thus questions of the effect of pressure on mineral equilibria and the regime of water and carbon dioxide in metamorphism required additional consideration. Of special importance for the study of iron-chert, carbonate, effusive, and intrusive rocks was preliminary theoretical examination of the particulars of metamorphism in conditions of a "deficit" of water both in the case of low P_{H_2O} and in "dry" systems closed to water. These questions were examined in detail in special papers, some of which have been published in English translation (Mel'nik and Yaroshchuk, 1970; Mel'nik, 1972c, 1976).

In all cases in the thermodynamic analysis we considered partial pressures of H_2O, CO_2, and other volatiles to be independent variables, if they were not related to one another by reactions. In addition the general conclusion was drawn that in thermodynamic calculations of metamorphic reactions it is impossible to assume different isotropic pressures on the solid phases and fluid. Lithostatic (nonhydrostatic) pressure or loading pressure has practically no effect on equilibrium in elastic deformation of rocks. Isotropic pressure equal to fluid pressure in the case of an excess of volatiles should be considered an equilibrium factor in actual natural processes.

In a study of the properties of real gases at high pressures (Mel'nik, 1972c, 1976, 1978) it was established that for CO_2 particularly substantial deviations from the properties of an ideal gas are observed, which is manifested in an increase in the fugacity ratio. Let us remember that fugacity (f) is the effective pressure which replaces real or measurable pressure in thermodynamic equations. Such a formal statement makes it possible to apply to real gases all the regularities ascertained for ideal gases. The deviation of the properties of a real gas from the ideal is characterized by the fugacity ratio γ, which is the ratio of the fugacity of the gas to its pressure:

$$\gamma = f/P \tag{4.3}$$

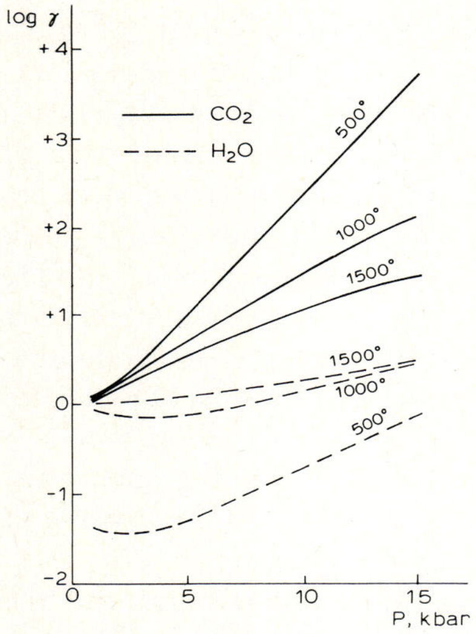

Fig. 68. Ratios of H_2O and CO_2 fugacities as a function of pressure, isotherms, °K (original data, Mel'nik, 1978).

The fugacity ratio of carbon dioxide, γ_{CO_2}, increases tens or hundreds or times as pressure increases (Fig. 68). Correspondingly f exceeds P, and the increase in pressure with a rise in temperature leads to a nonlinear increase in fugacity.

From comparison of the thermodynamic properties of H_2O and CO_2 it can be concluded that an increase in pressure leads not only to a proportional increase in partial pressures (P_{H_2O} and P_{CO_2}) in a water-carbon dioxide fluid, but also to a substantial (nonproportional) increase in P_{CO_2} compared to P_{H_2O}.

Thus Korzhinskiy's (1940, 1957) well known conclusion that there is a systematic increase in activity of carbon dioxide with depth finds new theoretical substantiation. It is to be expected that as pressure (depth) increases, if the temperature remains constant silicate mineral associations will be replaced by carbonate associations even in those cases where the composition of the water-carbon dioxide fluid ($m_{CO_2} : m_{H_2O}$ or $P_{CO_2} : P_{H_2O}$) remains unchanged.

Naturally the real processes of hydration and carbonation are complicated by variations in temperature and composition of the fluid, the effect of which is examined in more detail in the analysis of specific mineral equilibria.

The special features of CO_2 and H_2O at high pressures also lead to the fact that the *P-T* curves of reactions in which carbon dioxide takes part are characterized by a steep slope, in contrast to the gentle, sometimes horizontal or even reverse slopes of the *P-T* curves of reactions in which water takes part.

What has been discussed is valid if it is assumed that a mixture of real gases is ideal:

$$P_f = P_{CO_2} + P_{H_2O} = \frac{f_{CO_2}}{\gamma_{CO_2}} + \frac{f_{H_2O}}{\gamma_{H_2O}} \tag{4.4}$$

and the interaction of the individual gases with each other is not taken into account. Taking into account the deviation of a gaseous mixture from the ideal may make the calculations of the equilibrium parameters more precise but hardly alters the general conclusion that an increase in pressure should selectively affect the fugacity of the individual components of the fluid and consequently the particulars of deep mineragenesis. At the present time there is no doubt as to the "buffer" character of the mineral associations of iron-formations, determining the fugacity or partial pressure of oxygen in metamorphism. As has rightly been mentioned in several works (James and Howland, 1955; Eugster, 1961a, b; Sobolev, 1970), the main reason is the great "capacity" (oxidation-reduction capacity) of such rocks, made up of two iron minerals of constant composition and different degree of oxidation (for instance, hematite + magnetite). In our opinion, the concept of mineralogical "buffers" controlling the fugacity of the individual components in the fluid can feasibly be extended to other processes, in particular, to dehydration and decarbonation in normal progressive metamorphism.

Thus, in the dehydration of monomineralic rocks: hydrous silicate = anhydrous silicate + H_2O—the fugacity of water at a given temperature *T* is constant in the whole pile. The process of dehydration will go on irreversibly only in the case of a decrease in P_{H_2O} due to removal of water from the pile by diffusion along a concentration gradient or filtration along a pressure gradient. A decrease in P_{H_2O} leads to a corresponding drop in pressure of water, a shift of equilibrium to the right, and decomposition of some part of the hydrous silicate. Conversely, an increase in P_{H_2O} leads to hydration of the anhydrous silicate. Thus the association hydrous silicate + anhydrous silicate is a buffer and keeps the pressure of water within certain limits. A similar phenomenon should occur in the metamorphism of carbonate rocks with the formation of oxides or anhydrous silicates—for corresponding amounts of original and final products, the capacity of the buffer turns out to be sufficient to regulate P_{CO_2} in the case where fluid is removed or introduced.

In the metamorphic processes of dehydration and decarbonation of uniform thick sequences, filtration along the pressure gradient evidently played the main role in removal of water and carbon dioxide. Diffussion was of local importance, mainly occurring at the contacts of rocks of different composition, for instance silicate and carbonate. The direction of the pressure gradients of the fluid phase was determined by the permeability of the rocks and distance to the nearest fracture or fracture zone. Thus, in the case of a sparse network of fractures removal of the fluid required greater drops in pressure than in the case of intensive jointing in zones of folding. The decrease in the necessary pressure gradient in such zones could have led to a drop in fluid pressure and P_{H_2O} or P_{CO_2} which in turn led to "intensification" of metamorphism or even to the appearance, in isothermal conditions, of "higher-temperature" mineral associations.

It is obvious that in the case of pressure gradients determined by the maximum P_f controlled by buffer reactions in the pile of rocks, and by the minimum P_f in fractures (in the case of open circulation $P_{f(min)} = P_{(hydr)}$ of the column of fluid), mechanical movement of the fluid was in one direction —from rock to fracture. For movement in the opposite direction—from fracture to rock—it was necessary to create a corresponding pressure gradient. Such phenomena, in addition to diffusion along the concentration gradient, presumably have occurred in hydrothermal metamorphism with typical reactions of hydration and carbonation. However, for normal progressive metamorphism it is hard to imagine a mechanical model in which a fluid with a strictly constant value of P_{H_2O} or m_{H_2O} is introduced from without into a uniform pile of rocks along a pressure gradient.

In the metamorphism of mixed rocks buffered exchange reactions of the type carbonate + quartz + H_2O = hydrous silicate + CO_2, regulating the proportion of water and carbon dioxide in the equilibrium fluid when T = const, are possible. Such reactions are of very great importance in low-temperature metamorphism of iron-formations. The introduction of a water-carbon dioxide fluid with varying proportions of CO_2 and H_2O into the rock constituting the buffer association leads to a shift of equilibrium and formation of either silicate or carbonate. For instance, in metamorphism of silicate-carbonate iron formations in greenschist facies conditions, introduction of fluid with fixed P_{H_2O} = const = 1000 bar would lead to irreversible replacement of the siderite + quartz association by grunerite in the case where $P_s = P_f$ from 1.0 to 3.5 kbar. The wide distribution of carbonates with hydrous silicates suggests the buffer character of such associations.

The course of metamorphism is often complicated by variable water content, inasmuch as the paths of metamorphism probably are different in conditions of an excess or deficiency of water, as in systems closed to water the production of the particular mineral associations depends on the water content. Study of the details of metamorphism in conditions of an excess or deficiency of water has made it possible to draw a number of general conclusions (Mel'nik and Yaroshchuk, 1970):

(1) In rocks in which metamorphism is accomplished with an excess of water (for instance, pelitic), the processes of dehydration occur strictly consecutively, and the metamorphic zoning of these rocks very faithfully reflects the variation in temperature conditions in progressive metamorphism.

(2) In rocks metamorphosed with a water deficiency, the mineral associations do not always faithfully reflect temperature variation. In these rocks, zoning similar to temperature zoning in the same rocks with excess water can arise in isothermal conditions just as a result of variable saturation of the pile by solutions. This is the reason for the variance in metamorphism of meta-igneous rocks and rocks of pelitic composition that occur together.

(3) Only solid phases with closely similar water content can be in thermodynamic equilibrium. Any association of solid phases not close in water content is metastable.

(4) The presence of metastable associations and frequent alternation of mineral parageneses belonging to different metamorphic facies may indicate variable water content and the presence of sectors of "dry" rocks.

Thermodynamics of metamorphism of cherty iron sediments

The particulars of the metamorphism of Precambrian BIF usually have been established on the basis of study of the mineral associations, interrelationships of minerals, sequence of mineral formation and paragenetic analysis. Thermodynamic calculations and experimental data have been used to a lesser extent.

Among the petrological works dealing with the metamorphism of various types of rocks of the BIF of the Ukrainian shield, one must first and foremost cite the fundamental work by Semenenko and his colleagues, summarized in the well known monograph "Petrography of the cherty iron-formations of the Ukrainian S.S.R." (Semenenko et al., 1956) and other works (Semenenko, 1953; Semenenko et al., 1967). Problems of the metamorphism of the Krivoy Rog BIF were examined by Svital'skiy et al. (1932), Kanibolotskiy (1946), Belevtsev (1957, 1959, 1962), and later by Pirogov and Pirogova (1969) and Fedorchenko (1969). Of recent works, the investigation of metamorphic zoning of the Krivoy Rog basin conducted by R. Belevtsev (1975), on the basis of data of a study of the parageneses of metaleptites, should be mentioned. Interesting works by Khodyush (1967a, b, 1969), Pirogov and Shtoda (1969), Kornilov (1969) are devoted to other areas of the Ukrainian shield. New summaries of data on the metamorphism of the BIF of the Ukraine are contained in the works by Strygin (1960, 1969) and Nikol'skiy and Kaukin (1969).

Metamorphism of the Precambrian rocks of the KMA has been studied in detail mainly through the investigations of Glagolev (1965).

Phase equilibria in metamorphic rocks of the BIF have been examined, taking into account oxidation in high-temperature petrogenesis and the distribution of elements between coexisting minerals, in works by Mueller (1960, 1961a, b), Kranck (1961), Gross (1961), Klein (1966a, b, 1973), Klein and Fink (1976), Bonnichsen (1969, 1975), French (1968, 1973), and others (Davis, 1966; Chakraborty and Taron, 1968; Seguin, 1971; Ayres, 1972).

Physicochemical analysis of the equilibria of iron-ore minerals (hematite, magnetite, and siderite) in metamorphism has repeatedly been undertaken on the basis of thermodynamic calculations using the standard tabulated constants (Mel'nik, 1964, 1966a, c; Kornilov, 1969), or on the basis of treatment of individual experimental data (Eugster, 1961a, b; Yoi, 1966; Seguin, 1968a, b). Certain particulars of mineral equilibria in which fayalite (Mel'nik and Yaroshchuk, 1966) and other silicates (Hsu, 1968; Fed'kin, 1971) take part have also been studied. The data obtained contributed to a deeper understanding of the role of individual minerals (graphite) and mineral associations (hematite + magnetite, magnetite + fayalite) in determining the conditions of metamorphism of iron cherts, in particular in creating a controlled oxidation-reduction setting (buffer associations fixing the fugacity of oxygen—f_{O_2}) (James and Howland, 1955; French, 1966; Mel'nik, 1969).

However, these physicochemical investigations were to a substantial extent approximate and incomplete, due to the lack of thermodynamic constants of many rock-forming minerals (amphiboles, micas, chlorites, etc.); to imprecision in the tabulated calculated constants (siderite, ferrosilite, fayalite, etc.); to the substantial discrepancies between theoretical and experimental equilibrium data; and to the lack of data on the properties of the fluid at high pressures. Moreover, many thermodynamic calculations were made by approximate methods, which led to substantial errors.

Thermodynamic analysis of mineral equilibria was first made by us (Mel'nik and Siroshtan, 1973) on the basis of the new system of thermodynamic constants of minerals that is consistent with experiments (Mel'nik, 1972b), which also includes important hydrous minerals like grunerite and minnesotaite. These data were subsequently amended (Mel'nik and Radchuk, 1977b), and supplemented by the thermodynamic constants of greenalite and of some "diagenetic" iron and magnesium minerals (see Appendix).

In addition, new experimental data were given on the conditions of formation of goethite, hematite, grunerite, and iron-magnesian amphiboles and pyroxenes.

All the calculations were carried out on a computer by a very precise procedure which took into account the effect of pressure on the solid phases (correction for ΔV) and on the fluid. The procedure used is described in a separate paper (Mel'nik and Radchuk, 1978), and the new values of the fugacity ratios of the main gases entering into the composition of natural fluids, at T up to 1500°K and P up to 15 kbar, are taken from our new summary (Mel'nik, 1978).

The purpose of the thermodynamic calculations was to plot a series of diagrams of mineral equilibria at high T and P and different variations in composition of the fluid for the most typical isochemical series of Precambrian BIF. In so doing it was fundamental to find the general regularities of metamorphism of idealized iron cherts (the system $FeO-Fe_2O_3-SiO_2-H_2O-CO_2$ with excess silica). The effect of additional components, mainly MgO, was analyzed only approximately on the basis of comparison of corresponding reactions in which pure iron and magnesium minerals take part. The particulars of metamorphism and metasomatism in which other components take part, in particular Al_2O_3, were not examined inasmuch as there are no reliable consistent thermodynamic constants of the minerals (garnets, chlorites, hornblende, etc.). On the basis of available data, it is not possible to calculate the precise distribution of the individual components between coexisting minerals in the case where solid solutions are formed. Experience with petrological investigations (Perchuk, 1970) has shown that study of the isomorphism of natural minerals so far gives more reliable information than thermodynamic methods.

Particulars of metamorphism of iron-formations of various types

As has already been mentioned, iron-rich rocks characterized by uniform chemical composition and very diverse mineral associations, textures, and structures are of decisive importance in the composition of the Precambrian BIF. A general feature of these rocks is similar silica and total iron contents; the other essential components occur in subordinate amounts, although some of them, such as MgO and Al_2O_3, greatly influenced mineral formation.

In general the proportion of ore (oxide and carbonate) and non-ore (silicate) iron minerals was already predetermined by the degree of oxidation of the iron at the time of the formation of the original sediment, inasmuch as at all stages of metamorphic transformation of these sediments oxygen remained an only slightly mobile component. The oxidation-reduction capacity of the iron-rich rocks was changed during metamorphism only when an appreciable amount of iron carbonate or free carbon was present in the composition of the oxide facies. The degree of oxidation of the iron, the presence of carbonates and silicates, and the silica content are the main criteria for classification of the BIF as a whole and of the individual bands and layers.

The classification of the rocks of the BIF into oxide (or ore), carbonate, and silicate that has been adopted corresponds formally to the primary sedimentary facies of iron-formation according to James (1954), if the relatively infrequent sulfide facies is not taken into account. However, these facies, or more correctly rock types, cannot be regarded as primary chemogenic sediments, the composition of which was completely predetermined by the physicochemical conditions of sedimentation. Processes of diagenesis and protometamorphism, largely controlling the formation of stable or kinetically resistant metastable mineral types also were of great importance.

The rock types distinguished hardly ever occur in pure form, for the oxide facies always contains a certain amount of carbonate in addition to hematite and magnetite, and the iron silicate rocks usually contain carbonate, sometimes magnetite. Sustained mineralogical composition is observed only in thin bands; on the whole, in conditions of low and medium degrees of metamorphism the pile of BIF looks like a "layer cake". The composition of highly metamorphosed rocks is more uniform, which can be explained both by the disappearance of carbonate rocks due to thermal dissociation and by more intensive diffusion.

Oxide iron-formations. These are represented by the hematite (micaceous iron), hematite-magnetite, and magnetite varieties. These rocks are richest in iron (32–42% Fe), contain practically no silicates except in the magnetite rocks, but have impurities of magnesian-iron carbonates. The distinguishing feature of the chemistry is the high degree of oxidation of the iron—the Fe_2O_3:FeO ratio ranges from 10–30 in the hematite varieties to 1 in the magnetite.

In the absence of reducing agents, occurring either in the sediments (organic compounds, free carbon) or introduced from without (H_2, CO, CH_4, and other volatiles), the oxidation-reduction capacity of the rocks was not changed and metamorphism led only to crystallization of minerals with larger grains and formation of more clearly manifested magnetite crystals. This process is often observed in petrographic investigations: around large grains of ore minerals there is a "shell" of disseminated material. Crystallization of quartz went on in parallel. Tiny quartz grains are usually heavily dusted with ore material (both magnetite and hematite). As the size of the grains increases the quartz becomes purer and ultimately almost completely free of impurities. Petrographic investigations show that in the oxide iron-formations of the BIF of the Ukrainian shield, KMA, and other regions of the U.S.S.R., metamorphosed in greenschist, epidote-amphibolite, and amphibolite facies conditions, hematite remains stable and is not replaced by magnetite or Fe^{2+} silicates. Klein (1973) has shown, by examples of BIF from the southern part of the Labrador trough (Newfoundland), the Lake

Superior district, the Hamersley formation (Western Australia), and the iron-ore belt in the state of Minas Gerais (Brazil), that the quartz-magnetite, quartz-hematite, and quartz-hematite-magnetite associations are preserved when metamorphism of higher rank is imposed. On this basis, oxygen is considered to be an inert component. With increasing rank of metamorphism, to the kyanite and sillimanite zones, there occurs only recrystallization of the iron oxides and cherty matter, which is accompanied by enlargement of mineral grains, especially of quartz.

In addition to the phenomena of isochemical crystallization, in the earliest stages of metamorphic transformation in hematite rocks the process of dehydration should occur:

goethite → hematite

Primary goethite is not encountered in metamorphic rocks.

In the presence of a reducing agent the sequence of metamorphic reactions is determined by thermodynamic and kinetic factors, and also by the nature of the reducing agent.

In the case of reduction by organic compounds, free carbon, or gaseous carbon monoxide, different courses of the process are possible depending on the partial pressure of carbon dioxide and temperature:

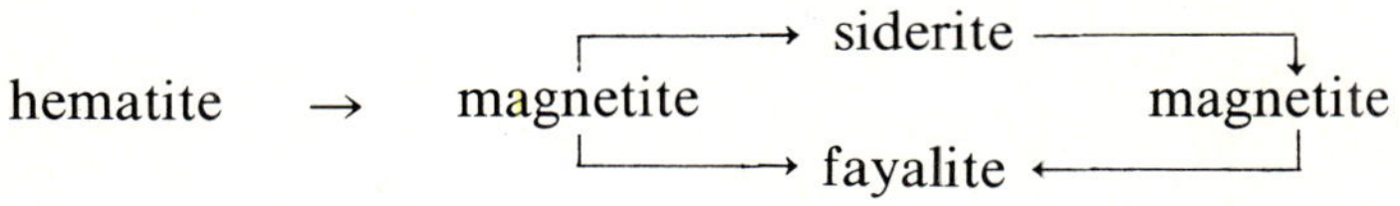

In the case of reduction by hydrogen, the appearance of hydrous silicates in the oxide sediments is possible:

hematite → magnetite → minnesotaite,

grunerite → fayalite

or silicates associated with carbonates in the case of a reducing agent of complex composition. An insufficiency of reducing agent leads to breaking the chain in places where ferric forms are converted to ferrous.

Actually, hematite practically never occurs in oxide BIF metamorphosed in granulite facies conditions, while magnetite is quite stable (French, 1973; Mel'nik and Siroshtan, 1973; Bonnichsen, 1975). If siderite or free carbon were absent in the original rocks, then the most likely reducing agent was gaseous hydrogen of deep origin—the most mobile component of the mantle fluid (see p. 87).

Silicate iron-formations are relatively rarely encountered in pure form; impurities of carbonates and magnetite are usual, hematite is rare. All transitions are observed between magnetite-oxide, carbonate, and silicate rocks.

As has already been mentioned, very interesting works have been published recently containing results of detailed study of highly metamorphosed BIF (Beukes, 1973; Bayley and James, 1973; French, 1973; Klein, 1973, 1974; Appel, 1974; Floran and Papike, 1975; Klein and Fink, 1976; Klein and Bricker, 1977). Investigation of highly dispersed iron silicates of variable composition and establishment of their relationships to other minerals is a very complicated matter. Most investigators are inclined to think that the main primary-diagenetic iron silicate was greenalite containing a certain amount of magnesium and impurities of aluminum. Stilpnomelane, chamosite, and ripidolite occur along with greenalite—formation of these protometamorphic minerals was furthered by a high content of aluminum in the original rocks, apparently caused by impurities of pelitic material in the chemogenic sediment. Minnesotaite probably is a secondary mineral, formed mainly at the expense of greenalite. French (1973) points out that in the Biwabik BIF in Minnesota, minnesotaite forms sheaf-like or acicular aggregates which replace greenalite either in the stage of diagenesis or in the lower stage of metamorphism. Stilpnomelane also is a secondary mineral replacing greenalite in this formation. Similar relationships of these silicates have been described in other works (Bayley and James, 1973; Appel, 1974). In the BIF occurring in the U.S.S.R. no such low-grade metamorphic associations have been found. Individual finds of minnesotaite in hydrothermal formations (Tanatar-Barash, 1965) have no direct bearing on the problem under consideration.

However, in most iron deposits, amphiboles of the cummingtonite-grunerite series, represented mainly by cummingtonite with an iron content of the order of 70–90% (i.e., grunerite), are widespread among the iron silicates in the more highly metamorphosed rocks. Two paths of formation of cummingtonite are possible—at the expense of silicates of minnesotaite type (Beukes, 1973; Garrels et al., 1973), or as a result of interaction of primary iron carbonate with silica (Semenenko et al., 1956; Klein, 1966; Khodyush, 1970).

Low-temperature metamorphism of silicate rocks is complicated by oxidation reactions in which free oxygen (French, 1973) or, what is more likely, goethite or hematite take part. In most cases the final product of such reactions is magnetite.

When intercalations of shale (pelite), which usually contains Al_2O_3, are present in iron formations, iron-magnesian chlorites or micas of stilpnomelane type predominate among the hydrous silicates. Such aluminum-bearing rocks, metamorphosed at higher temperatures or under unusual conditions (for instance, in dry systems), contain garnet of almandine type, sometimes hornblende.

High-temperature metamorphism of alumina-free silicate rocks leads to the formation of the fayalitic rocks sporadically encountered among the BIF of the Ukrainian shield and other areas (Polovko, 1956; Semenenko et al., 1956; Mel'nik and Yaroshchuk, 1966; Val'ter, 1969). The main rock-forming minerals are quartz and fayalite, constituting about 80–90%, and there also are ferro-hypersthene, garnet (almandine), and magnetite. In isolated cases the amount of magnetite increases and it becomes a major mineral. In chemical respects the fayalitic rocks are characterized by a low degree of oxidation of the iron ($FeO \gg Fe_2O_3$) and a low content of MgO, Al_2O_3, and other minor components. The main silicate—fayalite—has an iron content of the order of 94–96%. In rocks with a high MgO content, magnesian-iron pyroxene occurs instead of fayalite.

Metamorphism of silicate BIF containing no excess free carbon is characterized by successive dehydration reactions:

$$\xrightarrow{+\text{ hematite}} \text{magnetite}$$

$$\text{greenalite} \rightarrow \text{minnesotaite} \rightarrow \text{grunerite} \rightarrow \text{fayalite}$$

Carbonate iron-formations. These are widespread in slightly metamorphosed BIF, although thick uniform piles of such rocks are practically never encountered. Bands (up to 1–5 cm, in jaspilites up to 1–5 mm) of carbonate, silicate (shaly), and siliceous (chert or jasper) composition are more typical.

As has already been mentioned, pure iron carbonate—siderite—is rarely observed; a solid solution with magnesium of the order of 10–30%, more rarely up to 40–50%, is usual. A definite relationship is observed between the position of the rocks in the stratigraphic section, or more correctly, in the layer (horizon), and the composition of the carbonate, apparently caused by cyclic fluctuations in the oxidation-reduction conditions. In the opinion of most investigators of slightly metamorphosed BIF, siderite is a primary (sedimentary or early diagenetic) iron mineral. Thus, La Berge (1973) believes siderite is the least controversial primary iron mineral. Grains of siderite are chiefly of siltstone size, they often are part of granules, and they form regular or cross bedding.

The sequence of decarbonation processes is determined by the presence of iron oxides (hematite) and the water content of the fluids. In pure anhydrous siderite-chert rocks with an excess of quartz the following transformations are possible:

siderite → magnetite + graphite

↓

fayalite

siderite → fayalite

Oxide-carbonate iron-formations. Hematite-siderite rocks are relatively rarely encountered, which apparently is explained not by the exceptional nature of this association in sediments, but by metamorphic transformation into magnetite rocks at relatively low temperatures:

siderite + hematite → magnetite

There are enough petrographic observations confirming the reality of formation of magnetite at the expense of siderite by one of the schemes given (La Berge, 1964, 1973; Klein, 1973; French, 1973).

Magnetite-siderite rocks are very common in the section of the iron-ore sequence of the Krivbass (Krivoy Rog basin) and of numerous magnetic anomalies of the Ukrainian shield. Such rocks also occur in the KMA, in the Lake Superior and other areas of slightly metamorphosed BIF. The presence of excess magnetite does not affect further metamorphic transformations of siderite, which proceed as in pure carbonate iron-formations.

Silicate-carbonate iron-formations are the most widespread type, especially when they contain magnetite bands or disseminated magnetite in silicate and carbonate layers. Impurities of graphite are frequent in rocks in which shaly material predominates.

The sequence of metamorphic transformation may be different depending on the composition of the original sediment, the proportions of CO_2 and H_2O in the fluids, and the presence of reducing agents.

For pure iron silicate-carbonate rocks these transformations are most likely:

$$\text{greenalite + siderite} \rightarrow \text{minnesotaite + siderite} \rightarrow \begin{bmatrix} \text{grunerite + siderite} \\ \text{grunerite + magnetite} \\ \text{grunerite} \end{bmatrix} \rightarrow$$

$$\rightarrow \text{grunerite + fayalite} \rightarrow \text{fayalite}$$

The association hematite + iron silicate, encountered in sediments, is not characteristic of metamorphosed iron formations.

Klein (1973) points out that if there is not enough aluminum in the primary silicate associations, then their metamorphosed derivatives are indistinguishable from the derivatives of chert-carbonate varieties. Let us also note that the scheme of transformations given will become different if the rocks contain a certain amount of magnesium and calcium in addition to aluminum. In such rocks ortho- and clinopyroxenes and garnets form instead of fayalite (Semenenko et al., 1956; Mueller, 1960; Gross, 1961; Kranck, 1961; Glagolev, 1965; Klein, 1966; Chakraborty and Taron, 1968; Bonnichsen, 1969, 1975).

Thermodynamic analysis of the conditions of metamorphism

The conditions of metamorphism of BIF of different types are examined on the basis of thermodynamic calculations consistent with experimental investigations of phase equilibria at high temperatures and pressures.

Metamorphism of oxide iron-formations

In idealized form rocks of this type originally were iron hydroxide sediments: metastable $Fe(OH)_3$ or $FeOOH \cdot n\,H_2O$ and hydromagnetite—$Fe_3O_4 \cdot n\,H_2O$ or Fe_3O_4 (dispersed), transformed during diagenesis into stable crystalline goethite and magnetite.

Despite the substantial number of publications, the conditions of conversion of goethite into hematite were not determined precisely until recently. There was little experimental work related to investigations of the interrelationships of these minerals at high P and T and the data obtained were in poor agreement with each other; in particular, the discrepancies in determinations of the equilibrium temperature for P_{H_2O} up to 3 kbar reached 50–60°C (Smith and Kidd, 1949; Schmalz, 1959; Wefers, 1966a, b; Mel'nik, 1972b).

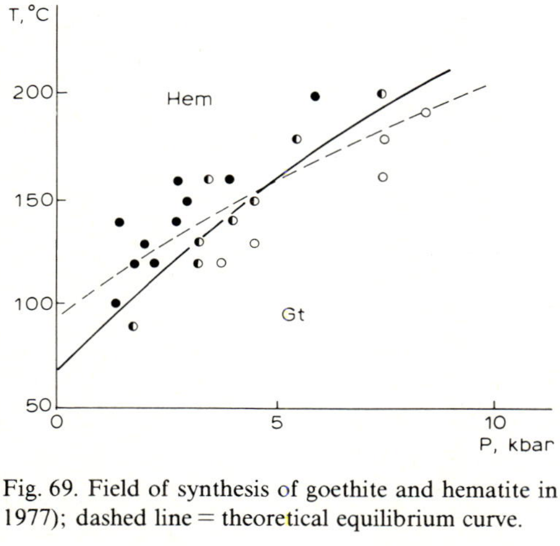

Fig. 69. Field of synthesis of goethite and hematite in hydrothermal conditions (Vorobyeva and Mel'nik, 1977); dashed line = theoretical equilibrium curve.

In this connection, we undertook an experimental investigation of the system Fe_2O_3–H_2O under hydrothermal conditions at $T = 100$–$200°C$ and P up to 9 kbar (Vorob'yeva and Mel'nik, 1977). The main experimental difficulties were related to the very low rate of crystallization and to the formation of metastable solid phases. Despite their considerable duration (up to 480 hours), mixtures of goethite and hematite rather than pure phases were obtained from the amorphous hydroxide in most of the experiments, especially in the region of the equilibrium curve. The diagram taking kinetic factors into account (Fig. 69) apparently is similar to the stability diagram of these minerals, which is confirmed by thermodynamic calculations of the dehydration reaction:

$$2\,FeOOH = Fe_2O_3 + H_2O \qquad (4.5)$$

using the constants given in the Appendix.

Thus the conversion of goethite to hematite is already accomplished before the beginning of regional metamorphism under conditions of epigenesis at T up to 150°C. This conclusion is confirmed by the data of isotope thermometry. Becker and Clayton (1976) believe that in the Hamersley area of Western Australia coarse-grained hematite could have formed from goethite at ~140°C. In the least metamorphosed of all the Middle Precambrian BIF of Biwabik, where the temperature of metamorphism outside the contact aureole of the Duluth massif was about 150°C (isotopic data of Perry et al., 1973), no primary goethite has been preserved.

Metamorphic transformation of hematite into magnetite by way of thermal dissociation:

$$6\,Fe_2O_3 = 4\,Fe_3O_4 + O_2 \qquad (4.6)$$

even at high temperatures, does not seem very likely, either in pure aqueous or in carbonic acid fluids. The equilibrium partial pressure of oxygen in greenschist facies conditions occurs at a level of 10^{-30} to 10^{-15} bar (there is practically no free O_2), in amphibolite facies conditions $P_{O_2} = 10^{-15}$ to 10^{-10} bar, and for granulite facies $P_{O_2} = 10^{-10}$ to 10^{-7} bar (traces of free O_2 are negligible). For reaction (4.6) to proceed to the right, oxygen must be removed from the rocks being metamorphosed by way of diffusion, or carried away in solution. Calculations show that to transform 1 mole of Fe_2O_3 (160 g) completely into Fe_3O_4 it is necessary to remove $\frac{1}{6}$ mole of O_2 (~5.3 g). If the process goes on in the P-T conditions of the granulite facies, for instance at 800°C, the metamorphic fluid carrying away oxygen would not contain more than one part O_2 per 10^9 parts of neutral volatiles ($\log f_{O_2} = -9$). To remove 5.3 g would require of the order of 5000 tons of fluid —a completely unrealistic figure for any geologic processes. Nor could diffusion over considerable distances lead to the removal of oxygen in appreciable amounts, at such low concentrations. Thus hematite should be taken to be a thermodynamically stable mineral in the conditions of all facies of metamorphism up to granulite.

The hematite + magnetite association, so widespread in iron-formations, is the buffer regulating the fugacity or partial pressure of oxygen, which remains constant at a fixed temperature (pressure has little effect on the equilibrium constant of reaction 4.6). The main condition governing the buffering properties— substantial predominance of the mass of the reacting solid substances over the mass of the volatile component— is fulfilled by the hematite + magnetite association at any realistic temperatures. The buffering mechanism consists of the fact that when a reducing agent is introduced into the rock from the outside and the P_{O_2} is lowered, the process of reduction of hematite begins, with the liberation of oxygen. This process goes on irreversibly until the original equilibrium value of P_{O_2} is restored. If an oxidant is introduced the reverse process will take place—oxidation of magnetite. As a result, in buffered systems the relative amount of the solid phases, for instance Fe_2O_3 and Fe_3O_4, may fluctuate widely and the only necessary condition is the simultaneous presence of the two minerals.

Thus metamorphic transformation of hematite into magnetite is possible only in the presence of a reducing agent (free carbon, gaseous CO, H_2, or CH_4), but it is expedient to consider the oxide-reducing reactions that take place when we analyze mineral equilibria in silicate and carbonate rocks.

Metamorphism of silicate iron-formations

It is presumed that at the end of deposition and diagenesis there is no siderite in rocks of this type, the stable Fe^{2+} silicate is a mineral of greenalite type, there is excess silica, ferric oxides possibly are present, and the fluid phase in the intergranular space consists mainly of water.

The sequence of dehydration reactions in the progressive metamorphism of silicate rocks depends on the presence of excess iron hydroxide in the original sediment, inasmuch as the lowest-temperature reaction is the transformation of the greenalite + hematite association into magnetite according to the reaction:

$$Fe_3Si_2O_5(OH)_4 + 3\ Fe_2O_3 = 3\ Fe_3O_4 + 2\ SiO_2 + 2\ H_2O \qquad (4.7)$$

the *P-T* curve of which, calculated for equilibrium of the crystalline minerals, falls in the 180–220°C interval for moderate pressures (Fig. 70).

Dehydration of pure greenalite begins at somewhat higher temperatures (200–220°C):

$$Fe_3Si_2O_5(OH)_4 + 2\ SiO_2 = Fe_3Si_4O_{10}(OH)_2 + H_2O \qquad (4.8)$$

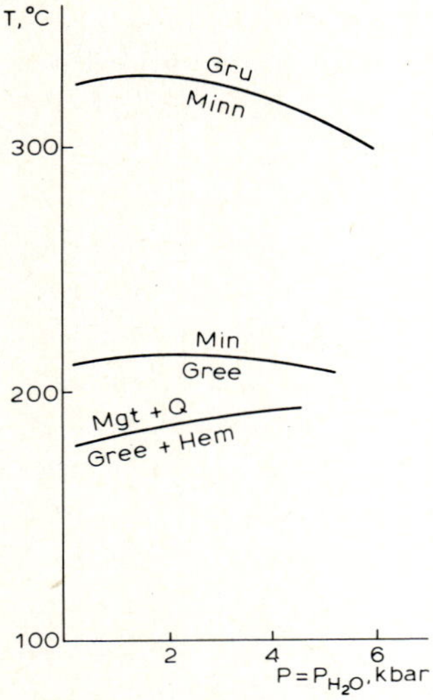

Fig. 70. *P-T* curves of reactions of dehydration of iron silicates under conditions of low-temperature metamorphism.

where minnesotaite, considered by many investigators to be an epigenetic or early metamorphic mineral, becomes stable. In the 200–300°C range the association minnesotaite + magnetite is finally formed. Iron chlorites of variable composition, stilpnomelane, and secondary magnetite are developed under the same conditions in rocks with a high content of Al and Mg.

With further progressive metamorphism minnesotaite is transformed into grunerite according to the dehydration reaction:

$$7\ Fe_3Si_4O_{10}(OH)_2 = 3\ Fe_7Si_8O_{22}(OH)_2 + 4\ SiO_2 + 4\ H_2O \qquad (4.9)$$

the *P-T* curve of which (see Fig. 70) falls in the 300–330°C interval at low and moderate pressures and is characterized by a reverse slope. It should be noted that that position of the curves cannot be considered to be reliably established inasmuch as the thermodynamic constants of greenalite and minnesotaite are based on insufficiently reliable or incomplete experiments (see Appendix). Still, the disappearance of minnesotaite apparently may indicate the presence of metamorphism in greenschist facies conditions. Judging from the natural mineral parageneses, grunerite is stable over a wide range of temperatures and pressures, but the conditions of its formation were not studied in detail until recently inasmuch as practically all attempts to synthesize pure iron amphibole were unsuccessful. Amphibole was obtained only in the experiments by Schürmann (1967), carried out with the

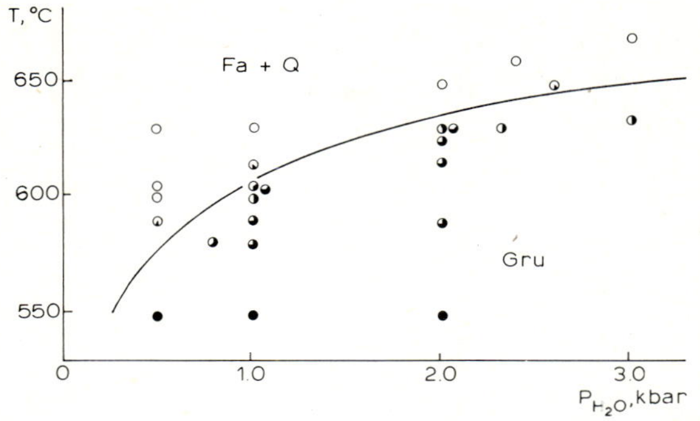

Fig. 71. Experimental *P-T* curve of the stability of grunerite (Mel'nik and Radchuk, 1977a).

addition of a CaO stabilizer in the 450–595°C range at $P_{H_2O} \approx 1$ kbar. We used these data (Mel'nik, 1971, 1972b) to calculate the approximate thermodynamic constants of grunerite and the *P-T* curve of the dehydration reaction. Later we conducted experiments as a result of which the curve of the upper stability limit of pure iron grunerite was obtained for the first time (Fig. 71) and its thermodynamic properties determined. At the same time and independently Forbes (1977) obtained similar results, although in his experiments oxygrunerite with a variable content of Fe^{3+} was obtained in addition to grunerite.

On the basis of the experimental and the fairly reliable theoretical data it can be considered that grunerite is quite stable beginning in the early stages

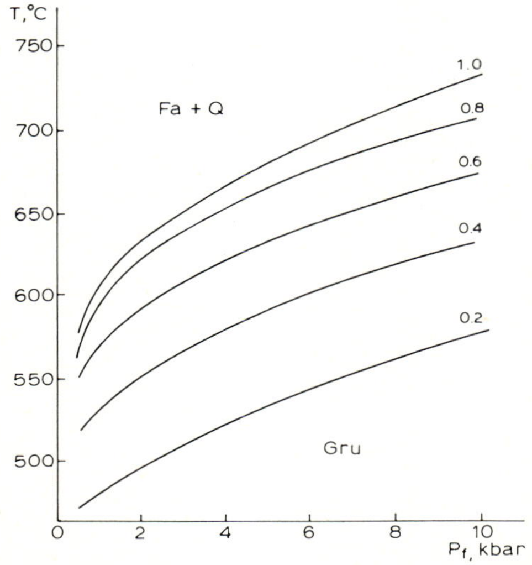

Fig. 72. Theoretical *P-T* curves of the stability of grunerite. Figures on curves indicate molar percent of water in the fluid (Mel'nik and Radchuk, 1977a).

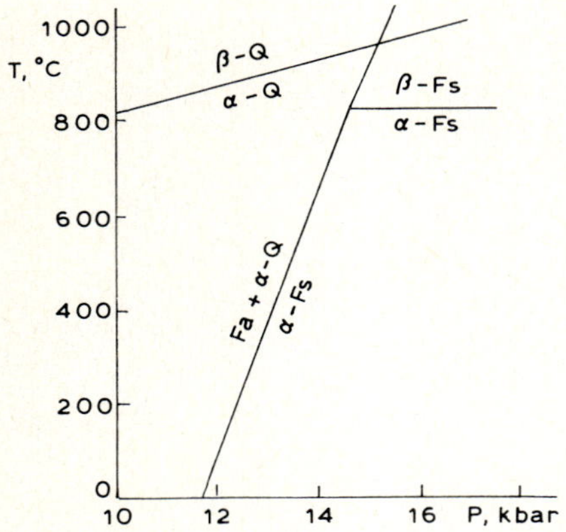

Fig. 73. *P-T* diagram of the stability of ferrosilite (Fs) in a "dry" system in the absence of volatile components (H_2O, CO_2 etc.).

of regional metamorphism, and is a typical mineral of iron cherts metamorphosed in both greenschist and amphibolite facies conditions. Grunerite does not decompose until the upper amphibolite facies at a temperature of 640–690°C (see Fig. 71):

$$2\ Fe_7Si_8O_{22}(OH)_2 = 7\ Fe_2SiO_4 + 9\ SiO_2 + 2\ H_2O \tag{4.10}$$

into the fayalite plus quartz association, which is stable in granulite facies

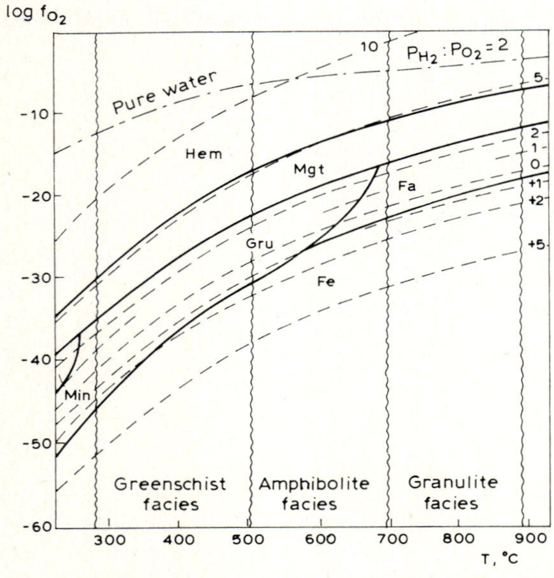

Fig. 74. Metamorphism of silicate iron-formations (log f_{O_2}-*T* diagram). Diagram for $P = P_f = \Sigma(P_{H_2O}, P_{H_2}, P_{O_2}) = 5$ kbar. Lines of equal $\log(f_{H_2}/f_{H_2O})$ are shown by dashed lines.

conditions. The temperature of this transition may be lowered considerably if the metamorphic fluid contains other volatile components besides water (Fig. 72), the most likely one being an admixture of CO_2.

At pressures below 13–15 kbar iron pyroxene—ferrosilite—is unstable at any temperature, which is confirmed by thermodynamic study of the solid-phase reaction (Fig. 73):

$$2\ FeSiO_3 = Fe_2SiO_4 + SiO_2 \tag{4.11}$$

However, in the case of an admixture of magnesium (more than 10–15%) the reaction goes in the opposite direction.

The general diagram of mineral equilibria in coordinates of $\log f_{O_2}$–T for pressures of 5000 bar is presented in Fig. 74. From the diagram it follows that magnetite equilibria according to the reactions:

$$6\ Fe_7Si_8O_{22}(OH)_2 + 7\ O_2 = 14\ Fe_3O_4 + 48\ SiO_2 + 6\ H_2O \tag{4.12}$$

$$3\ Fe_2SiO_4 + O_2 = 2\ Fe_3O_4 + 3\ SiO_2 \tag{4.13}$$

also buffer, controlling the oxygen fugacity when T is constant and $P_s \approx P_{H_2O}$. Tentative boundaries between the facies are entered on this same diagram. The thermodynamic confirmation of the conclusion that the association hematite-plus-ferrous iron silicates is not in equilibrium in the P-T conditions of any metamorphic facies is very important.

Inasmuch as when the fluid consists essentially of water only the products of decomposition of water are present in it (of course, not counting the neutral gases), diagrams in coordinates of f_{H_2O}–f_{H_2}–T can be plotted. Diagrams in coordinates of $\log f_{H_2O}$–$\log f_{H_2}$–T and especially isothermal sections of them also prove convenient in practice (Fig. 75). More detailed sections are given in Fig. 76 for temperatures and pressures corresponding approximately to the conditions of metamorphism. Isobars of total fluid pressure and isobars of $\log f_{O_2}$ are entered in the diagrams. The succession of mineral associations is clearly traceable.

Directly by way of computer calculations or graphically using the diagrams, one can easily determine the composition of the fluid in equilibrium with the mineral association being investigated, in particular its hydrogen content. For instance, for $P_f = 5000$ bar and $T = 627°C$ (amphibolite facies) the fluid in equilibrium with the grunerite + magnetite + quartz association is characterized by the composition: $\log f_{H_2O} = +3.52$ and $\log f_{H_2} = +0.90$ (Fig. 76), which corresponds to $f_{H_2O} \approx 3290$ bar and $f_{H_2} \approx 8$ bar; taking into account the fugacity ratios ($\gamma_{H_2O} = 0.658$ and $\gamma_{H_2} = 2.70$), we find the partial pressures: $P_{H_2O} \approx 5000$ bar and $P_{H_2} \approx 3$ bar; the ratio of partial pressures or molar amounts, $H_2 : H_2O$, has an insignificant value—$6 \cdot 10^{-4}$ or 0.06%

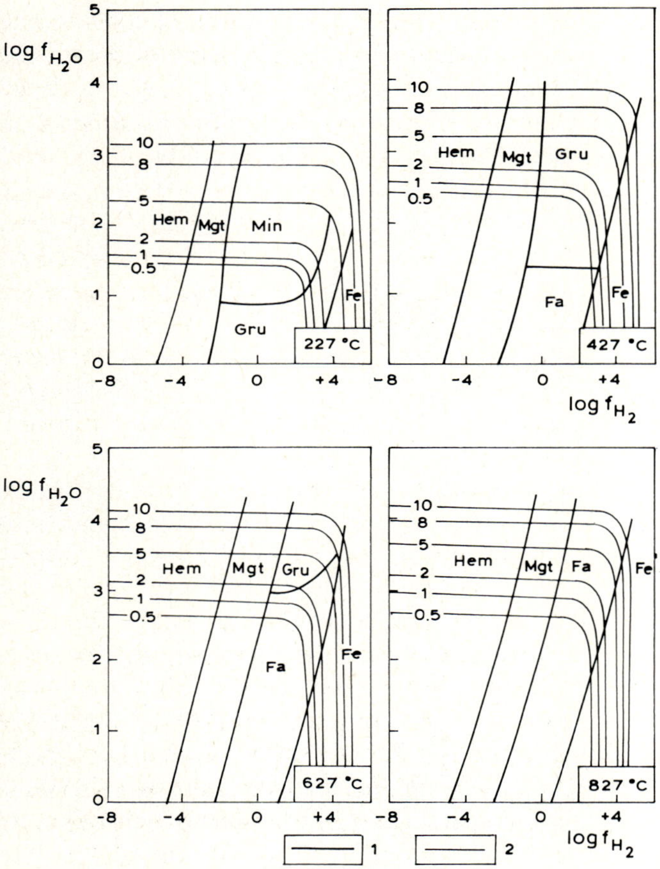

Fig. 75. Diagram of phase equilibria in the system Fe–Si–H_2–O_2 (isothermal ↔ sections in coordinates of log f_{H_2O}-logf_{H_2}): *1* = phase boundaries; *2* = isobars of total fluid pressure ($P_f = P_{H_2O} + P_{H_2}$, in kbar).

hydrogen. The value of log f_{O_2} is close to 18, i.e. there is practically no oxygen in the mixture.

Pure water is an oxidant with respect to ferrous iron silicates and if substantial amounts of aqueous solutions are introduced, partial or complete replacement of anhydrous minerals by magnetite is possible according to the reactions:

$$3\ Fe_7Si_8O_{22}(OH)_2 + 4\ H_2O = 7\ Fe_3O_4 + 24\ SiO_2 + 7\ H_2 \quad (4.14)$$

$$3\ Fe_2SiO_4 + 2\ H_2O = 2\ Fe_3O_4 + 3\ SiO_2 + 2\ H_2 \quad (4.15)$$

According to the equation of reaction (4.14), to replace grunerite by

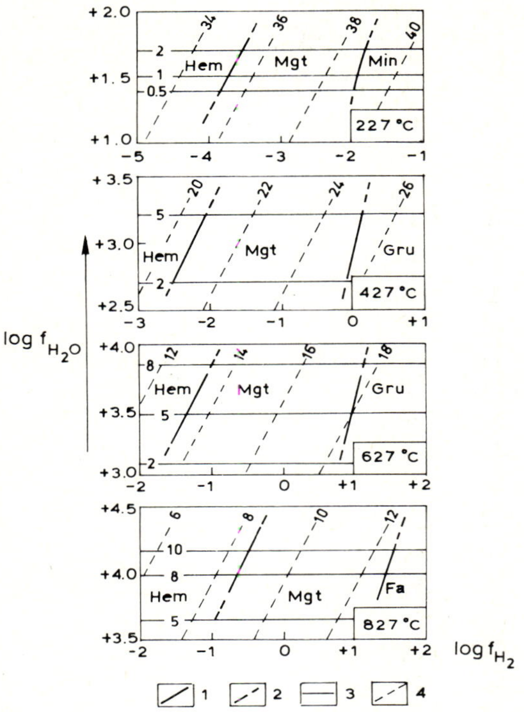

Fig. 76. Diagrams of mineral equilibria in silicate iron-formations in the absence of carbon dioxide (isothermal sections): *1* = actual boundaries of stability fields of minerals; *2* = boundaries unrealistic under the given conditions; *3* = isobars of fluid pressure ($P_f = P_{H_2O} + P_{H_2}$, in kbar); *4* = isobars of log f_{O_2}.

magnetite it is necessary to remove 14 g of H_2 for every 3005 g of $Fe_7Si_8O_{22}(OH)_2$, or 4.7 mg of hydrogen per 1 g of silicate. "Pure" water with an undisturbed molar ratio of $H_2:O_2 = 2$ contains practically no free hydrogen (at 627°C and 5 kbar, $\sim n \cdot 10^{-8}\%$ H_2). For the oxidation process to go the other way, the water removed from the reaction zone must contain less than 0.06% hydrogen, for instance 0.04–0.05%, or 40–50 mg H_2 per 100 g H_2O. From this it follows that it takes about 10–12 g of H_2O to remove 4.7 mg of H_2. Thus, to oxidize 1 g of grunerite to magnetite under conditions of amphibolite facies metamorphism requires the introduction of approximately 10–12 g of "pure water"—an amount quite acceptable for hydrothermal activity. However, in normal regional metamorphism oxidation of Fe^{2+} silicates by water can hardly be of great ore-forming importance, inasmuch as the $H_2:O_2$ ratio in the metamorphic fluids introduced from the underlying or adjacent silicate rocks will be shifted.

To oxidize magnetite to hematite according to the reaction:

$$2\,Fe_3O_4 + H_2O = 3\,Fe_2O_3 + H_2 \qquad (4.16)$$

under conditions of metamorphism (hypogene martitization variant) tremendous amounts of "pure" water (at 627°C and 5 kbar, of the order of 1500 g H_2O per 1 g Fe_3O_4) are needed, because of which this process could be only of local importance.

Metamorphism of carbonate iron-formations

A typical feature of carbonate rocks is the siderite-plus-quartz association, and usually also the presence of iron oxides, both magnetite and hematite; it is assumed that there is no Fe^{2+} in hydrous silicates.

The character of metamorphism of carbonate BIF depends to a certain extent on the presence of hematite, inasmuch as in that case the reaction:

$$FeCO_3 + Fe_2O_3 = Fe_3O_4 + CO_2 \qquad (4.17)$$

becomes possible at relatively low temperatures—300°C at $P_{CO_2} = 2000$ bar (Fig. 77).

Siderite + hematite + magnetite equilibrium depends to a considerable extent on pressure, inasmuch as the slope of the *P-T* curves of decarbonation reactions are much steeper than those of dehydration reactions. It is presumed that the fluid consists mainly of carbon dioxide. Therefore the reaction in question does not strictly determine the lower temperature limit of the greenschist facies, and siderite plus hematite are often retained alongside grunerite all the way to temperatures of the order of 380–420°C at $P_{CO_2} = 5$–7 kbar.

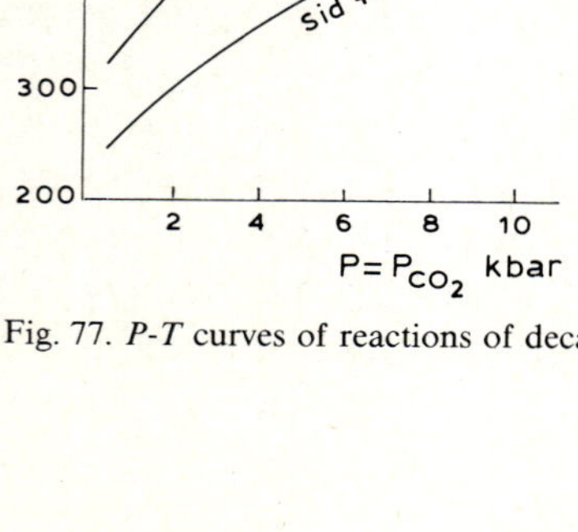

Fig. 77. *P-T* curves of reactions of decarbonation of the minerals of iron-formations.

We examined earlier the particulars of metamorphism of carbonate iron-formations with the formation of magnetite on the basis of thermodynamic calculations (Mel'nik, 1964, 1966c, 1969) and some experimental data (Yepatko and Mel'nik, 1965; Mel'nik, 1966a; Mel'nik and Kucher, 1967). Special attention was paid to siderite as one of the main original minerals, which dissociated thermally to yield magnetite even without the addition of an oxidant. Iron carbonates were no less important as natural reducing agents furthering the metamorphic transformation of iron hydroxides and hematite into magnetite.

In the works cited the two main schemes of metamorphic transformation of siderite into magnetite at high temperatures were studied theoretically:

(a) In environments with little or no water according to the thermal dissociation reaction:

$$3\ FeCO_3 = Fe_3O_4 + 2\ CO_2 + CO \tag{4.18}$$

(b) In aqueous environments (in the presence of a substantial amount of water in the rocks being metamorphosed or with the introduction of solutions from external sources) according to the reaction of siderite oxidation by water:

$$3\ FeCO_3 + H_2O = Fe_3O_4 + 3\ CO_2 + H_2 \tag{4.19}$$

The latter reaction would occur at lower temperatures than the former (by about 80–100°C).

Both reactions were accomplished experimentally. It actually turned out (Mel'nik, 1966a, c) that in a stream of water vapor the dissociation of siderite to magnetite with liberation of carbon dioxide and hydrogen takes place at a lower temperature (by 60–80°C) than in a stream of nitrogen with liberation of CO_2 and CO. The experimental works of other authors (Baykov and Tumarev, 1937; Berg and Buzdov, 1961) have shown that in a neutral atmosphere or in the presence of CO_2, siderite dissociates according to reaction (4.18) with the formation of CO.

In all the experiments and thermodynamic calculations, it was assumed that carbon monoxide was stable at the time of decomposition of iron carbonate and was removed from the rocks being metamorphosed. Nor were possible secondary reactions in the case of oxidation of $FeCO_3$ by water taken into account—the result of the reactions would be the formation of some CO, CH_4 and more complex hydrocarbons. Such assumptions are valid in examining the results of short-term experiments in which metastable carbon monoxide arises and does not undergo subsequent decomposition. However, in long-term natural processes the establishment of complete equilibrium between all the components of the gas phase, with the decomposition of metastable compounds, cannot be ruled out.

In the absence of water the gas phase in equilibrium with the siderite + magnetite association likewise will consist of CO_2 and CO, but with a CO_2:CO ratio different from the extreme value of 2:1 established by us earlier, due to the secondary reaction:

$$2\,CO = C + CO_2 \tag{4.20}$$

In the presence of water several secondary reactions between the gases are possible:

$$CO + H_2O = CO_2 + H_2 \tag{4.21}$$

$$CO + 3\,H_2 = CH_4 + H_2O \tag{4.22}$$

or between the gases and carbon (graphite):

$$C + 2\,H_2 = CH_4 \tag{4.23}$$

$$C + H_2O = CO + H_2 \tag{4.24}$$

As a result of these exchange reactions, the gas phase in equilibrium with the siderite + magnetite association will be a complex mixture of H_2O, CO_2, CO, H_2, and CH_4, the relative amounts of which depend on temperature, total pressure, presence or absence of graphite (carbonaceous matter), and oxygen fugacity.

Experimental investigations of the thermal decomposition of siderite in equilibrium with graphite under a total pressure up to 2000 bar showed that in more prolonged experiments reaction (4.20) apparently also occurs, accompanying the dissociation reaction (4.18). As a result the CO_2:CO ratio in the equilibrium gas mixture is determined by the buffer reaction of CO breakdown. In our experiments on thermal decomposition of natural siderite in a neutral atmosphere at $P = 1$ bar and $T = 380–460°C$, the CO_2:CO ratio ranged from 6:1 to 3:1 and never reached the 2:1 ratio needed for reaction (4.18).

Consequently, if dissociation of siderite takes place at relatively low temperatures, the process goes on fairly slowly and no rapid removal of the gaseous dissociation products from the reaction zone is observed, the decomposition of siderite may be accompanied by breakdown of CO. Then the overall process can be expressed by the equation:

$$6\,FeCO_3 = 2\,Fe_3O_4 + 5\,CO_2 + C \tag{4.25}$$

with the oxidation reaction of the graphite that is formed proceeding in

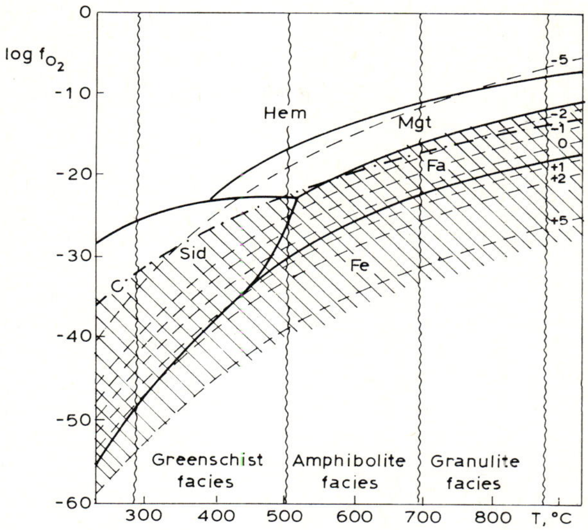

Fig. 78. Metamorphism of carbonate BIF (diagram of log f_{O_2}-T). Diagram for $P = P_f = \Sigma(P_{CO_2}, P_{CO}, P_{O_2}) = 5$ kbar. Isolines of log(f_{CO}/f_{CO_2}) are shown by dashed lines. The field of metastability below the graphite line is hatched.

parallel from right to left. On the diagram (Fig. 78) for $P_{CO_2} = 5000$ bar the temperature range of 390–510°C corresponds to the phase boundary of bivariant equilibrium of siderite + magnetite.

As was shown earlier (French, 1966; Yoi, 1966; Mel'nik, 1969) graphite, newly formed or originally present in the rock, is an oxygen buffer regulating f_{O_2} and f_{CO} in a fluid of carbonic acid composition. The line of graphite stability divides the diagram (see Fig. 78) into two parts. Only minerals whose stability fields are intersected by this line—siderite, magnetite, and fayalite—can be in equilibrium with graphite. Mineral associations occurring in the shaded field cannot exist because it is physically impossible to create such low values of f_{O_2} in carbonate rocks in the presence of a fluid containing CO_2 and CO. Associations outside the shaded field are stable only in the absence of graphite.

Similar observations should be taken into account in examining the isothermal sections of the diagram in coordinates of log f_{CO_2}–log f_{CO} for the temperatures of the various metamorphic facies (see Figs. 79, 80).

The petrological implications that follow from analysis of the *P-T* curves (see Fig. 77) and diagrams are quite obvious:

(1) Hematite cannot exist in equilibrium with graphite at any temperature. In greenschist facies conditions hematite would be transformed into siderite:

$$2\ Fe_2O_3 + C + 3\ CO_2 \rightarrow 4\ FeCO_3 \tag{4.26}$$

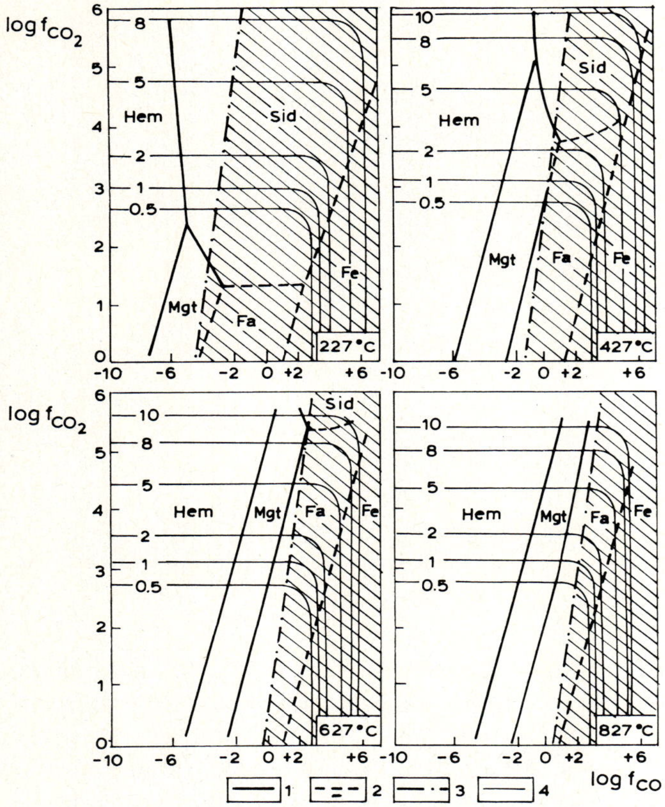

Fig. 79. Diagrams of phase equilibria in the system Fe–Si–C–O_2 (isothermal sections in coordinates of log f_{CO_2}-log f_{CO}): *1* = stable boundaries; *2* = metastable boundaries; *3* = line of graphite stability; *4* = isobars of fluid pressure ($P_f = P_{CO_2} + P_{CO}$, in kbar). Field of metastable states is hatched.

or into magnetite:

$$6\ Fe_2O_3 + C \rightarrow 4\ Fe_3O_4 + CO_2 \qquad (4.27)$$

According to thermodynamic calculations, these reactions proceed irreversibily to the right. Under conditions of the amphibolite and granulite facies only reduction according to reaction (4.27) is possible.

(2) Siderite is stable up to the temperatures of the beginning of the amphibolite facies.

(3) Equilibrium conversion of siderite into fayalite according to the reaction:

$$2\ FeCO_3 + SiO_2 = Fe_2SiO_4 + 2\ CO_2 \qquad (4.28)$$

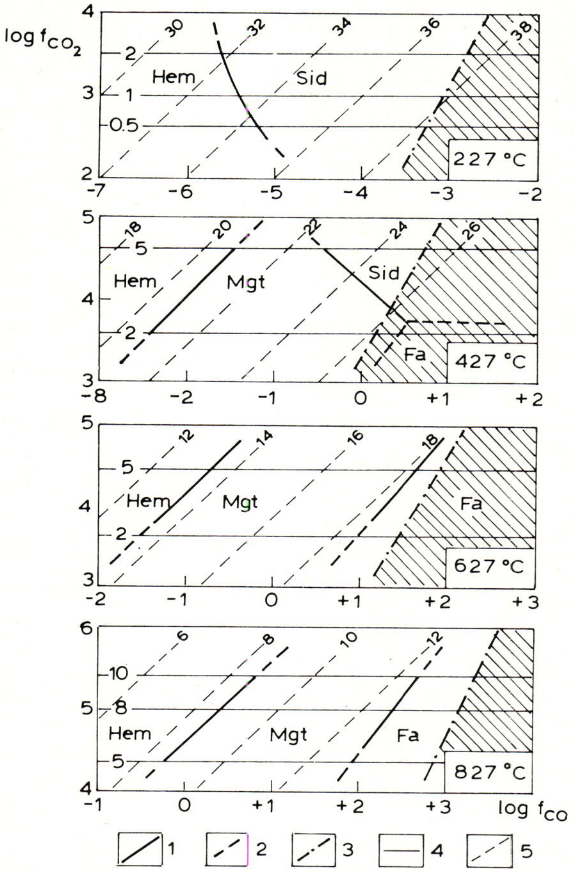

Fig. 80. Diagrams of mineral equilibria in carbonate iron-formations in the absence of water (isothermal sections): *1* = actual boundaries of stability fields of minerals; *2* = boundaries that are metastable or unrealistic under the given conditions; *3* = line of graphite stability; *4* = isobars of fluid pressure ($P_f = P_{CO_2} + P_{CO}$, in kbar); *5* = isobars of log f_{O_2}. Field of metastable states is hatched.

is impossible at pressures up to 6–8 kbar, because the stability fields of these minerals are separated by the magnetite field along the graphite line at any temperatures.

Therefore in the region of pressures not exceeding 6–8 kbar dissociation proceeds according to reaction (4.25), and subsequently the independent process of reduction of magnetite to fayalite takes place.

(4) At temperatures above 500–600°C the magnetite-plus-graphite association becomes unstable (Figs. 79, 80) due to the reaction:

$$2\,Fe_3O_4 + 3\,SiO_2 + C = 3\,Fe_2SiO_4 + CO_2 \qquad (4.29)$$

(5) There is a point of invariant equilibrium of siderite + fayalite + magnetite + quartz + graphite + fluid which fixes $T = 600°C$ and $P_f \approx 6-8$ kbar.

(6) In carbonate iron-formations metamorphosed under conditions of the amphibolite and granulite facies ($> 600°C$), graphite exists stably only in association with fayalite.

The compositions of the equilibrium fluids can also be found from the diagrams (see Figs. 79, 80). Inasmuch as the graphite field decreases appreciably as temperature rises, the carbon monoxide content in the fluid in equilibrium with graphite should increase. As an example let us trace the change in composition of the fluid on passing from the greenschist to the granulite facies, assuming the pressure to be constant at 5000 bar.

At 427°C graphite is in equilibrium with siderite (see Fig. 80), the fluid contains 99.99% CO_2 and 0.01% CO ($\log f_{CO_2} = +4.58$ and $\log f_{CO} = +0.58$, which for $\gamma_{CO_2} = 7.56$ and $\gamma_{CO} = 10.96$ corresponds to $P_{CO_2} \approx 5000$ bar and $P_{CO} \approx 0.5$ bar). At 627°C graphite is in equilibrium with fayalite, the composition of the fluid changes toward an increase in carbon monoxide content—99.75% CO_2 and 0.25% CO. At 827°C graphite also is in equilibrium with fayalite and with a fluid composed of 97.7% CO_2 and 2.3% CO.

Thermodynamic calculations show that carbon dioxide, like water, theoretically can serve as an oxidant of silicates of divalent iron:

$$3\,Fe_2SiO_4 + 2\,CO_2 = 2\,Fe_3O_4 + 3\,SiO_2 + 2\,CO \qquad (4.30)$$

and magnetite:

$$2\,Fe_3O_4 + CO_2 = 3\,Fe_2O_3 + CO \qquad (4.31)$$

However, the amount of "pure" carbon dioxide needed for oxidation without disturbing the ratio of the volatile products of dissociation ($CO:O_2 = 2$) would be still greater than the amount of water in analogous reactions.

In comparable conditions of the amphibolite facies ($T = 627°C$; $P_f = 5$ kbar), introduction of 80–100 g of CO_2 is needed to oxidize 1 g of fayalite, and of the order of 12–15 kg of CO_2 to oxidize 1 g of magnetite.

Therefore the formation of magnetite in that way could hardly be of essential importance in the metamorphism of iron-formations, and martitization is still less likely. However, in deposits of other genetic types, for instance skarn deposits, oxidation of iron silicates to magnetite at the contact with large masses of carbonate rocks (dolomite, magnesite) can be considered an ore-forming process. The last conclusion is still feasible because the carbon dioxide released in the dissociation of carbonates probably had an undisturbed $CO:O_2$ ratio.

Metamorphism of oxide-carbonate iron-formations

It is assumed that in such rocks oxides—hematite and magnetite—were essential minerals as well as siderite. The association of siderite with magnetite and fayalite in equilibrium with graphite in reducing conditions has already been examined; now it remains to analyze in detail the particulars of metamorphism in oxidizing conditions in equilibrium with a hematite-magnetite buffer.

From the diagrams (see Figs. 79, 80) it has already been seen that at high CO_2 pressures magnetite is unstable and breaks down to siderite + hematite, due to reaction (4.17) proceeding from right to left. This process is a

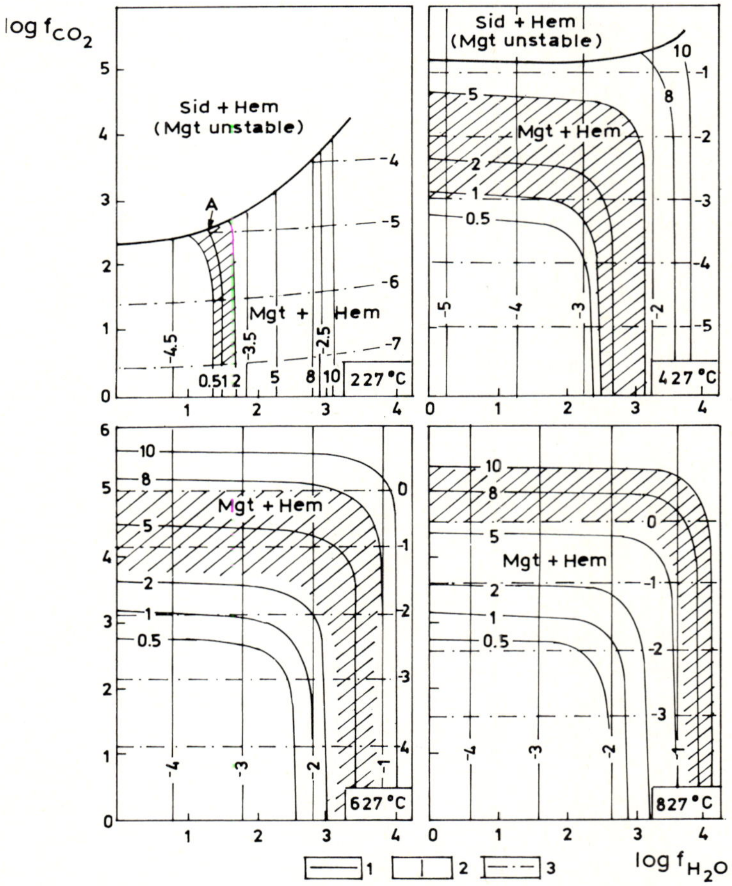

Fig. 81. Diagrams of mineral equilibria in oxide-carbonate iron-formations (isothermal sections in coordinates of log f_{CO_2}-log f_{H_2O}, magnetite-hematite buffer): *1* = isobars of total fluid pressure ($P_f = P_{CO_2} + P_{H_2O}$); *2* = vertical thin straight lines—isobars of log f_{H_2}; *3* = isobars of log f_{CO}. Fields of actual pressures in metamorphism are hatched.

relatively low-temperature one and is possible only under greenschist facies conditions, being bounded by the Mgt + Sid + Hem curve. The "hypogene" martitization of magnetite was studied in detail by Kornilov (1969) on the Krivoy Rog and Belozerka deposits and is theoretically substantiated by thermodynamic calculations.

However, it should be kept in mind that decomposition of magnetite is not a metamorphic process proper, but reflects the manifestation of carbon dioxide metasomatism, in which the introduction of corresponding amounts of CO_2 from external sources is necessary. In aqueous and water-carbon dioxide fluids magnetite is quite stable in association with hematite both at low and at high temperatures (Fig. 81). Magnetite becomes unstable only at a certain CO_2 content in the fluid, controlled from without, i.e. in the case of "completely mobile" behavior of carbon dioxide. For instance, at 227°C and $P_f = 1000$ bar, magnetite is stable if the CO_2 content in the fluid does not exceed 35% (see Fig. 81, *A*).

If there is no constant influx of fluid of a certain composition, decomposition of magnetite ceases. The limiting case is a "dry" system closed to CO_2. By analogy with systems closed to water, in such a system with constant pressure ($P_s = P_f =$ const) the fluid phase disappears entirely, and the Mgt + Sid + Hem association (system Fe–C–O) becomes bivariant and can exist stably below the *P-T* curve (see Fig. 77) in the stability field of the Sid + Hem (+fluid) association. From these considerations the Mgt + Sid + Hem association cannot be used to judge the low-temperature limit of mineral formation; the upper limit is fixed quite definitely inasmuch as removal of CO_2 begins at $P_f \geqslant P_s$ and the reaction proceeds irreversibly to the right. The extensive occurrence of magnetite in oxide-carbonate iron-formations of low-rank metamorphism apparently indicates the absence of equilibrium or even a deficiency of CO_2 and special "dry" conditions.

Metamorphism of silicate-carbonate iron-formations

As has already been mentioned, a substantial part of iron-formations consists not of pure silicate or carbonate varieties, but of mixed banded silicate-carbonate cherty formations with or without magnetite. The essential minerals are silicates of divalent iron and siderite, and a certain amount of graphite usually occurs, acting as an oxygen buffer. Graphite might be formed as a result of metamorphism of carbonaceous organic matter in the sediment, or it can arise in the course of conjugate oxidation-reduction reactions (4.18, 4.25) of dissociation of carbonates with subsequent breakdown of CO according to reaction (4.20).

Therefore for an analysis of mineral equilibria it is necessary to plot sections of the diagrams along the line of graphite stability (see Figs. 79, 80). If the fugacities of CO_2 and H_2O are used as independent variables such a section, for constant temperature, will be a plane (surface) on every point of which the presence of graphite is implied, controlling the oxygen fugacity according to the reaction:

$$C + O_2 = CO_2 \tag{4.32}$$

This provides the possibility of entering the isobars of oxygen fugacity on the diagram, and according to reaction (4.20) also the isobars of carbon monoxide fugacity. Finally, knowing f_{O_2} and f_{H_2O}, the fugacity of hydrogen, f_{H_2}, can be determined at every point on the diagram according to the reaction of decomposition of water:

$$2\,H_2O = 2\,H_2 + O_2 \tag{4.33}$$

and then that of methane, f_{CH_4}, according to reaction (4.23).

The diagrams obtained make it possible to obtain a detailed analysis of mineral equilibria and to determine the content of any of the six volatile components of the fluid at a given T and P.

Before the beginning of regional metamorphism ($T < 250$–$280°C$) only greenalite or minnesotaite and siderite, the stability fields of which are determined by the reaction:

$$3\,FeCO_3 + 4\,SiO_2 + H_2O = Fe_3Si_4O_{10}(OH)_2 + CO_2 \tag{4.34}$$

can exist stably in equilibrium with graphite and the fluid at high pressures (< 500 bar). Therefore there is reason to suppose that the finding of crystalline magnetite in the iron cherts of slightly metamorphosed sequences containing graphite indicates either the absence of equilibrium or "dry" conditions, when there was no fluid at all or $P_f \ll P_s$. The appearance of a fluid of any composition in magnetite-graphite rocks leads to the irreversible reaction of replacement of magnetite by siderite when the content of $CO_2 \geqslant 4$–13% (for 227°C and $P_f = 0.5$–2.0 kbar), or by minnesotaite in the case of a lower carbon dioxide content:

$$2\,Fe_3O_4 + 8\,SiO_2 + 2\,H_2O + C = 2\,Fe_3Si_4O_{10}(OH)_2 + CO_2 \tag{4.35}$$

We note that the phenomena of replacement of magnetite by siderite and more rarely by silicates, observed in iron cherts of low rank metamorphism, probably were caused by conjugate reactions in which graphite (carbonaceous matter) and fluids of complex composition took part directly or indirectly.

At $T > 300$–$330°C$ the widely occurring amphibole grunerite appears instead of the rare minnesotaite; the interrelationships of grunerite and siderite are governed by the well known reaction:

$$7\,FeCO_3 + 8\,SiO_2 + H_2O = Fe_7Si_8O_{22}(OH)_2 + 7\,CO_2 \qquad (4.36)$$

The equilibrium of siderite + grunerite (+ quartz) + graphite occupies a broad field in greenschist *P-T* conditions (see Fig. 82), but on passing to the amphibolite facies ($T = 550$–600°C) the association theoretically can be retained only in the case of very high fluid pressure.

In the stability field of this association the formation of grunerite depends on the ratio of CO_2 and H_2O in the fluid, rather than on temperature. For instance, at $T = 427$°C and $P = 5000$ bar the siderite + grunerite association is in equilibrium with a fluid of the composition $P_{H_2O} \approx 2660$ bar and

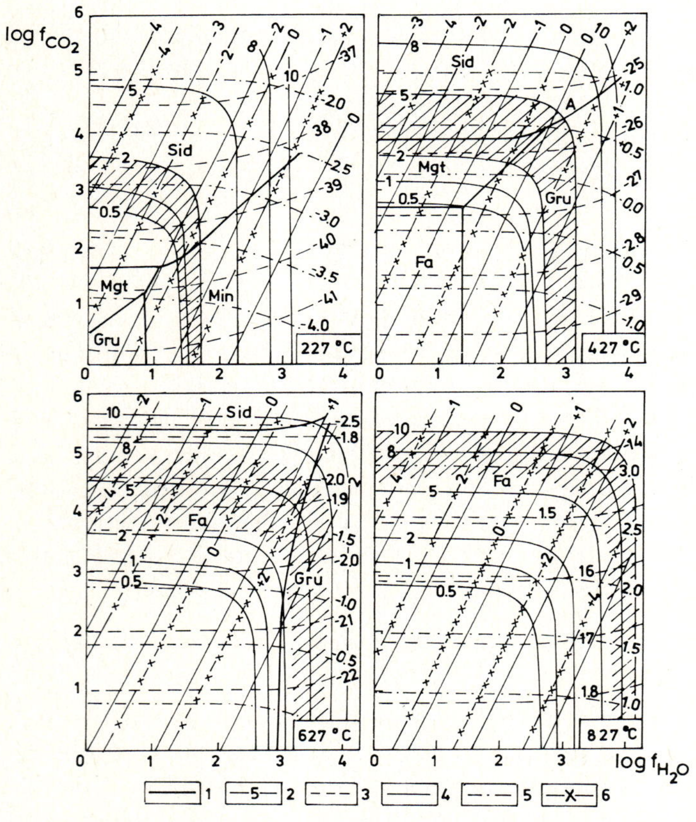

Fig. 82. Diagrams of mineral equilibria in silicate-carbonate iron-formations (isothermal sections in coordinates of log f_{CO_2}-log f_{H_2O}, graphite buffer). Fields of actual pressures in metamorphism are hatched: *1* = phase boundaries; *2* = isobars of fluid pressure, in kbar; isobars of partial fugacity (log f_i, in bar); *3* = O_2; *4* = H_2; *5* = CO; *6* = CH_4.

TABLE X
Composition of fluid in equilibrium with the siderite + grunerite + graphite + quartz association at 700°K (427°C) and $P_s = P_f = 5$ kbar

Component (i)	f_i (bar)	γ_i	P_i (bar)	m_i (%)
H_2O	795.0	0.299	2659.0	53.18
CO_2	17683.0	7.56	2339.0	46.78
CH_4	16.3	10.51	1.55	0.03
H_2	1.1	3.64	0.30	$6 \cdot 10^{-3}$
CO	2.6	10.96	0.24	$5 \cdot 10^{-3}$
O_2	$3{,}4 \cdot 10^{-26}$	6.48	$5.3 \cdot 10^{-27}$	–
Total	–	–	5000.0	100.0

$P_{CO_2} \approx 2340$ bar (Table X), which corresponds to $P_{H_2O} : P_{CO_2} = 1.14$.

A change in this ratio as a result of introduction of water or CO_2 leads to the formation of grunerite or siderite, respectively. Thus the main reason for the formation of grunerite is the presence in the rocks being metamorphosed of an amount of water sufficient to maintain a high P_{H_2O} and $P_{H_2O} : P_{CO_2}$ ratio when water is consumed in reaction (4.36).

The diagrams given in Fig. 82 contain complete information on the composition and evolution of the metamorphic fluid, but they are not clear enough. Therefore we, together with Radchuk (1977b, 1978), subsequently plotted several diagrams in which the composition of the fluid is expressed in molar fractions of CO_2 and H_2O. The improved thermodynamic constants of the minerals and fugacity ratios of the gases were used (Mel'nik, 1978).

As has already been mentioned, the effect of carbon dioxide on the

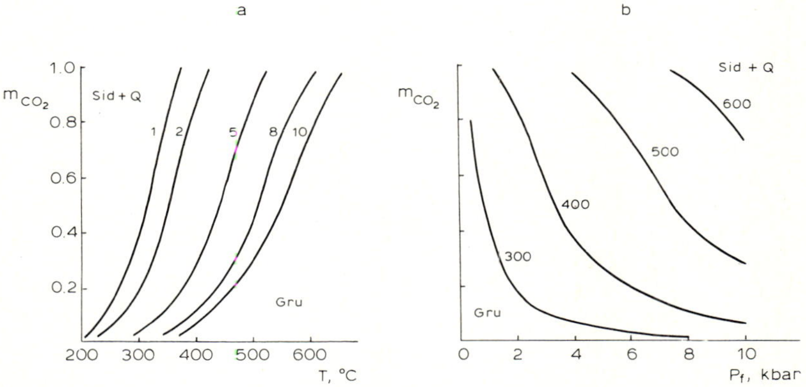

Fig. 83. Stability of grunerite in water-carbonic acid fluids (equilibrium Sid + Q + Gru). a. Isobars; figures on curves designate $P_f = P_{H_2O} + P_{CO_2}$. b. Isotherms, figures on curves designate T^0C (Mel'nik and Radchuk, 1977a). *Gru* = grunerite; *Q* = quartz; *Sid* = siderite.

stability of grunerite is of dual character. At high temperatures in the field of fayalite stability CO_2 behaves as an inert component, taking no part in the reactions but lowering the temperature of decomposition of amphibole, as is readily seen from a comparison of the *P-T* curves in Fig. 72. At lower temperatures (300–600°C) the stability of grunerite is determined by reaction (4.36), which plays the most important role in the formation of this amphibole in progressive metamorphism of carbonate BIF. The computer-calculated isobars and isotherms of this reaction are presented in diagrams (Fig. 83). The effect of all the independent variables (*T*, *P*, composition of the fluid) on this equilibrium is readily seen here. The lower temperature limit of stability of the amphibole is lowered (and its stability field correspondingly enlarged) as the CO_2 content in the fluid is reduced and the total fluid pressure decreases. Or in other words, the lower the total pressure and the higher the water content in the fluid, the lower the temperature at which siderite-quartz rocks are metamorphosed with the formation of grunerite. Let us consider the effect of each of the equilibrium parameters separately.

When the composition of the fluid is constant, for instance at $m_{CO_2} = 0.5$ or $m_{CO_2} = m_{H_2O}$ (Fig. 83a), the conversion of the siderite + quartz associa-

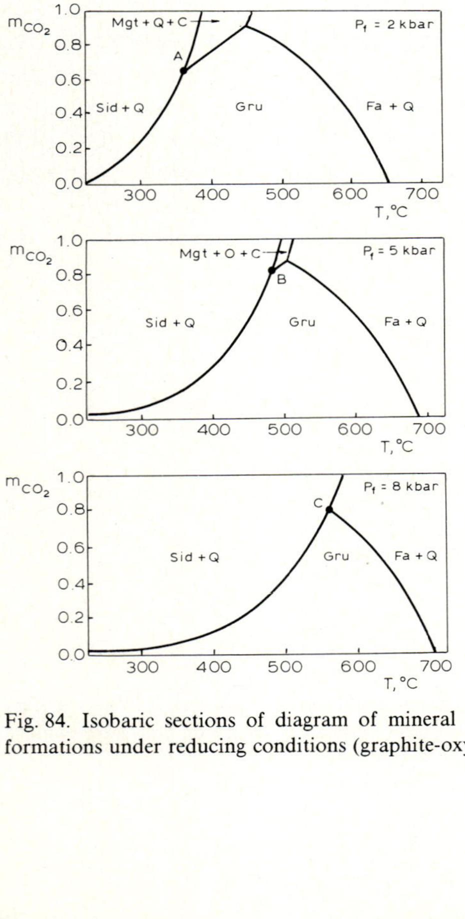

Fig. 84. Isobaric sections of diagram of mineral equilibria in metamorphosed silicate-carbonate iron-formations under reducing conditions (graphite-oxygen buffer).

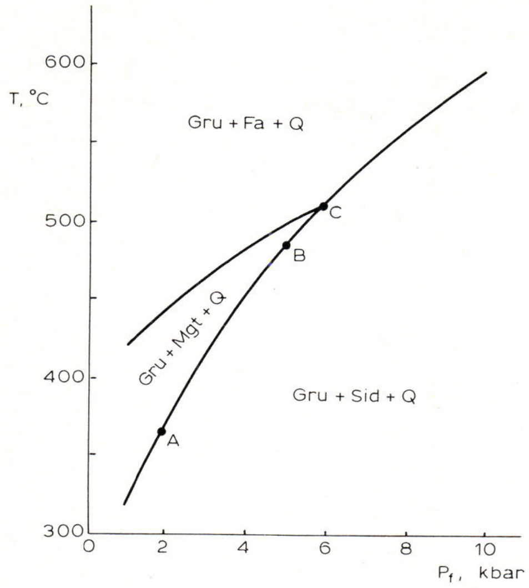

Fig. 85. *P-T* curves of monovariant equilibria of decarbonation and dehydration reactions in the metamorphism of silicate-carbonate iron-formations containing graphite.

tion to grunerite is determined by the curve of monovariant equilibrium which passes through the points: $T = 315°C$ for $P_f = 1$ kbar; $T = 445°C$ for $P_f = 5$ kbar; and $T = 555°C$ for $P_f = 10$ kbar.

When the pressure is constant, for instance at $P_f = 5$ kbar, the conversion from an essentially aqueous fluid ($m_{CO_2} = 0.2$ or 80% H_2O) to a carbon dioxide fluid ($m_{CO_2} = 0.8$ or 20% H_2O), raises the equilibrium temperature from 380°C to 485°C.

On the isobaric and isothermal sections of the diagrams (Fig. 84) it is seen that at low and medium pressures ($P_f < 5$–6 kbar) grunerite is in equilibrium with magnetite:

$$14\,Fe_3O_4 + 48\,SiO_2 + 7\,C + 6\,H_2O = 6\,Fe_7Si_8O_{22}(OH)_2 + 7\,CO_2 \qquad (4.37)$$

The equilibrium of siderite + grunerite + magnetite + graphite + quartz is monovariant and at pressures of 3–6 kbar corresponds to a temperature of the order of 400–500°C (Fig. 85).

However, this equilibrium fixes only the upper temperature limit fairly definitely. In the absence of fluid in "dry" systems or when $P_f \ll P_s$, the siderite + grunerite + magnetite association can exist stably at higher temperatures.

At higher temperatures and in the presence of a fluid, bivariant grunerite + magnetite equilibrium occurs, which is superseded above 450–510°C by grunerite + fayalite equilibrium, and above 600–650°C by fayalite alone.

We note that the conjugate values of temperature and pressure in Fig. 85 correspond to a fluid of a certain composition. Thus, the siderite + grunerite + magnetite + quartz + graphite association at $P = 2$ kbar and 370°C (point A in Fig. 85) is in equilibrium with a fluid containing ~ 70% CO_2 and ~ 30% H_2O, and at $P = 5$ kbar and 485°C (point B) the fluid contains ~ 83% CO_2 and ~ 17% H_2O. At the point of invariant equilibrium (point C) where $P = 6$ kbar at $T = 510$°C and $m_{CO_2} = 0.85$ (85% CO_2 and 15% H_2O) the association siderite + grunerite + fayalite + magnetite + quartz + graphite is stable. It is quite understandable that a combination of such "rigid" limits is not very likely in natural processes and this mineral association has not been found in metamorphosed BIF.

From comparison of the monovariant P-T curves and diagrams it is seen that in fluids of complex composition the temperatures of equilibria in which only one volatile takes part are somewhat lower ("dilution" effect) and fayalite becomes stable in amphibolite facies conditions.

Figure 86 presents another diagram of mineral equilibria in coordinates of log f_{O_2}-T for a fluid of constant composition ($m_{CO_2} = m_{H_2O} = 0.5$) which demonstrates the systematic change in mineral associations as the oxidation-reduction conditions change. Here too the line of graphite stability is entered as a dashed line. To the right of this line graphite is oxidized, and to the left it occurs in the field of synthesis of graphite from the fluid phase in sharply reducing conditions, which hardly arise in normal regional metamorphism.

When the diagrams are examined, the broad field of siderite stability, which increases sharply as m_{CO_2} and total pressure (P_f) increase, is striking. Metamorphism of the siderite-quartz association can go in two directions,

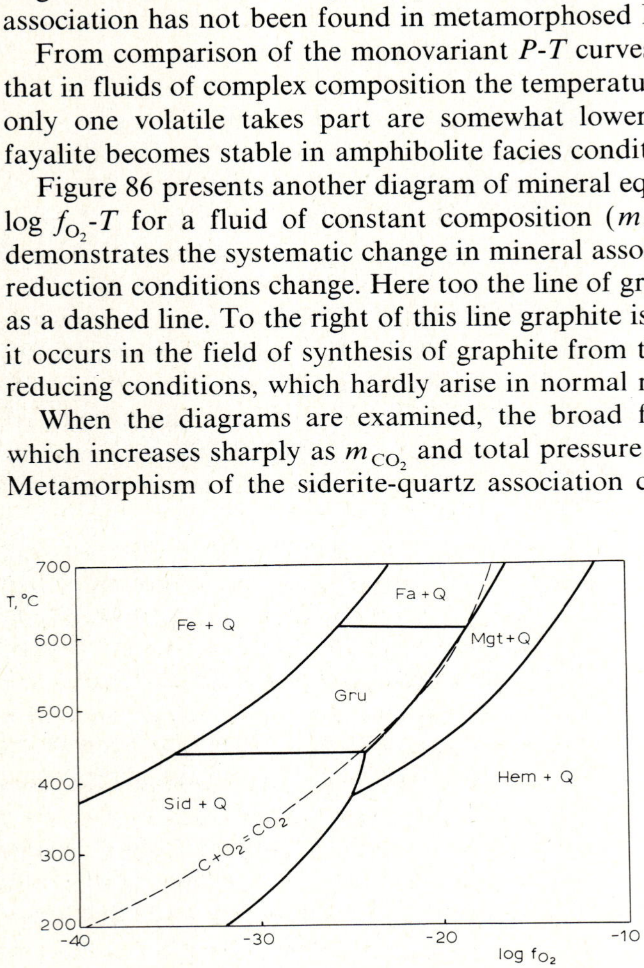

Fig. 86. Isobaric section of diagram of mineral equilibria in metamorphosed BIF of complex composition ($P = 5$ kbar, $P_{CO_2} = P_{H_2O} = 0.5$).

which can arbitrarily be called "rock-forming" and "ore-forming". In the first case grunerite transforms according to reaction (4.37), and in the second magnetite forms according to reaction (4.25).

Formation of grunerite is possible in a wide range of temperatures (300–560°C at $P_f = 2$–8 kbar) and variations in composition of the fluid (m_{CO_2} from 0.05 to 0.85), and a fluid of a certain composition (buffer association) corresponds to conjugate values of T and P. We note that according to thermodynamic data, formation of grunerite at the expense of siderite according to reaction (4.36) begins at the same temperature as the dehydration of minnesotaite (300–320°C) and should be selective. Numerous petrographic observations confirm this conclusion—primarily the process involves the zone of contact between siderite-quartz and silicate layers, where substantial amounts of water are liberated in the metamorphism of minnesotaite or chlorites. In the P-T conditions of the greenschist facies ($P = 2$–5 kbar) reaction (4.36) proceeds toward formation of grunerite, if the water content in the fluid reaches 75–90% ($m_{CO_2} = 0.10$–0.25). Acicular crystals and aggregates of grunerite can often be seen in thin section forming a reaction rim at the contact between schist and siderite-quartz layers, where diffusion of H_2O and CO_2 occurs. As the temperature rises the intensity of amphibolitization of sideritic rocks increases, and the process involves not only the marginal but also the central parts of the carbonate sequence, which is explained both by kinetic (increase in rate of diffusion) and thermodynamic factors (decrease in H_2O content of the fluid to 15–25%).

The ore-forming process—transformation of siderite into magnetite in reducing conditions—begins at higher temperatures than the formation of grunerite (370–510°C at $P_f = 2$–6 kbar) and, the most important point occurs only if the H_2O content of the water-carbon dioxide fluid is low enough ($m_{CO_2} = 0.7$–1.0). The temperature range of the stability of magnetite in association with graphite is relatively narrow—about 60–70°C at $P_f = 2$ kbar and 10–15°C at $P_f = 5$ kbar. In more oxidizing conditions the field of magnetite formation and stability is considerably expanded.

From the data given it follows that for any variations of f_{O_2} and fluid composition, siderite cannot be retained at temperatures above 510–530°C and its disappearance indicates the beginning of metamorphism in epidote-amphibolite facies conditions. In this case, in interbedded schist-carbonate sequences, magnetite can be formed either in the absence of excess quartz or in the inner parts of sideritic sequences, where the introduction of water was limited.

From the diagrams that have been plotted, information can be obtained on the composition of the fluids in equilibrium with iron-ore minerals over wide ranges of temperature (200–850°C) and pressure ($P_f = 0.5$–10 kbar), and the evolution of the composition of the fluids as a function of tempera-

TABLE XI
Calculated variations in composition of fluid in metamorphism of iron-formations in equilibrium with graphite buffer

T (°C)	P_f (kbar)	Mineral association	Composition of fluid, in mol.%				
			H_2O	CO_2	CH_4	H_2	CO
227	0.5– 2.0	Sid + Min + Q	80–95	20– 5	~0.1	<0.001	<0.001
427	2.0– 3.0	Mgt + Gru + Q	5–20	25–80	~0.02	~0.01	~0.02
427	3.0– 5.0	Sid + Gru + Q	20–50	80–50	~0.03	~0.01	~0.01
627	3.0– 7.0	Gru + Fa + Q	75–50	20–50	5.0–0.1	0.6–0.1	~0.2
827	7.0–10.0	Fa + Q	~10	~90	~0.01	~0.05	~1.0
827	7.0–10.0	Fa + Q	~85	~10	10.0–5.0	~1.0	~0.5

ture, pressure, and ratio of water to carbon dioxide and the sequences of mineral associations can be traced. The physical state of the fluid (liquid or gas) does not affect the determination of the composition, inasmuch as the actual properties (deviation from the ideal—γ) are taken into account for all the volatile components.

Table XI gives the limits of possible fluctuations of the composition of the fluid in equilibrium with graphite and with typical iron mineral associations in the conditions of the main facies of metamorphism. For $T = 827°C$, where strict limits cannot be set on the composition of the fluid in equilibrium with just the one iron mineral (fayalite), two extreme variants are arbitrarily taken: in one H_2O predominates in the fluid ($CO_2/[H_2O + CO_2] = 10$), and in the other CO_2 ($H_2O/[H_2O + CO_2] = 10$). From the table it is seen that, at all T and P, water and carbon dioxide predominate in the fluid; the content of other volatiles at low temperatures is negligible, and at high temperatures

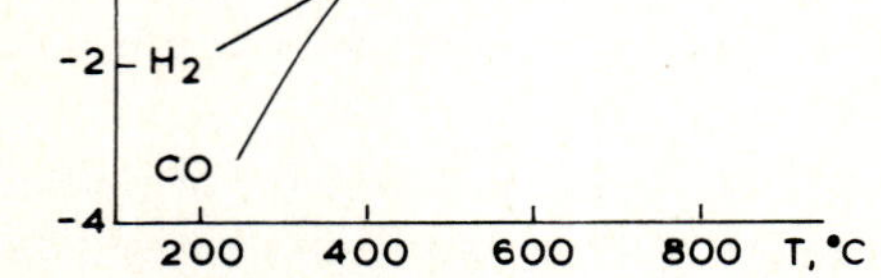

Fig. 87. Maximum partial pressures of volatile components $P_{i(max)}$ in the metamorphism of carbonate-silicate iron-formations containing graphite.

may reach 5–10%. These relationships are seen more clearly in a graph (Fig. 87) where the evolution of the maximum possible partial pressures of volatiles in equilibrium with iron minerals and graphite is shown as a function of temperature (at pressures corresponding to the respective metamorphic facies).

The systematic increase in the content of reducing agents (CH_4, H_2, CO) with rise in temperature with methane predominant in high-temperature fluids, is noteworthy. A dependence of P_{CH_4} on P_{H_2} and P_{H_2O} and of P_{CO} on P_{CO_2} is also observed. At 800–850°C the methane content in aqueous fluids (85–95% H_2O) may reach 5–10%, and of hydrogen 1–3%, while in carbon dioxide fluids (90–95% CO_2) carbon monoxide is sharply predominant over other volatiles (1–3% CO).

The established regularities in the evolution of the composition of fluids also explains some particulars of the metamorphism of iron-formations of other types. In particular, the frequently observed disappearance of hematite in oxide rocks metamorphosed in amphibolite and granulite facies conditions can be caused by reduction by volatiles, especially by easily diffused hydrogen. Acceleration of diffusion with rise in temperature and increase in concentration gradients of volatiles leads to the appearance of more uniform mineral associations in place of formerly extremely variegated banded sequences.

Effect of solid solution of magnesium on the stability of the minerals of iron-formations

It is known that solid solutions of magnesium, typical of silicates and carbonates of iron formations, would shift the equilibrium of dehydration and decarbonation reactions into a higher temperature region (Dobretsov and Ponomareva, 1964; Perchuk, 1968). In this case the distribution of Mg^{2+} and Fe^{2+} between coexisting minerals, like that of other cations, is a function of temperature (Bartholomé, 1962; Grant, 1963; O'Hara, 1963; Perchuk, 1970). The temperature dependence of the distribution of Mg has been derived for many mineral pairs on the basis of use of various geologic thermometers (Sen and Rege, 1966; Marks, 1968). Theoretical (Ramberg and De Vore, 1951; Kretz, 1961; Ghose, 1962; Grover and Orville, 1969; Saxena, 1969a, b) and experimental (Larimer, 1968; Medaris, 1969) investigations are also known.

Here we will not go into the question of giving a definitive analysis of the effect of Mg on mineral equilibria, which would include calculation of the cation distribution between coexisting minerals. That problem cannot be solved today by thermodynamic methods even if the consistent constants of the minerals are used, inasmuch as the properties of real solid solutions are

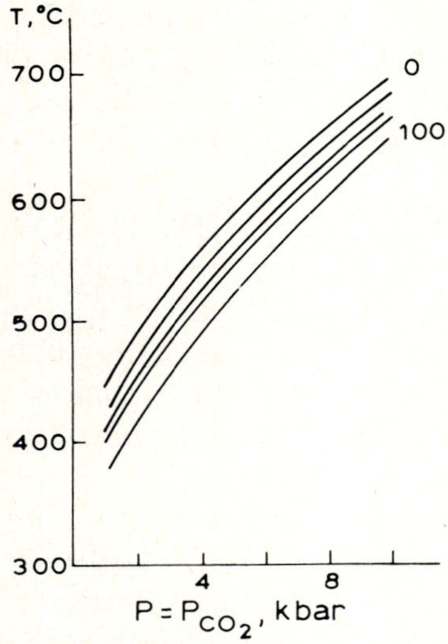

Fig. 88. Stability of carbonates of the isomorphous series siderite-magnesite in rocks with excess silica (upper temperature limits). Figures on curves indicate iron content (in mol.%); curves are drawn through every 20% Fe.

quite inadequately known. Marakushev (1968) has pointed out that the lack of agreement between the theoretical temperature dependences in the pairs clinopyroxene-orthopyroxene and olivine-orthopyroxene and the actual distribution in coexisting minerals is explained precisely by the inapplicability of the laws of very simple solutions to such systems, especially in the temperature range below 1000°C.

However, some general relationships, particularly for the rock type in question with an excess of silica, can be found by thermodynamic calculations or graphic methods. For instance, comparing the stability of siderite and magnesite, the *P-T* curves of which differ by 50–60°C (Fig. 88), it can be presumed that the temperatures of dissociation of carbonates of intermediate composition with the formation of the corresponding Fe-Mg silicates will fall in a given temperature range. For ideal solid solutions the dependence of temperature of dissociation on composition is almost linear in first approximation. Sufficiently reliable data based on experiments (Lindsley et al., 1968; Akimoto and Syono, 1970) have been obtained for the common mineral paragenesis olivine + pyroxene + quartz (Kurepin, 1970). The new consistent constants make it possible to determine more precisely the position of the *P-T* curves of this solid phase equilibrium and also to delineate the fields of development of cummingtonite in aqueous fluids and of Fe-Mg carbonates in carbonate fluids.

Stability of magnesian-iron carbonates

Figure 88 shows the *P-T* curves of the decarbonation reaction:

$$(Fe,Mg)CO_3 + SiO_2 = (Fe,Mg)SiO_3 + CO_2 \quad (4.38)$$

In the case of a high iron content in the carbonate, olivine is formed instead of orthopyroxene. A 10% increase in magnesium content leads to a 5–8° shift of the *P-T* curve into the higher temperature region. The shift is relatively small, but it plays a definite role in the metamorphic redistribution of iron among the minerals, particularly in reactions in which iron oxides are formed in the absence of water:

$$6(Fe_m,Mg_{1-m})(CO_3) = 6(Fe_{n-m}Mg_{1-n})(CO_3)_{1-m} + 2\,mFe_3O_4 + mC + 5\,mCO_2 \quad (4.39)$$

As a result the formation of magnetite is accompanied by an increase in the magnesium content of the carbonate. An analogous phenomenon can also occur in the oxidation of carbonates.

Metamorphism of carbonate iron-formations in the presence of aqueous fluids leads to the formation of cummingtonite. When equilibrium is reached, reactions of the type:

$$7(Mg_n,Fe_{1-n})(CO_3) + 8(1-n)SiO_2 + (1-n)H_2O = 7(Mg_{n-m},Fe_m)(CO_3)_n + (Mg_m,Fe_{1-m-n})_7[Si_8O_{22}(OH)_2]_{1-n} + 7(1-n)CO_2 \quad (4.40)$$

can also be accompanied by redistribution of magnesium and iron, as a result of which the magnesium content of the carbonate coexisting with amphibole increases as the temperature rises or the water content in the fluid increases ($H_2O:CO_2$ ratio).

Stability of amphiboles

On the basis of experimental investigations of equilibria at fluid pressure ~ 1000 atm (Hinrichsen, 1967; Schürmann, 1967) it is presumed that the stability fields of isomorphous series of Fe-Mg amphiboles—orthorhombic anthophyllite and monoclinic cummingtonite—overlap (Fig. 89). However, the experiments cited are not completely comparable inasmuch as cummingtonite could be synthesized only after addition of 2% CaO and equilibrium was not always attained, which we had already pointed out earlier (Mel'nik, 1972b). Therefore the existence of two amphiboles of the same

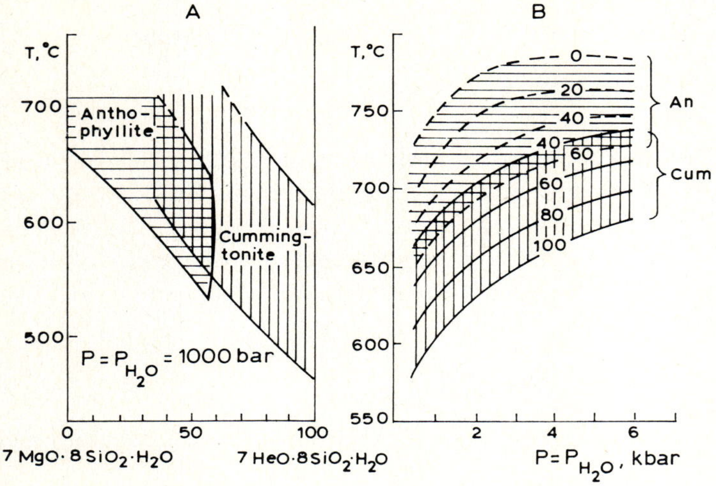

Fig. 89. Comparison of stability fields of synthetic anthophyllite and cummingtonite. Cross-hatched region shows T-x field in which anthophyllite and cummingtonite with the same Mg : Fe ratio could exist.

composition in the region of overlap at equal T and P is metastable.

Comparison of the calculated P-T curves of the thermal stability of grunerite, anthophyllite, and tremolite shows a systematic increase in dehydration temperature on passing from iron to magnesian amphiboles and beyond to magnesian-calcic (Fig. 90). Obviously, substitutions of Mg and Ca

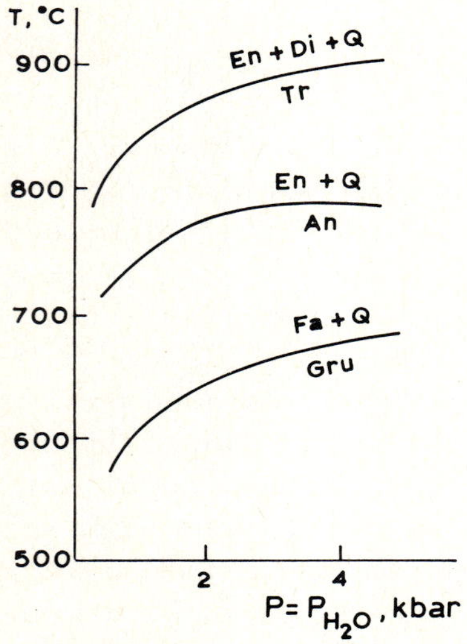

Fig. 90. Calculated P-T curves of dehydration of Fe-, Mg-, and Ca-amphiboles in rocks with excess silica (upper temperature limits).

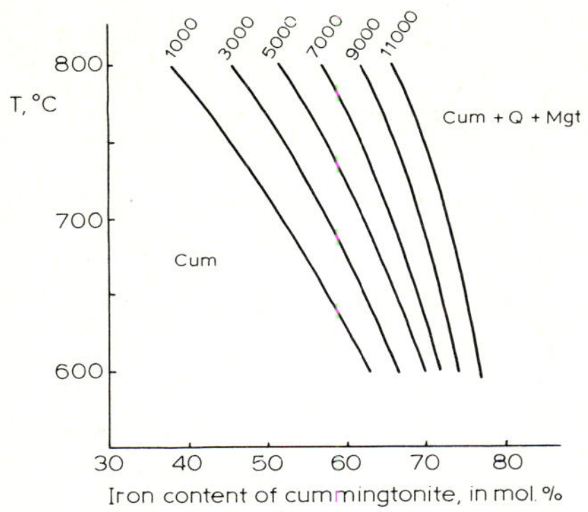

Fig. 91. Calculated isobars (P, in bar) of the compositions of cummingtonite in the association Cum + Mgt + α-Q (Fonarev et al., 1976).

would correspondingly increase the stability of Fe amphiboles.

Figure 89 shows schematic *P-T* curves of the upper temperature limit of stability of Mg-Fe amphiboles. The region of overlap here is arbitrarily retained. From the experimental data it follows that the iron content of cummingtonite does not fall below 30–40%, while ferroanthophyllite does not contain more than 60–70% iron molecule. Therefore the interpolation was not strictly linear, and the *P-T* curves obtained may contain an error of the order of ± 5–$15°$. From these approximate formulations it follows that a 10% increase in the magnesium content of cummingtonite would cause a 10–15° rise in dehydration temperature. As in the carbonate reactions, the product of decomposition will be olivine in the case of a high iron content and orthopyroxene in the case of an iron content below 70–90%.

Recently (Fonarev et al., 1976) experimental data were obtained on the stability of magnesian-iron cummingtonite in more oxidizing conditions (Ni/NiO oxygen buffer). On the basis of these data the equilibrium composition of cummingtonite in association with magnetite and quartz was calculated for various temperatures and pressures (Fig. 91). The upper temperature limits of cummingtonite stability given on this diagram are too high inasmuch as the field of orthopyroxene stability occurs at $T \geqslant 750°C$.

Stability of pyroxenes and olivines

Pure iron orthopyroxene (α-$FeSiO_3$) is unstable at the pressures presumably existing in metamorphic processes, due to breakdown into the stable association fayalite + quartz (see Fig. 75):

$$2\ FeSiO_3 \rightarrow Fe_2SiO_4 + SiO_2; \quad \Delta V = +3.26\ cm^3/mol \qquad (4.41)$$

For pure magnesian enstatite (α-$MgSiO_3$) the relationship is the opposite:

$$2\,MgSiO_3 \leftarrow Mg_2SiO_4 + SiO_2; \quad \Delta V = +3.89\ cm^3/mol \qquad (4.42)$$

Both these reactions are accompanied by an increase in volume, and an increase in pressure should shift the equilibrium to the left, toward pyroxene stability. As a result, at very high temperatures $FeSiO_3$ becomes stable, and at lower pressures magnesian-iron pyroxenes, the maximum iron content of which is a function of pressure, are stable.

Under equilibrium conditions the change in isobaric-isothermal potential ($\Delta G_{T,P}$) of the exchange reaction:

$$2(Fe,Mg)SiO_3 = (Fe,Mg)_2SiO_4 + SiO_2 \qquad (4.43)$$

should be zero. Approximate calculation of the equilibrium states of pyroxenes is possible if it is assumed that the $\Delta G_{T,P}$ of reaction (4.43) is a linear function of the ratio of Mg and Fe at T-P-const.

For instance, at $T = 1000°K$ and $P = 1$ bar, $\Delta G = -1.10$ kcal/mol for the reaction of decomposition of $FeSiO_3$, and for the reaction of decomposition of $MgSiO_3$ (4.42), $\Delta G = +1.33$ kcal/mol. Linear interpolation gives $\Delta G = 0.00$ for pyroxene containing $\sim 55\%$ of the iron molecule.

At $T = 1000°K$ and $P = 10$ kbar, the ΔG values of the reaction of decomposition of the pure end members are -0.30 and $+2.28$ kcal/mol, respectively, which corresponds to stability of a pyroxene with $\sim 88\%$ $FeSiO_3$.

The P-T curves of the stability of pyroxene with maximum iron content, in association with olivine and quartz, calculated in that way are given in Fig. 92. We note that the diagram obtained is based entirely on the consistent thermodynamic constants and differs somewhat from the diagram

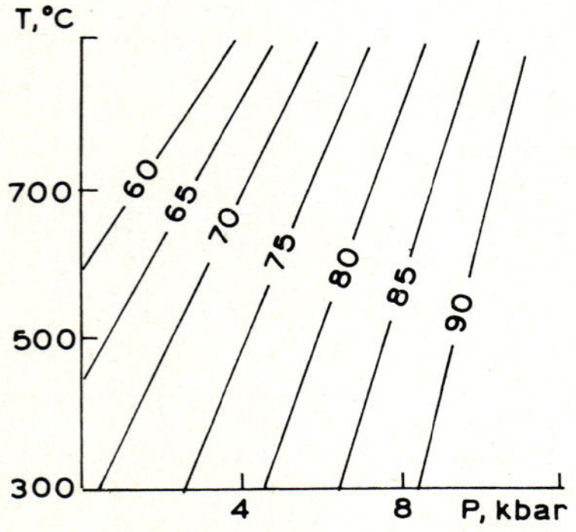

Fig. 92. Equilibrium of olivine with pyroxene in rocks with excess silica in the absence of fluid. Figures on curves indicate maximum iron content of orthopyroxene, in mol.%.

plotted by Kurepin (1970) taking into account low-pressure experimental data. The reason for the discrepancy apparently lies both in the approximate nature of the calculations (the properties of real solid solutions were not taken into account) and in the error in the initial thermodynamic constants of the silicates.

The association orthopyroxene + olivine + quartz is of great interest for geologic barometry, especially in the region of relatively high temperatures and pressures where the slope of the curves (to the temperature axis) is slight and equilibrium can be established rather rapidly. Pyroxenes containing 80–88% $FeSiO_3$ occurring in metamorphosed iron-formations (Semenenko et al., 1956; Val'ter, 1969; Kravchenko, 1971) crystallized at a pressure of the order of 7–10 kbar.

In the low-temperature region pyroxene and olivine can be retained only in "dry" rocks containing no volatiles. In the presence of a fluid phase the appearance of cummingtonite should be expected in the case of "excess" water, or carbonates in the case of "excess" carbon dioxide, which happens in hydrothermal metasomatism or even in progressive metamorphism of silicate or carbonate iron formations, respectively.

The coincidence of the stability diagrams of Fe-Mg pyroxenes (see Fig. 92) and amphiboles (see Fig. 93) makes it possible to plot a schematic diagram of mineral equilibria for $P_s \approx P_{H_2O}$ (Fig. 93A) which is divided into two parts by the line of cummingtonite decomposition. With an "excess" of water, only amphiboles are stable below this line, and orthopyroxene plus olivine and quartz above it.

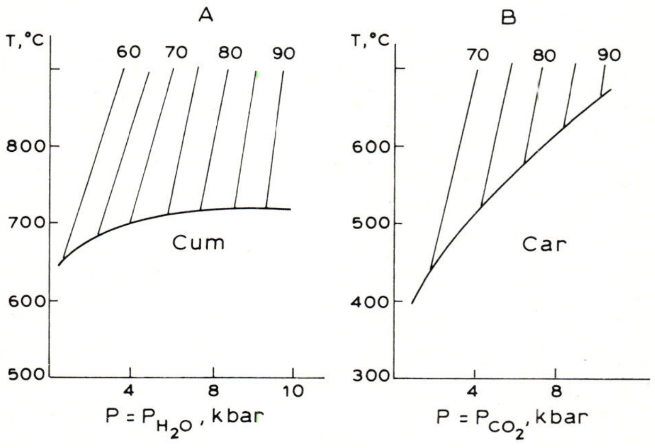

Fig. 93. Equilibrium of magnesian-iron minerals in rocks with excess silica. A. In silicate iron-formations (aqueous fluid, Cum + Px + Ol + Q association). B. In carbonate iron-rich rocks (carbonic acid fluid, Car + Px + Ol + Q association). Figures indicate maximum iron content of orthopyroxene in association with quartz and olivine.

The association pyroxene + olivine + cummingtonite + quartz, often encountered in highly metamorphosed BIF, is also very important for judging the thermodynamic parameters of mineralization. This association fixes the temperature at 700–720°C, which depends little on pressure and iron content of the silicates, while pressure can be estimated fairly precisely from the iron content of orthopyroxene. A decrease in partial pressure of water in conjunction with "mechanical" equilibrium ($P_s = P_f$) due to "dilution" of the fluid by other volatiles, for instance carbon dioxide, can lead to some shift of the *P-T* curve into the lower-temperature region. However, the amount of such a shift, especially at high pressures (8–10 kbar) cannot be significant because Fe-Mg carbonates which are stable in carbon dioxide fluids already appear at 610–650°C (Fig. 93b), and their paragenesis with anhydrous silicates is not typical of iron formations.

When olivine and pyroxene are oxidized the same regularities are observed as in the oxidation of carbonates. In particular, it was established in a study of the rich and unique olivine-magnetite iron ores of the Volodarsk deposit (Ukrainian shield) that the magnesium content of the silicates increases as the magnetite content of the ore increases (Mel'nik and Yaroshchuk, 1966). In this case the most likely oxidants may be water and carbon dioxide, the main components of the fluids causing metasomatic reworking of the olivine- and pyroxene-bearing iron formations.

Some particulars of low-temperature metamorphism of BIF

An intimate and frequent alternation of bands of different composition and conditions of formation is typical of most cherty iron-formations. Sometimes in the space of a few centimeters bands of siderite-hematite-magnetite are succeeded by grunerite-magnetite, and silicate bands containing graphite alternate with hematite-magnetite bands containing no layer silicates. Detailed study of such rocks provides grounds for presuming that special equilibrium conditions were often fulfilled in spatially very limited sectors (mosaic or local equilibrium), which were characterized by a certain composition of the fluid in the intergranular space. Closely adjacent areas with other mineral associations contained a fluid of quite different composition. Evidently at low temperatures diffusion was limited not only in the solid phases, but also in the fluid phase, which is also confirmed by study of the distribution of oxygen isotopes in the ore minerals. Therefore individual places in the rocks being metamorphosed can be regarded as closed systems, and the cherty iron-formation as a whole as a multitude of closed systems. These systems differed not only in composition of the fluid, but also in the content of individual volatiles.

Thus, in siderite or siderite-magnetite bands there apparently was no water at all, the system was "dry" and the main component of the fluid was carbon dioxide, while in adjacent silicate (schist) bands water occurred in excess. Local diffusion of water from schist to carbonate layers led to formation of grunerite, and diffusion of carbon dioxide in the opposite direction to the formation of carbonates in the schists.

Similar ideas have also been expressed recently by other investigators of the BIF. Klein (1973) pointed out that in the Labrador Trough some parts of the section of the BIF, of quartz-carbonate composition, may have been involved in reactions which produced silicates of the metamorphic series, while in other places only recrystallization of the quartz-carbonate associations with or without magnetite occurred. In the author's opinion this indicates that the chemical potential of CO_2 varied from place to place in the course of metamorphism and, consequently, CO_2 could not have been an ideally mobile component at any moment of time. Perry et al. (1973) believe that in the BIF of Biwabik, Minnesota, individual bands behaved as isolated subsystems and that different minerals formed in each band depending on the composition of the original unmetamorphosed rocks.

In amphibolite-facies conditions a tendency toward equilization of fluid compositions and a decrease in the variety of mineral associations is noted.

Analysis of the metamorphic conditions of the individual rock types of the BIF makes it possible to establish that they belong to certain metamorphic facies and indicates the existence of definite metamorphic zoning.

For instance, in the Ukrainian iron-ore province such zoning is traced along the strike of the BIF of the Krivoy Rog–Kremenchug structural-facies zone. In the central part of the Krivoy Rog basin rocks belonging to the greenschist facies are developed. Here mineral associations including siderite, hematite, graphite, and chlorites are substantially developed. Toward the north and south along the strike hematite and siderite gradually disappear, and cummingtonite is more common. In these areas the typical association is magnetite + cummingtonite (plus garnet), reflecting *P-T* conditions of the amphibolite facies. In turn, northward along the strike this facies grades into the granulite facies, represented by fayalitic rocks. Then a change to amphibolite and greenschist facies again is observed in the Kremenchug basin.

Thus the conditions of metamorphic transformation of iron-formations could have varied widely, from low- to high-temperature, even within short distances.

Metamorphism of iron-formations and ore formation

The rocks of the BIF are low-grade iron ores whose value is determined mainly by their magnetite content. The formation of magnetite apparently occurred in the period of diagenesis and in the initial stages of metamorphism as a result of reduction of hematite by carbon or siderite, and under the conditions of passing to the amphibolite facies, due to thermal dissociation of siderite as well. Thus low- and medium-temperature metamorphism helped to enhance the quality of the iron ores to a considerable extent. However, the introduction of water played a negative role in these stages of metamorphism, leading to the formation of silicates such as grunerite instead of magnetite. In amphibolite- and granulite-facies conditions an excess of graphite also led to formation of silicates—grunerite or fayalite—at the expense of magnetite, contributing to some impoverishment of the ores. The introduction of substantial amounts of pure water, and in isolated cases of carbon dioxide, could have led to the inverse process—oxidation of silicates to magnetite.

The formation of high-grade secondary hypogene iron ores due to removal of silica and some redistribution of iron is a special problem which has been examined by Belevtsev (1959). The physico-chemical aspects of ore formation are in need of special study.

Stability of the rocks of the BIF under the conditions of ultrametamorphism and granitization

By ultrametamorphism we usually mean regional metamorphism in deep folded zones, accompanied by migmatization, granitization, and formation of pegmatites. The final result of granitization of an initially heterogeneous pile usually is a relatively uniform pile of granitic or granidioritic composition, which sometimes contains remnants of the more stable original rocks.

The dissimilar behavior of different rocks toward granitization has not yet found a satisfactory explanation. Most investigators believe that the closer the chemical composition of the original rocks to the composition of granite, the more intensive their granitization (Backlund, 1949; Sudovikov, 1955; Nikol'skiy, 1970). In other works attention is paid to the physical properties and tectonic preparation of the rocks (Yeliseyev, 1959) or their ability to diffuse (Read, 1949). On the basis of geologic field observations, Sudovikov (1950) drew the general conclusion that the difference in the resistance of rocks to the granitization process is governed by the dissimilar stability of the minerals constituting the rock, the granularity, the closeness of the overall composition to granite, and the presence of minerals that react easily with introduced substances. However, the role of each of the factors enumerated is not well enough known.

Such a situation can be explained not only by the complexity and differences in the interpretation of observed natural factors, but also by the inadequate development of the physicochemical side of the theory of ultrametamorphism and granitization. If we proceed from general principles of chemical thermodynamics, then the most important factor which mainly determined the resistance of the rocks is their equilibrium or nonequilibrium under the physicochemical conditions that arose in the zones of ultrametamorphism. There are very few data for a rigorous thermodynamic analysis and plotting of diagrams—practically nothing is known about the properties and composition of the "permagmatic" (Korzhinskiy, 1952) solutions causing granitization. Experimental modeling (Winkler, 1961) has demonstrated the theoretical possibility of obtaining granitoid melts from sedimentary rocks in the case of an excess of water, but these data are still insufficient for quantitative estimation of the effect of numerous independent factors on the resistance of various rocks. Therefore an analysis of such complex phenomena must be limited to general physicochemical considerations. In particular, to explain the unequal stability of rocks we developed ideas concerning differences in the energy state and energy anisotropy of the pile being granitized. The main conclusions obtained will be given here briefly; more detailed physicochemical substantiation and geologic data are given in papers published both in Russian (Mel'nik, 1962b, 1966b) and in English translation (Mel'nik, 1965).

As a whole, a pile of deposits being granitized can be pictured as a complex system consisting of many systems which exist at different energy levels. In the course of establishing equilibrium under corresponding conditions some of the systems release energy and others receive it. On the basis of these considerations all rocks can be divided into two large groups—those actively and those passively granitized.

Argillaceous and mixed rocks containing a substantial amount of clay components belong to the actively granitized rocks. Acutely nonequilibrium clastic rocks (conglomerates, siltstones) and uncrystallized effusives would be somewhat less actively granitized. Argillaceous rocks are of special importance among the rocks of this group, as storers of water, which is given off before or during the process of granitization and can actively affect anhydrous rocks. Complex reactions between minerals leading to the establishment of chemical equilibrium and equalization of energy anisotropy always occur in the presence of and with the participation of a fluid phase.

Chemogenic sedimentary rocks (carbonate, ferruginous), some uniform clastic rocks (sandstones, quartzites), and all holocrystalline magmatic rocks, especially granitoids, must be assigned to the passively granitized rocks. Basic and ultrabasic rocks capable of more active granitization under the influence of a substantial amount of aqueous fluids can be distinguished in this group. Therefore, a fairly fine alternation of such rocks with water-bearing sedimentary rocks (even passive iron cherts) should be an association less resistant to granitization than each rock individually.

The vast fields of migmatites and granites in the Ukrainian shield were formed chiefly on a substrate of sedimentary rocks, now represented in remnants by paragneisses and schists. At the same time numerous layers or lenses of little-altered or unaltered metabasites are found nearly everywhere, interlayered with migmatite-granite material or preserved in it in the form of inclusions. Numerous inclusions of passively granitized iron-formations are widespread among the migmatites west of the Krivoy Rog, in the areas of the Orekhov–Pavlograd zone and in the Pri-Azov area. However, it should be mentioned that cases are known of rather intensive alteration of the rocks of the BIF, especially when they do not constitute thick uniform piles but are interlayered with paraschists.

Chapter 5

On the origin of the ore minerals of the Precambrian iron-formations

Introduction

Clarification of the genesis of the main ore minerals (magnetite, hematite, and siderite) is closely related to the solution of the fundamental problem—the origin and conditions of formation of the BIF. If we limit ourselves only to hypotheses in which the iron-formations are interpreted as chemogenic sediments, then it turns out that there are numerous variations in the interpretation of the original composition of these sediments and of the explanation of the conditions of formation of the main ore mineral—magnetite. The main points of modern hypotheses can be summarized in three groups.

(1) The original sediment, deposited from colloidal solution, was chiefly of hydroxide (hematite) composition. Magnetite and siderite are the products of reduction of hematite during diagenesis or metamorphism (Yurk and Shnykov, 1961). Buried organic matter, in part also preserved in the metamorphosed rocks, is usually regarded as the possible reducing agent (Strakhov, 1960; Plaksenko, 1966).

(2) The original sediment was precipitated from ionic solution in an oxygen-free environment and had a carbonate (siderite or sideroplesite) composition (Svital'skiy et al., 1932; Gershoyg, 1949; Tochilin, 1952). La Berge (1973) believes that siderite was one of the main primary minerals of the BIF. During metamorphic transformation of the original sedimentary or diagenetic siderite, magnetite was produced (Mel'nik, 1964, 1966c, 1969; Beukes, 1973; French, 1973).

(3) The original sediment had a complex composition which was determined by fluctuations in *Eh*, pH, P_{CO_2}, etc. in the sedimentary basin. Magnetite and hematite are believed to be primary sedimentary or diagenetic minerals, essentially unaltered in metamorphism (James, 1954; Huber, 1960; Garrels et al., 1973; Klein, 1973; Klein and Fink, 1976; Drever, 1974).

Use of the methods of nuclear geochemistry seems to be very promising in solving the complex problems of the origin of the ore minerals. Detailed study of the chemistry of iron-formations and the minerals constituting them also can give some new information on the origin of the magnetite. Free carbon is of particular importance as the main reducing agent of oxide sediments in diagenesis and metamorphism, therefore the distribution of C_{org} in the various types of rocks, individual bands, and minerals was studied specially.

Variations in the oxygen isotopic composition of the ore minerals and genesis of magnetite

Preliminary results of study of the distribution of oxygen isotopes in the ore minerals of the iron cherts, carried out together with Lugovaya (Mel'nik and Lugovaya, 1972) have been discussed in several papers and articles. Here these data will be summarized and systematized. Questions of the procedure of the investigations have been discussed in sufficient detail in the publications cited and will not be considered here.

The first isotope-geochemical investigations (Dole and Slobod, 1940) already showed that $^{16}O/^{18}O$ is not the same in iron ores of different genetic types. According to the data of determinations of the oxygen isotopic composition of the Krivoy Rog iron ores, made using the flotation method by Tanater and later by Lebedev and Petrov, the oxygen of unoxidized high-grade ores and iron-rich rocks is enriched in ^{18}O compared to ores oxidized in the supergene zone. However, no determinations were made for the individual minerals. Vinogradov and Dontsova (1952) suggested, on the basis of a study of the oxide minerals, that these minerals should differ in oxygen isotopic composition depending on their origin. This suggestion was confirmed by subsequent work. Dontsova (1964) established that the oxygen of the magnetite of skarns has a high ^{18}O content, while iron oxides of sedimentary origin are virtually not enriched in the heavy isotope compared to river water. This important conclusion was later substantiated both geologically (Clayton and Epstein, 1958) and experimentally. O'Neil and Clayton (1963) established that magnetite in equilibrium with water at low temperatures is only slightly enriched in ^{18}O, but quartz concentrates the heavy isotope maximally (the coefficient of isotopic fractionation $\alpha \approx +35\%$); an intermediate value of α is characteristic of carbonates. According to the data of James and Clayton (1962), in natural hematite-magnetite associations $\delta^{18}O$ is about the same in both phases.

Thus it is to be expected that in the case of sedimentary or diagenetic formation of magnetite and hematite of iron-formations the oxygen isotopic ratios of these two minerals will be similar and will correspond approximately to the $\delta^{18}O$ of the water of the sea basin. Iron carbonates (siderite, sideroplesite) formed in the same way would be substantially enriched in the heavy isotope (up to 15–20‰).

The final distribution of oxygen isotopes between coexisting minerals in metamorphosed iron-formations was determined by a number of factors: composition of the original sediment, attainment of isotopic equilibrium in metamorphism, temperature, constancy of the total oxygen isotopic composition of the rock, relative abundance of the various oxygen-containing minerals, composition and amount of intergranular fluid. In progressive metamorphism the coefficient of oxygen isotope fractionation between iron oxides and quartz decreases as temperature increases (Perry and Bonnichsen, 1966a, b; Garlick and Epstein, 1967; Taylor, 1970), which theoretically should lead to an increase in the $\delta^{18}O$ of magnetite. In highly metamorphosed rocks such an effect actually is observed at the contact with intrusions (Perry and Bonnichsen, 1966a, b), but in slightly and moderately metamorphosed iron-formations there is a decrease in the $\delta^{18}O$ of quartz while the $\delta^{18}O$ of magnetite yields only small fluctuations around the average value (Clayton and Epstein, 1958). It is not precluded that oxygen isotopic equilibrium was limited in such rocks (Anderson, 1967; Belevtsev et al., 1969; Becker and Clayton, 1976); original differences in the oxygen isotopic composition were retained in the metamorphic minerals also. In reality, the oxygen isotopic composition of magnetite and hematite in the rocks of the BIF of the Lake Superior district varies even within the same metamorphic zone, which cannot be satisfactorily explained by certain differences in temperature of crystallization and retrograde changes. The magnetite of the BIF of Quebec has a high ^{18}O content, as do other minerals there, compared to the corresponding minerals of the Lake Superior district (Sharma et al., 1965; Fig. 94). For the magnetite and hematite of the iron

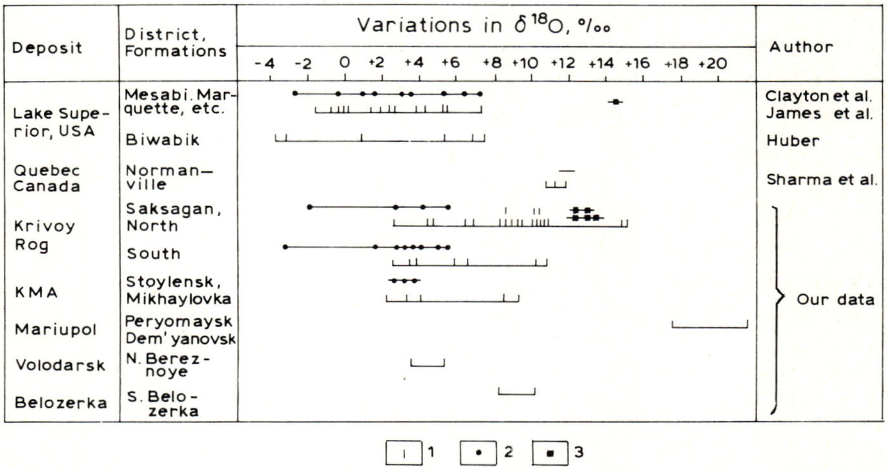

Fig. 94. Variations in the oxygen isotopic composition of the ore minerals of Precambrian BIF: *1* = magnetite; *2* = hematite; *3* = siderite.

formations of the Middle suite of the Krivoy Rog group, which belongs to the same metamorphic facies and crystallized in about the same temperature conditions, a still greater spread of $\delta^{18}O$ than could have been caused by the effect of steep gradient of the temperature of crystallization (Perry and Bonnichsen, 1966) has been established (Belevtsev et al., 1969).

Comparing the data in Fig. 94, it can be suggested that the spread of $\delta^{18}O$ of the ore minerals of Precambrian BIF was caused by substantial differences in the original oxygen isotopic composition, which makes it possible to use the results obtained to reconstruct the conditions of sedimentation and determine the composition of the iron sediments and the origin of the main ore mineral, magnetite.

Isotopic data

Figures 94 and 95 give the results of determination of the oxygen isotopic composition of the ore minerals of the iron formations and schists of various districts of the Ukrainian shield and KMA. Analysis of these data has shown that in the Krivoy Rog no clear-cut relationship is observed between the isotopic composition of the magnetite and the stratigraphic position of the rocks in the section of the middle suite of the Saksagan group, although the magnetites of individual districts, for instance the YuGOK and TsGOK *, differ somewhat. Rocks of the same metamorphic rank differ considerably in $\delta^{18}O$ both in the Krivoy Rog and KMA (greenschist facies) and in the Volodarsk and Mangush districts (granulite facies), and in the latter case the differences are so considerable ($\Delta\delta^{18}O$ of magnetites more than 12‰) that they suggest fundamental genetic differences in the very similar iron-formations of these two districts.

A quite definite dependence of the oxygen isotopic composition of the ore minerals on the characteristic mineral parageneses is observed for the Krivoy Rog and KMA (see Fig. 95). The values obtained for $\delta^{18}O$ fall into four groups.

(1) Very low $\delta^{18}O$ values are established in the hematite and hematite-magnetite rocks which occur in some ore horizons of the Krivoy Rog and KMA, constituting individual members and bands in magnetite rocks proper. Variations in the $\delta^{18}O$ of hematite fall within +2.6–5.5‰ for the Saksagan district, +1.3–5.5‰ for the southern districts of Krivoy Rog and +2.9–3.6‰ for the KMA. The lowest values ($\delta^{18}O = 1.3$–3.0‰) in this group of parageneses are those for hematite from rocks containing no silicates (chlorites,

* YuGOK stands for South Mining and Beneficiating Combine, TsGOK for Central Mining and Beneficiating Combine.

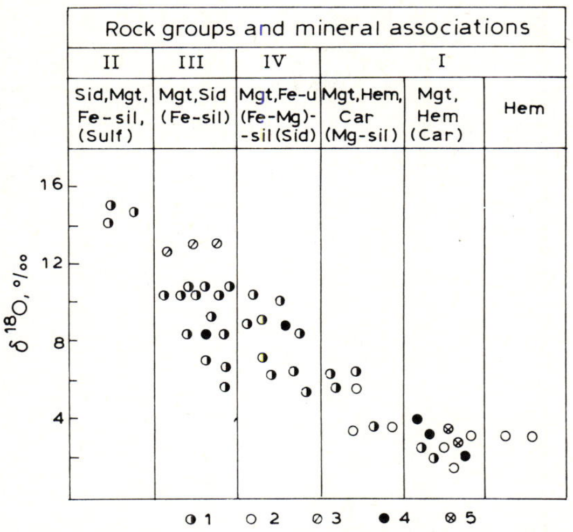

Fig. 95. Distribution of oxygen isotopes in relation to mineral associations of iron-formations and schists. Krivoy Rog: *1* = magnetite; *2* = hematite; *3* = siderite. KMA: *4* = magnetite; *5* = hematite.

amphiboles) and poor in carbonates (sideroplesite, pistomesite, dolomite). An extremely low content of free carbon and a low content of carbon dioxide are typical of all the hematite-magnetite rocks. The variations in the $\delta^{18}O$ of magnetite associated with hematite amount to +2.6–6.8‰ for the Saksagan district, +2.5–64‰ for the southern districts of Krivoy Rog, and +2.1–4.0‰ for the KMA. The fairly close correspondence of the oxygen isotopic ratios of hematite and magnetite (Fig. 96) is noteworthy. In all cases but one, the oxygen of the magnetite is somewhat enriched in ^{18}O compared to that of hematite. We note that for Lake Superior there is the opposite

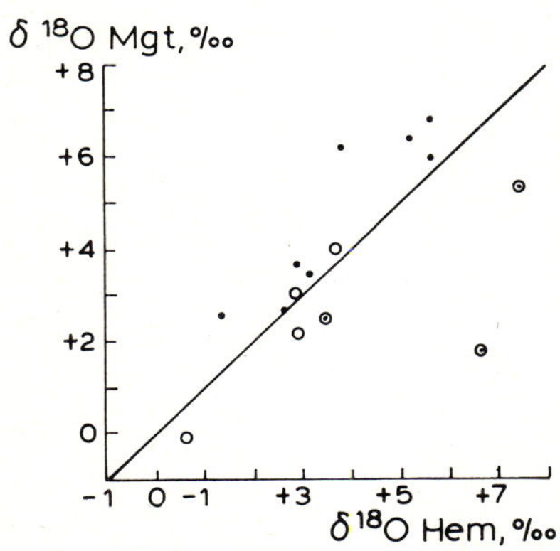

Fig. 96. Ratios of isotopic compositions of oxygen of magnetite and hematite: *1* = Krivoy Rog; *2* = KMA; *3* = Lake Superior.

relationship, and the correspondence of the $\delta^{18}O$ of associated oxides is much poorer. However, new measurements made by Becker and Clayton (1976) for the BIF of the Hamersley district in Western Australia have shown that the oxygen of magnetite is much heavier ($\delta^{18}O = +4.62$–$4.75‰$) than that of hematite ($\delta^{18}O = -1.87$ to $-2.19‰$). It is suggested that the magnetite possibly was formed at the expense of hematite.

(2) The highest values of $\delta^{18}O$ ($+14.8$–$15.3‰$) are found in magnetite from low-grade siderite BIF and some schists of the Krivoy Rog. These rocks underlie the ore iron-formations of the YuGOK and are the transitional sequence between the schists (K_2^{4s}) and the iron horizons (K_2^{4i}). Of the ore minerals, siderite (sideroplesite) predominates, magnetite occurs in subordinate amount, silicates are represented by chlorites and more rarely by biotite, and there is no hematite. In the schists frequently interbedded with the BIF, very fine-grained silicates predominate, sometimes carbonates and quartz; the principal ore mineral often is magnetite, crystals of which form relatively large porphyroblasts in the schist fabric. An admixture of syngenetic iron sulfides (pyrite, pyrrhotite, marcasite), disseminated along the layering, is typical of both the low-grade and barren rocks of the BIF and of the schists. These data suggest that a reducing environment was maintained at all stages of formation of the rocks of this group. Such high values of $\delta^{18}O$ in magnetite from Precambrian BIF were obtained for the first time; the maximum values established previously in the rocks of Quebec (Sharma et al., 1965) did not exceed $+11.9‰$. This apparently is explained by inadequate study of rocks of this type rather than by their exceptional nature, inasmuch as chiefly hematite-magnetite and silicate-magnetite rocks, which are characterized by lower $\delta^{18}O$, were studied in the works cited (James and Clayton, 1962; O'Neil and Clayton, 1963; Perry and Bonnichsen, 1966a, b).

(3) A very numerous group of $\delta^{18}O$ values for magnetites from iron-formations falls in a narrow interval of $+9.4$–$10.8‰$, with an average value of $10.2‰$ (of 8 determinations). Typical magnetite-siderite (sideroplesite) BIF ores with predominant magnetite, low-grade rocks in which carbonate predominates over magnetite, and also bands of siderite rocks in barren schist sequences fall into this group. Common to the rocks of this group are the obligatory presence of iron carbonate as an ore-forming mineral, the absence of hematite, and a low content of silicates in the BIF proper. The C_{org} content in the siderite-magnetite rocks is at the same level as in the hematite-magnetite rocks, but in the iron-rich rocks interbedded with schists and containing an appreciable admixture of silicates of chlorite type, the C_{org} content is somewhat higher. There is no correlation between the $\delta^{18}O$ of magnetites and the relative content of magnetite and siderite. Comparison of the $\delta^{18}O$ of associated magnetite and siderite (sideroplesite) is interesting. Isotopic data obtained for the first time for syngenetic iron carbonates from

Precambrian BIF (Mel'nik and Lugovaya, 1972) give a $\delta^{18}O$ value for total (combined) oxygen between +12.3 and +12.9‰, which is 2.3 to 2.6‰ higher than in the associated magnetites. For siderites of other genetic types we obtained values of $\delta^{18}O =$ +12.3‰ for a siderite vein with sulfides from the Rudňany deposit in Czechoslovakia, +14.4‰ for siderite from the Bakal iron deposit in the South Urals, and +15‰ for siderite from a polysulfide deposit in Central Asia. According to the data of Degens (1967), hydrothermal-vein siderites have $\delta^{18}O$ between 18.5 and 20.7‰. Becker and Clayton (1976) cite similar values of $\delta^{18}O$ (+20.12 to +21.20‰) for siderite occurring in association with quartz in the BIF of the Hamersley district of Western Australia. The differences in $\delta^{18}O$ of sedimentary-metamorphic and hydrothermal siderite are explained by differences in the oxygen isotopic composition of the aqueous solutions (O'Neil and Clayton, 1963).

(4) And finally, the fourth group includes schists of diverse composition which contain disseminated magnetite, and magnetite-silicate rocks. The $\delta^{18}O$ value for the magnetite of the schists falls between +6.4 and +10.2‰, averaging 8.1; for the silicate facies of the BIF it is somewhat lower than in the siderite rocks. Common to the rocks of this group is the presence of magnesian-iron silicates and aluminosilicates plus a variable content of magnetite and quartz; there is no hematite, and carbonates are usually of secondary importance. High C_{org} contents are typical of the schists, but the ore mineral is chiefly magnetite; the CO_2 content is relatively low. There is no connection whatever between the $\delta^{18}O$ of magnetite and the C_{org} content. The $\delta^{18}O$ values of the magnetite in the rocks of this group are comparable to the very "heavy" oxygen of Lake Superior magnetites.

The magnetite from the pyroxene-magnetite rocks of the Mangush district in the Azov area differs from all the samples studied from the Krivoy Rog and KMA and other districts as well. Such high values of $\delta^{18}O$ (+17.7 to +21.6‰) were previously found only in the metasomatically altered iron-rich rocks and ores of North Krivoy Rog. Subsequent investigations of metasomatic rocks yielded values of $\delta^{18}O$ of up to 20‰ for the magnetite in the aegirine-magnetite rocks at Krivoy Rog and up to 16.0‰ in the alkalized carbonate-magnetite-hematite rocks of the Mikhaylovka deposit of the KMA. As has already been mentioned, the iron-formations of the Volodarsk district, which are analogous in composition, rank of metamorphism, and apparently in stratigraphic position, are characterized by "lighter" oxygen in the magnetites ($\delta^{18}O =$ +3.7 to +5.3‰), which is comparable to the hematite-magnetite rocks of Krivoy Rog and the KMA. However, small bodies of rich metasomatic iron ores containing sulfide mineralization (Mel'nik and Yaroshchuk, 1966) which occur in this same Volodarsk district gave $\delta^{18}O$ values of +13.3‰ and +20.1‰ for the magnetite.

Interpretation of the isotopic data

The data obtained indicate that the oxygen of the original primary iron-rich rocks is retained in the ore minerals of metamorphic BIF. Variations in the $\delta^{18}O$ of magnetite even over relatively short distances—from tens to a few meters— proved to be very substantial: 5–12‰ at Krivoy Rog and up to 4.6‰ in the KMA. More detailed study showed that even in adjacent bands, at a distance of a few centimeters, the oxygen isotopic ratios of magnetites differ by 1–2‰. Thus, in one hand specimen where bands (0.5–3.0 cm thick) of low-grade siderite are interbedded with bands of dark biotite-chlorite schist with disseminated carbonate and magnetite, the $\delta^{18}O$ of the magnetite in a quartz-siderite band is about +9.7‰ and in the adjacent schist, +8.4‰. The $\delta^{18}O$ of magnetite in bands 1–2 cm thick from the Republic Mine, Michigan (Anderson, 1967) was +1.7‰ in a hematite-magnetite, +1.7‰ in a quartz, and +2.3‰ in a magnetite band.

According to the data of Dontsova (1970), such variations must indicate the absence of equilibrium oxygen exchange during metamorphism. For the rocks of the Krivoy Rog group, characterized by frequent alternation of iron-rich rocks and schists, the latter could have been barriers preventing circulation of solutions and isotope exchange.

The wide range of $\delta^{18}O$ of magnetite and hematite from Krivoy Rog and the KMA cannot be explained if the hypothesis of magmatic or metasomatic replacement is adopted (Knyazev et al., 1969), or that of "hydrothermal" deposition (Tochilin, 1969; Gusel'nikov, 1972). Deposition of the ore material in the form of oxides and subsequent metamorphism of the oxide sediment likewise does not lead to substantial fractionation of oxygen isotopes in the magnetites of iron-formations, inasmuch as differences in the $\delta^{18}O$ of hematite and magnetite are insignificant at any temperatures.

On the basis of the isotopic data obtained, the hypothesis of chemogenic deposition of cherty iron sediments of complex composition in sea basins seems to be most applicable. Only by such a process was it possible for compounds to form which already differed sharply in oxygen isotopic composition—oxides (goethite or hematite, magnetite) and carbonates (siderite, sideroplesite)—in the original iron sediment.

The oxygen isotopic composition in the primary sedimentary or diagenetic hematite and magnetite would not be significantly different and would be close to the oxygen isotopic composition of the water. Carbonate oxygen is substantially enriched in ^{18}O.

The actual isotopic ratios in associated hematite and magnetite (see Fig. 96) and siderite (see Fig. 94) are in good agreement with this hypothesis. In reality, the $\delta^{18}O$ of hematite and magnetite are similar and vary within +1.3 to +6.8‰, and siderite is enriched in the heavy isotope—$\delta^{18}O$ is up to

+12.9‰. The usually established variations in absolute values of $\delta^{18}O$ of coexisting hematite and magnetite could be caused by various factors, the most probable of which is the extent of equalization (depth of isotopic exchange) of the oxygen of iron oxides and the oxygen of carbonates and quartz in metamorphism. Apparently this also explains the variations in $\Delta\delta^{18}O$ between iron oxides.

The high $\delta^{18}O$ values of magnetite from siderite rocks are explained by metamorphic formation of the magnetite in these rocks by decomposition of carbonate. The physicochemical particulars of this process were examined by us earlier on the basis of thermodynamic calculations and experiments. Isotopic fractionation between the magnetite being formed and the fluid phase in the conditions of metamorphism is not known, but according to the data of Dontsova (1964), Anderson (1967), and others it can be presumed that magnetite formed in reactions with carbonates or with CO_2 will contain oxygen enriched in ^{18}O. Possibly enrichment of magnetite in the heavy isotope was furthered by a reducing environment of mineralization, and by the presence of sulfides and sometimes of an excess of free carbon (graphitite), which acted as an oxygen buffer in metamorphic processes.

In siderite-magnetite, magnetite, and silicate-magnetite rocks (groups 3 and 4) with intermediate $\delta^{18}O$ values of magnetite (+6.4 to +10.9‰), complex reactions in which primary oxides, carbonates, and silicates take part could have occurred during metamorphism. However, the reaction of siderite with hematite with liberation of CO_2, typical of the greenschist facies, apparently was of principal importance in the redistribution of oxygen isotopes. As a result of this process the oxygen of the magnetite that was formed acquired an "intermediate" composition between the original hematite with low $\delta^{18}O$ and carbonate with high $\delta^{18}O$; a high content of the heavy isotope in magnetite may indicate that "siderite" oxygen predominates in it.

In the formation of magnetite as a result of reduction of hematite during diagenesis or metamorphism by carbon or organic remains present in the sediment, some impoverishment in ^{18}G would be expected in the magnetite obtained compared to the original hematite, inasmuch as at low and moderate temperatures the liberated CO_2 is enriched in the heavy oxygen isotope. In the Krivoy Rog rocks not one case has been established where the ^{18}O content of the magnetite is lower than that of the hematite, which indicates that reduction of that type is of little importance.

In other districts, for instance in the KMA, reduction of oxide sediment to magnetite and siderite by organic matter probably was of great ore-forming importance (Plaksenko, 1966). Isotopic data on this district are still insufficient for definite conclusions: in two out of three analyses (see Fig. 96) the $\delta^{18}O$ of coexisting oxides are similar, and in one the magnetite oxygen is actually "lighter" than that of hematite. Depletion of magnetite in ^{18}O compared to hematite is typical of the Lake Superior district, while in the BIF of the Hamersley district of Western Australia the magnetite is substantially enriched in the heavy oxygen isotope (by more than 6‰) compared to hematite.

Relationship of the occurrence of types of iron minerals to the content of carbon and some other elements

Investigations of the geochemistry of carbon in recent and ancient sedimentary rocks (Strakhov, 1960; Stashchuk et al., 1964) have demonstrated the special role of this element in the formation of iron minerals. The data obtained by S. Sidorenko (1971) for Precambrian sequences and especially by Plaksenko (1966) for the BIF of the KMA stimulated study of the distribution of carbon and some other elements in the iron-rich rocks, schists, and ore minerals of the Krivoy Rog deposits to a substantial extent.

Figure 97 shows the distribution of C_{free}, CO_2, Al_2O_3, TiO_2, S_{sulf}, and also Fe_{total} and the $Fe^{2+}:Fe^{3+}$ ratio in the most typical clastic (schists) and chemogenic (BIF) rocks of the Middle suite of the Krivoy Rog group. These

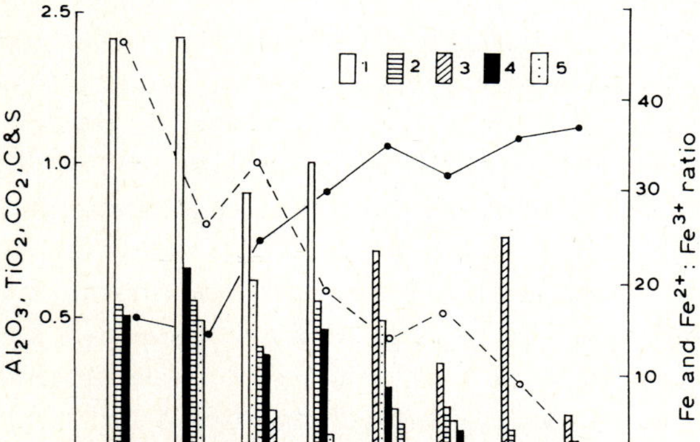

Fig. 97. Content of some components of schists and iron-formations of the Middle suite of the Krivoy Rog group: *a* = various schists, containing no oxides, carbonates, or sulfides of iron; *b* = black slates containing sulfide impurities; *c* = schists with disseminated magnetite and sulfides; *d* = schists with disseminated magnetite (large crystals—porphyroblasts); *e* = magnetite-silicate-carbonate rocks with minor sulfides; *f* = silicate and silicate-carbonate iron-rich rocks with magnetite; *g* = siderite and siderite-magnetite rocks; *h* = hematite and hematite-magnetite rocks; (iron-mica jaspilites).
1 = $Al_2O_3(\times 10^{-1})$; *2* = TiO_2; *3* = $CO_2(\times 10^{-1})$; *4* = C; *5* = $S_{sulfide}$; solid line = total iron content (Fe_{met}); dashed line = ratio of $Fe^{2+}:Fe^{3+}(\times 10)$. Plotted from data of 96 chemical analyses.

two groups of rocks are fairly clearly differentiated on the basis of content of the typical detrital elements Al_2O_3 and TiO_2, and also of free carbon. The magnetite-silicate-carbonate rocks with an admixture of sulfides occupy an intermediate position.

It is known that the iron mineral type is related to the content of C_{free} or "organic matter" (Strakhov, 1960). However, in the rocks studied this relationship is manifested only roughly.

In schists with a high content of C_{free} no hematite—a common mineral in iron-formations devoid of free carbon—is found. The other iron minerals (magnetite, siderite, silicates, sulfides) occur both in various schists and in the iron-rich rocks. A relationship between the presence of certain minerals and C_{free} content is not always manifested, not by far. For instance, in the dark chlorite-biotite-quartz schists typical of the Middle suite of the Krivoy Rog group ($C_{free} \sim 0.5\%$) there frequently is no siderite, and the ore mineral is magnetite, forming typical large idiomorphic crystals (porphyroblasts). At the same time siderite predominates in the widespread carbonate BIF with low carbon content ($C_{free} \sim 0.03\%$).

On the average, the carbon content is at a level of 0.5% for the group of schists studied (in the schist horizons of the middle part of the Saksagan district, 0.38% C_{free}). However, as is seen from Fig. 97, variations in C_{free} are not reflected directly in the appearance of sulfides or carbonates in the schists or in the $Fe^{2+}:Fe^{3+}$ ratio. It can be presumed that the formation of this or that iron mineral (besides the rock-forming silicates) in the schists was not determined by the amount of C_{free} observed now, but by other factors—the presence of reactive forms of sulfur, concentrations of dissolved carbon dioxide, physical properties of the sediments, and kinetic effects (rate of percolation and diffusion), and possibly by microbiological activity. As a result, the pH and *Eh* values and concentrations of ions which controlled mineral formation were determined by the complex interrelationships of the independent factors enumerated and could have varied considerably even over short distances. This could have occurred only after deposition of the sediments. In reducing conditions iron hydroxides, if they were present in the original sediments, could not be retained and underwent appropriate diagenetic transformations (as in recent iron-rich sediments). It is not ruled out that the magnetite porphyroblasts in the schists represent iron hydroxide concretions that were reduced during diagenesis and recrystallized during metamorphism.

The insignificant content of siderite in the schists of the clastic layers and bands themselves is noteworthy. It is believed that in the case of a high carbon content this mineral would arise in the series hematite (goethite) → magnetite → silicate → carbonate → sulfide. The omission of carbonate from that series (in the presence of silicates and sulfides) indicates the low probability that $FeCO_3$ is formed exclusively by reduction of the primary hydroxide $Fe(OH)_3$ by organic matter (carbon) in diagenesis or metamorphism.

Study of the second group of rocks—the BIF proper—confirms the suggestions made. Regardless of their composition, purely chemogenic sediments are practically devoid of free carbon, traces of which appear only after the addition of a clastic component: the correlation of the content of C_{free} with those of Al_2O_3 and TiO_2 is quite clear (see Fig. 97e–h). Carbonate and magnetite-carbonate rocks are little different from the hematite and hematite-magnetite varieties in content of C_{free} (on the average 0.03 and 0.01%, respectively). There is no clear correlation between C_{free} content and carbonates—pure siderite low-grade rocks ($CO_2 > 10\%$ with $C_{free} < 0.02\%$) and hematite-magnetite rocks with impurities of silicates ($CO_2 < 1$–2% with $C_{free} = 0.03$–0.05%) are encountered. As the content of Al_2O_3 and TiO_2 increases, a transitional group of chemogenic-clastic rocks of very variegated composition arises—magnetite-silicate-carbonate, often with an admixture of sulfides; the C_{free} content on them also is intermediate (0.2–0.3%).

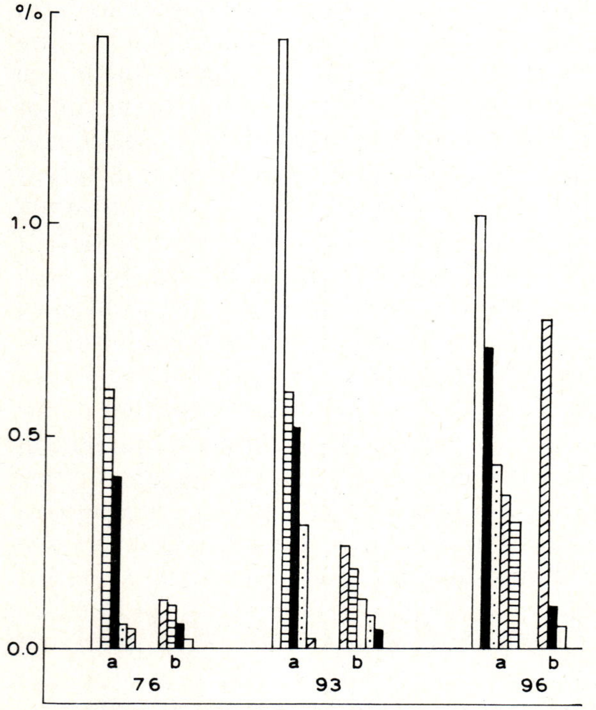

Fig. 98. Distribution of components in interbedded schist (clastic) and cherty iron (chemogenic) layers: a = schists with disseminated idiomorphic magnetite crystals (porphyroblasts); b = somewhat sideritic bands; figures = sample numbers. For symbols, see Fig. 97.

Detailed study of the mixed rocks, including layer-by-layer chemical analyses, has shown that in the chemogenic quartz-carbonate bands siderite usually predominates over magnetite and silicates ($C_{free} = 0.04–0.10\%$), but magnetite crystals occur in the adjacent schist bands notwithstanding the high carbon content (C_{free} up to 0.7%) (Fig. 98).

From the data given it can be concluded that the formation of a substantial amount of magnetite and especially of siderite in chemogenic iron sediments via diagenetic reduction of the hydroxide $Fe(OH)_3$ by buried organic matter is not very likely. In typical iron formations we find no appreciable traces of residual carbon retained after diagenesis and metamorphism. If it is assumed that this carbon existed originally and subsequently was completely used up in the reduction of iron oxides (Fe_2O_3 and Fe_3O_4), it then remains unclear why carbon was retained in the schist bands although the magnetite was not reduced. The hypothesis of a different activity (reactivity) of carbon in schists and quartz-carbonate rocks situated next to one another seems much too artificial.

Study of the composition of monomineralic fractions of the ore minerals showed that magnetite contains higher amounts of MgO and MnO than hematite. These differences are not great, but the appearance of some excess magnesium and manganese in the composition of the magnetite in the case where it formed by reduction of hematite requires explanation. An analogous conclusion can be drawn from the data on the distribution of certain "trace" elements, in particular Ni, Ti and especially Ge (Grigor'yev and Zelenov, 1965). We note that both magnetite and hematite contain practically no free carbon.

The geochemical data given make no claim to completeness and do not give the whole idea of the distribution of elements between all the rock types and between the various coexisting minerals. However, the basic conclusion that the ore minerals, primarily magnetite, are of complex polygenetic origin is consistent with the physicochemical and isotopic data.

As a result, generalized genetic groups of ore minerals can be designated.

(1) *Primary sedimentary: hematite (goethite) and magnetite* in hematite and hematite-magnetite jaspilites; *siderite* in siderite and siderite-magnetite rocks, barren and low-grade BIF bands in schist sequences. Only amorphous $Fe(OH)_3$ hydroxides and $FeCO_{3d}$ carbonates should be taken as strictly primary-sedimentary. Primary dispersed magnetite Fe_3O_{4d} probably arose when flocculent hydroxide sediments passed through reducing zones with low *Eh* values, or in unconsolidated sediment.

(2) *Diagenetic: magnetite* in the schists of the BIF and mixed chemogenic-clastic rocks, in part *siderite* in schists; *sulfides* in schists and carbonate-silicate BIF. The formation of diagenetic ore minerals was accompanied by reduction of iron and the appearance of silicates such as greenalite or even of

complex ferro-ferri-hydrochlorites. Considerable fluctuations in *Eh* and pH because of interaction of internal (carbon) and external (atmosphere and ocean) buffer systems produced the diversity of mineral forms and frequent changes in paragenetic mineral associations. In purely chemogenic iron sediments the diagenetic transformations were not accompanied by substantial changes in the oxidation-reduction state due to the great buffering capacity of the sediment itself and the low carbon (organic matter) content.

(3) *Metamorphic: magnetite* in magnetite and carbonate rocks. The main non-ore minerals in the silicate rocks belong to this group: minnesotaite, grunerite (cummingtonite), pyroxenes, and olivines; in part they are due to reduction of magnetite by excess free carbon at fairly high temperatures, but mainly they represent the products of metamorphic dehydration and decarbonation of silicate and carbonate sediments.

The derivatives of various secondary stages of mineral formation are not included here: hematites of specular type in veins of alpine type and in high-grade ores, siderite veins; martites, goethites, and carbonates from the oxidation zone.

Conclusion

On the basis of comparison and critical examination of the geologic, geochemical, isotopic, experimental, and theoretical thermodynamic data that have been given, general conclusions can be drawn.

(1) Typical BIF are metamorphosed chemogenic cherty iron sediments, deposition of which was typical only of the Precambrian and ceased in subsequent geologic epochs. The formation of these rocks reflects a certain stage in the irreversible process of evolution of the Earth's crust, atmosphere, hydrosphere and biosphere. The period of intensive deposition apparently was preceded by a long period of accumulation of dissolved iron and silica in the waters of the original sedimentary basins.

(2) For most BIF of the world a spatial and genetic relationship to the processes of volcanism is postulated. The nature of that relationship may be different depending on the proximity of the volcanic sources to the region of deposition of the sediments. The oldest iron formations, closely associated with volcanic rocks, were deposited from supersaturated solutions formed when acid thermal waters of submarine sulfataric springs mixed with ocean waters saturated with free carbon dioxide but devoid of oxygen. The main reasons for chemical deposition (geochemical barriers) were gradients of the pH (from 0–1 before to 5–6 after mixing), temperature (from 150–100 to 5–25°C) and, possibly, of concentration of carbon dioxide (or P_{CO_2}) and electrolytes. The iron was deposited directly from ionic solutions in the form of carbonates $FeCO_3$ or $(Fe,Mg)CO_3$ due to an increase in pH at constant P_{CO_2} (buffer system—atmosphere + ocean). Silica first was converted in part from ionic to colloidal (geochemical barrier—temperature gradient) and then the colloids coagulated when the pH was increased to 5–6, or under the influence of electrolytes (Na^+, Fe^{2+}, Mg^{2+}). The formation of the banding of iron-formations is explained either by cyclicity of volcanic activity and different rate of deposition of iron and silica (kinetic and sedimentational effects), or by sinusoidal variations of the pH due to sorption and desorption of hydrogen ions by dispersed particles, or by constant deposition of silica and periodic deposition of iron. In the last case the periodicity can be explained by fluctuations in P_{CO_2} as a result of complex conjugate processes, including oxidation of methane and free carbon due to the activity of the first organisms. It is not ruled out that very simple organisms interacted directly with Fe^{2+} ions, causing oxidation and deposition of iron in the form of hydroxide or hydromagnetite. Thus, iron cherts of this type were de-

posited in the immediate vicinity of submarine volcanic sources and had a chert-carbonate composition. Magnetite and iron silicates arose chiefly in diagenesis and metamorphism, the quantitative relationships between these minerals being controlled not only by temperature and pressure, but also by the regime of water and carbon dioxide in the fluids.

(3) Maximum accumulation of iron cherts occurred in the Proterozoic (Middle Precambrian). The relationship of the formations of this group to volcanism is believed to be remote, a volcanogenic source for the silica and most of the iron (fumarole fields) and a normal sedimentary mode of deposition of the iron cherts in relatively shallow marine water bodies being presumed. The transport of iron and silica to considerable distances and subsequent deposition from low-temperature aqueous solutions evidently were governed by various factors not directly related to each other. Only a favorable conjunction of many factors in limited parts of the Earth's surface and in certain periods of geologic evolution led to profound differentiation and chemical precipitation of cherty iron sediments uncontaminated by clastic material. Silica was transported in ionic-colloidal solutions, the colloidal component of which was deposited when the pH and electrolyte concentration increased. Coagulation began as soon as acid thermal waters mixed with surface river waters, but maximum chemogenic deposition occurred at some distance from the shore and did not coincide spatially with the maximum accumulation of clastic argillaceous sediments. Deposition of iron began after a period of accumulation, but the main factor was spasmodic variation in the oxidizing-reducing environment (*Eh* gradient).

(4) On the basis of thermodynamic analysis of the evolution of the ancient atmosphere and hydrosphere, the sequence of oxidation reactions occurring when oxygen was introduced and P_{O_2} increased is ascertained:

$$NH_3(N_2) \rightarrow CH_4(C) \rightarrow C(CO_2) \rightarrow S(SO_4^{2-}) \rightarrow Fe^{2+}(Fe^{3+})$$

In the early stages of development of the Earth's crust the only source of free oxygen was the processes of photochemical decomposition of water vapor in the upper layers of the atmosphere with simultaneous dissipation of hydrogen. However, calculations show that the amount of oxygen needed to oxidize the original atmosphere is such that only biochemical processes could have provided it. Very simple plants (of the type of the bluegreen algae), the remains of which have also been found in the metamorphosed rocks of the oldest BIF, apparently had already appeared in the Early Archean, with its oxygen-free atmosphere containing ammonia and methane. Thus large-scale deposition of iron in the form of hydroxide when there was a change in *Eh* did not become possible until after ammonia, carbon, and sulfur were oxidized, when nitrogen predominated in the atmosphere, with a minor amount of carbon dioxide. Free oxygen in equilibrium with the atmosphere did not appear until after complete oxidation of ferrous iron compounds in the hydrosphere and on land, in the interval of 1.7–1.2 b.y. ago. Before the beginning of deposition of iron hydroxides or simultaneously with it, buffered carbonate and carbonate-silicate equilibria controlling pH and possibly P_{CO_2} came to play a greater and greater role in the primitive ocean. The result of the appearance of internal buffer systems in the sedimentary basins was a decrease in P_{CO_2} and an increase in pH. When iron-magnesian carbonates in equilibrium with silicates were deposited, the pH values were within 6 to 7 and P_{CO_2} within 0.001 to 0.1 bar. For amorphous sediments the corresponding pH values were somewhat lower and P_{CO_2} higher. Data on the ratios of the stable isotopes of oxygen, carbon, and sulfur in Precambrian sedimentary rocks are consistent with this scheme of evolution.

(5) The evolution of the carbon dioxide atmosphere and deposition of iron oxide apparently were directly or indirectly interrelated. There is reason to suppose that substantial colonies of phytoplankton were developed in limited parts of the ancient water bodies, at a certain depth controlled by the thickness of the water layer shielding the organisms from ultraviolet radiation, and at an optimal distance from the shoreline. An indirect relationship of "oases" of the first life to the regions of volcanic activity, both terrestrial and submarine, from which the necessary components—in particular silica—were introduced is not ruled out. Periodic "flare-ups" of intensive "blooming" of phytoplankton in Precambrian conditions did not lead to release of the oxygen produced into the atmosphere, but rather to oxidation of Fe^{2+} to Fe^{3+} directly in the water layer and subsequent deposition of practically insoluble hydroxides; in this case the formation of primary magnetite was of no great importance. Cyclic (seasonal) deposition of iron compounds in relatively shallow basins against a background of continuous deposition of amorphous silica explains the nature of the banding and the diversity of textures, structures, and mineral associations of the BIF. Periodicity might also have been manifested in the deposition of silica, caused by fluctuations in the temperature of the basin (microrhythms) and intensity of fumarolic activity (macrorhythms).

(6) Diagenetic processes played a large part in the formation of the original mineral parageneses of iron cherts and in the stabilization of the "oxidation-reduction capacity". Quantitative energetic analysis made for the first time for the transformation of amorphous and highly dispersed sediments into crystalline minerals showed that these processes are accompanied by substantial changes in the isobaric-isothermal potential (up to 50 cal/g) and decrease in solubility (by 2–3 orders). Freshly precipitated iron hydroxide containing free oxygen, coming from the near-surface parts of the basin (zone of development of phytoplankton) into oxygen-free deep zones, was

partly or completely reduced, forming magnetite, carbonates (siderite, sideroplesite), silicates (minnesotaite, chlorites), or sulfides depending on the *Eh*, presence of reactive forms of silica, carbon, and sulfur, and ratios of the rates of the respective reactions. The oxidizing-reducing environment in the chemogenic iron-chert sediments was controlled by conjugate reactions in the atmosphere and hydrosphere (external buffer), and in clastic and clastic-chemogenic sediments also by buried organic remains, mainly dead phytoplankton (internal buffer). However, the primary-sedimentary association iron hydroxides + dispersed siderite + amorphous silica was mainly retained, fostered by the high partial pressures of carbon dioxide and relatively low concentrations of sulfur characteristic of the Precambrian. Diagenesis of such sediments amounts to crystallization, decrease in solubility, and formation of a thermodynamically stable mineral association—goethite + siderite + quartz. In clastic sediments iron hydroxides were not retained, but were reduced to magnetite and silicates.

(7) After diagenesis, cherty iron-formations were characterized by frequent variations in mineral associations of different degree of oxidation and different oxidation-reduction capacity, variable water and carbon dioxide contents, and uneven distribution of the respective minerals (oxides, carbonates, and silicates). Clastic, chiefly pelitic, rocks and also various volcanics are encountered in the make-up of the formations in question, along with the typical chemogenic iron-rich and carbonate rocks. On the whole, both macro- and micro-anisotropy of composition and properties are typical of the rocks of the BIF. The amount of anisotropy decreases as metamorphic reworking is intensified. Some theoretical problems, including improvement of methods of thermodynamic calculations and plotting of diagrams of mineral equilibria at high T and P on the basis of a new system of consistent constants, were specially worked out for physicochemical analysis of the metamorphism of complex heterogeneous piles; the effect of unequal pressures on equilibria in which a fluid phase takes part was examined; the particulars of metamorphism in systems closed to water were investigated; the buffering properties of oxide, silicate, and carbonate mineral associations as possible regulators of P_{O_2}, P_{H_2O}, and P_{CO_2} in progressive regional metamorphism were discussed. A new approach to estimating and predicting the properties of water and carbon dioxide at high pressures made it possible to find a better-grounded explanation for the stability of carbonate associations.

(8) The conditions of metamorphism of BIF of different original composition are examined on the basis of thermodynamic calculations and some experimental data. In oxide iron-formations the conversion of goethite to hematite was accomplished even before the beginning of regional metamorphism; the *P-T* curve of the dehydration reaction falls within 110° and

160°C for pressure up to 5 kbar. Metamorphic transformation of hematite into magnetite by thermal dissociation and removal of free oxygen is not very likely even at high temperatures, either in purely aqueous or in carbon dioxide fluids. In the absence of a reducing agent, hematite should be taken to be a thermodynamically stable mineral in all metamorphic facies up to the granulite facies. The hematite + magnetite association is a buffer regulating the fugacity of oxygen. In silicate rocks the lowest-temperature transformation is that of the greenalite + hematite association into magnetite (180–200°C). Dehydration of pure greenalite begins at somewhat higher temperatures (200–220°C), where minnesotaite becomes stable. In the 200–300°C range the minnesotaite + magnetite association finally is formed. Transformation of minnesotaite into grunerite (300–330°C) indicates the beginning of metamorphism in greenschist facies conditions. At 640–690°C and corresponding pressure, grunerite decomposes into the association of fayalite plus quartz, which is stable in granulite-facies conditions. The pyroxene + olivine + cummingtonite + quartz association in highly metamorphosed magnesian-iron rocks fixes $T = 700–720°C$, which depends little on pressure and iron content of the silicates. In carbonate rocks the hematite + siderite association is transformed into magnetite at $T = 300–400°C$, at 400–550°C pure $FeCO_3$ decomposes into the stable magnetite-plus-graphite association. At a temperature above 500–600°C the latter association also becomes unstable, transforming into stable fayalite with an excess of silica. Equilibrium conversion of siderite into fayalite at low pressures is impossible; the stability fields of these minerals are separated by the stability field of magnetite plus graphite, which acts as an oxygen buffer. In anhydrous carbonate rocks, invariant equilibrium of siderite + fayalite + magnetite + quartz + graphite + fluid exists, fixing $T = 600°C$ and $P_f \approx P_{CO_2} = 6–8$ kbar. In the metamorphism of silicate-carbonate rocks, only minnesotaite and siderite are in equilibrium with graphite before the beginning of regional metamorphism, but above 300–330°C the common association of grunerite + siderite is formed (in actual rocks, cummingtonite + sideroplesite). In the stability field of these minerals the appearance of grunerite depends on the ratio of CO_2 and H_2O in the fluid rather than on temperature or pressure. The main reason for the formation of this amphibole is a substantial amount of water, which is typical of shaly and shaly-cherty strata. The equilibrium of siderite + grunerite + magnetite + (graphite + quartz) in water-carbon dioxide fluids is monovariant and corresponds to a temperature of the order of 400–500°C. At higher temperatures in the presence of a fluid and an excess of graphite and quartz, the bivariant equilibrium of grunerite + magnetite occurs, which above 450–510°C is replaced by the equilibrium of grunerite + fayalite, and above 600–650°C by fayalite alone.

(9) Thermodynamic analysis of mineral equilibria in iron cherts has shown

that all the rock types are characterized by certain common features—the presence of nonequilibrium associations at low ranks of metamorphism, variable composition of the fluid even within individual layers and bands, the existence of "dry" sectors or zones, "closed" to fluids in general or only to individual volatile components (H_2O and CO_2).

(10) Diagrams of mineral equilibria in metamorphosed iron cherts made it possible to obtain sufficiently complete information on the evolution of composition of the fluid in progressive regional metamorphism. It has been established that in the rocks studied, the fluid phase consists of H_2O and CO_2, not counting the inert gases, at any realistic temperatures and pressures. The content of other volatiles is negligible at low temperatures (200–500°C), but at high temperatures (>700°C) it may reach 5–10%. The systematic increase in the content of reducing agents (CH_4, H_2, CO) in silicate-carbonate iron-formations with increasing temperature is noteworthy: at 800–850°C the methane content in aqueous fluids (85–95% H_2O) may reach 5–10%, and of hydrogen 1–3%, while in carbonic acid fluids (90–95% CO_2) carbon monoxide is sharply predominant over other volatiles (1–3% CO). Acceleration of diffusion of the reducing agent with increasing temperature and increasing gradients of concentration (fugacity) leads to the formation of more uniform mineral associations in place of formerly extremely variegated banded sequences. In particular, the frequently observed disappearance of hematite in oxide rocks metamorphosed in amphibolite and granulite facies conditions may be due to reduction by volatiles introduced from silicate-carbonate rocks, mainly highly mobile hydrogen.

(11) Iron-formations are widespread low-grade ores whose value is determined mainly by the content of magnetite, which is easily extracted by magnetic separation. Isotopic and geochemical investigations have made it possible to presume that the formation of magnetite occurred at various stages of formation of the sediment and during its subsequent diagenesis and metamorphism. In oxide iron-formations dispersed magnetite probably arose in the original sediment during its diagenesis, or even when flocculent hydroxide particles passed through reducing zones with low *Eh* values. Periodic fluctuations in *Eh*, caused by external chemical and biochemical processes, brought about deposition of iron in the form of goethite or magnetite. In carbonate-oxide rocks the formation of most of the magnetite only began upon metamorphism as a result of reduction of hematite by siderite in greenschist facies conditions, and also upon thermal dissociation of siderite under conditions grading into the amphibolite facies. In silicate BIF magnetite arose as a stable mineral in low-temperature (180–200°C) metamorphism of hematite + greenalite associations. Thus, low- and medium -temperature metamorphism helped to enhance the quality of the iron ores to a considerable extent. However, introduction of water during these stages

of metamorphism played a negative part, leading to formation of silicates of grunerite type instead of magnetite. In amphibolite and granulite facies conditions an excess of graphite also leads to the formation of silicates—grunerite or fayalite—at the expense of magnetite, contributing to some impoverishment of the ores. But the introduction of a substantial amount of pure water, and in isolated cases of carbon dioxide, could lead to the opposite process—oxidation of silicates to magnetite.

Appendix

System of consistent thermodynamic constants of minerals, sediments and ions

At the present time, thermodynamic methods are finding wider and wider application in ascertaining the physicochemical conditions under which natural processes take place. However, the use of thermodynamic analysis in geology has been considerably restricted due to the lack of reliable and sufficiently precise thermodynamic constants for many of the rock-forming minerals. Even new summaries (Karpov et al., 1968; Robie and Waldbaum, 1968; Wagman et al., 1968; Naumov et al., 1971) contain hardly any information on the thermodynamic properties of most of the silicates, including such important groups of minerals as the amphiboles, micas, chlorites, etc. Many constants have been insufficiently precisely determined, and calculations lead to inadmissible discrepancies with respect not only to experimental, but also to petrological data. Holland (1965) reports that the tabulated thermodynamic data on minerals, in particular on siderite, are erroneous. The uncertainties in ΔG^0 values obtained by thermochemical methods are usually an order of magnitude higher than needed to determine equilibrium T with a precision of $\pm 10°C$. At the same time, very precise thermodynamic constants of minerals can be obtained from reliable experimental phase diagrams. Therefore in some works, experimental data are given along with thermodynamic calculations, and the experimental data or even the thermodynamic constants of individual minerals are calculated and corrected on the basis of the experimental results (Marakushev, 1968; Langmuir, 1969; Karpov et al., 1971). It is obvious that agreement between thermodynamic calculations and experimental determinations of the equilibrium parameters will make it possible to greatly broaden the possibilities of physicochemical analysis of mineral formation in a wide range of T and P.

We have worked out a sufficiently precise and reliable system of consistent thermodynamic constants (Mel'nik, 1972) especially for analysis of the conditions of formation of iron ores. In this work, in addition to the constants of crystalline minerals, data on the stability and thermodynamic properties of the original finely dispersed amorphous or cryptocrystalline iron, magnesium, and silica sediments were systematized for the first time. Such sediments are metastable solid phases which in nature are converted into stable crystalline minerals during diagenesis and low-rank metamorphism. Later, the constants of some iron minerals were corrected (Mel'nik and Radchuk, 1977b), in connection with which the respective grounds for choosing the recommended values are given here. In addition, the thermodynamic constants of such important primary iron silicates as minnesotaite and greenalite were also calculated. Inasmuch as the constants of the magnesium minerals and sediments varied insignificantly, only the tabulated values of ΔG_f^0, the substantiation of which can be found in the principal work (Mel'nik, 1972b), are given here.

In this work, the main criterion determining the suitability of thermodynamic constants for analysis of mineral equilibria is correspondence of the theoretical calculations with experimental investigations. Only in the case of such correspondence are the tabulated constants taken to be consistent and recommended for use. If essential discrepancies between calculation and experiment are found, for instance in determination of the position of the *P-T* curves of heterogeneous equilibria or the solubility of stable minerals, then preference usually is given to reliable experimental data and the recommended thermodynamic constants are correspondingly corrected. If thermodynamic determinations for any compound are incomplete or altogether lacking, then ΔG_f^0 and other constants were calculated directly from the experimental data. Therefore the reliability of the experimental investigations of mineral equilibria, especially at high temperatures and pressures, is a fundamental problem in basing the choice and calculation of the recommended constants. The use of new experimental methods (long-term experiments at high *P* and *T*, measurements of dissociation vapor pressure, measurement of e.m.f. in cells with solid electrolyte at high *T*, etc.) substantially broadens the possibilities of direct determination of the equilibrium parameters and enhances the reliability of the experimental data.

From the scheme of reactions given in Fig. 99 it follows that in calculating the thermodynamic constants (Tables XII and XIII) of any intermediate or final compound, the recommended constants of the initial compounds and the standard (tabulated) constants of reference substances were used. For instance, in the case of fayalite—Fe_2SiO_4—the accepted constants of the

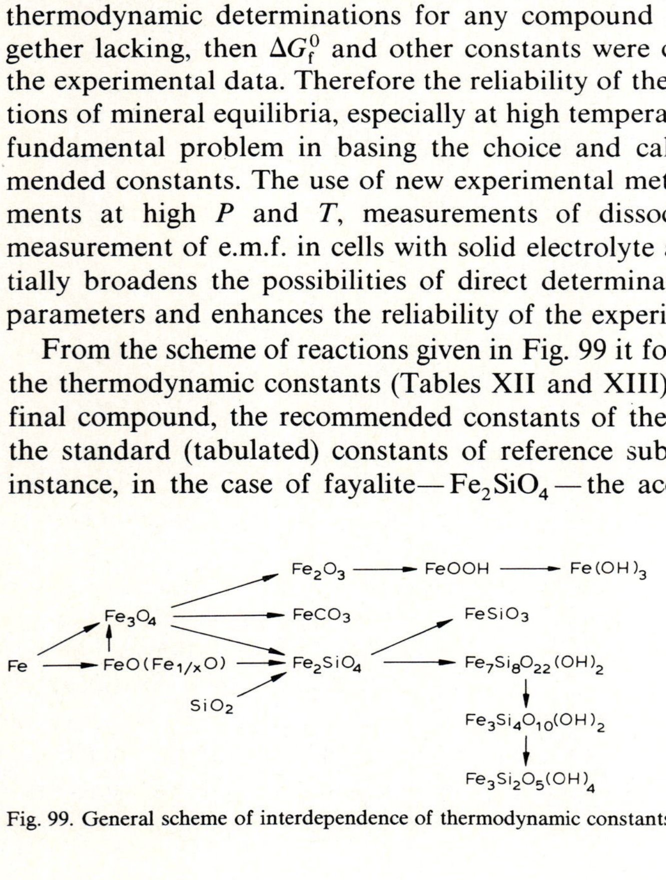

Fig. 99. General scheme of interdependence of thermodynamic constants of the iron minerals.

TABLE XII

Isobaric-isothermic potentials of formation ($-\Delta G^0_T$, kcal/mol, from elements) of some minerals of iron-formations

Mineral	Temperature (°K)										
	298	400	500	600	700	800	900	1000	1100	1200	Ref.
Troilite	24.22	24.26	24.35	24.50	24.66	26.00	24.63	23.25	21.84	20.42	4
Pyrite	38.30	37.31	35.99	34.50	32.91	33.66	28.97	24.30	–	–	4
Wüstite, stoichiometric	58.10	56.42	54.83	53.28	51.75	50.25	48.75	47.23	45.72	44.20	1, 2
Wüstite, nonstoichiometric	58.11	56.44	54.85	53.32	51.80	50.31	48.82	47.32	45.80	44.29	1, 2
Magnetite	240.61	232.32	224.38	216.63	209.09	201.75	194.63	187.60	180.55	173.50	1, 2
Hematite	176.52	169.92	163.56	157.34	151.24	145.24	139.30	133.46	127.63	121.80	1, 2
Goethite	116.76	111.00	105.37	99.75	–	–	–	–	–	–	1, 2
Siderite	162.02	156.08	150.32	144.62	139.00	133.47	127.89	–	–	–	1, 2
Fayalite	329.57	321.08	313.66	305.91	398.24	290.61	282.99	275.39	267.78	260.19	1, 2
Ferrosilite	266.67	260.19	254.28	248.22	242.21	236.25	230.31	224.40	218.51	212.63	1, 2
Grunerite	2135.4	2083.7	2033.0	1982.3	1931.6	1880.9	1830.2	1779.5	1728.8	1675.1	1, 2
Minnesotaite	1069.2	1040.8	1013.0	985.2	957.4	–	–	–	–	–	1
Greenalite	717.2	693.4	670.1	646.8	623.5	–	–	–	–	–	1
Periclase	136.09	133.45	130.87	128.29	125.73	123.19	120.64	117.92	115.15	112.39	4
Brucite	200.05	192.58	185.24	177.89	170.56	163.26	156.01	158.61	–	–	3
Magnesite	246.11	239.30	232.66	226.08	219.54	213.07	206.65	200.13	193.60	187.16	3
Forsterite	491.94	482.20	472.70	463.13	453.65	444.23	434.83	425.12	415.31	405.57	4
Enstatite	349.39	342.29	335.31	328.35	321.41	314.50	307.61	300.56	393.49	286.46	4
Anthophyllite	2717.87	2658.90	2601.16	2543.44	2485.95	2428.68	2371.47	2313.35	2254.89	–	3
Talc	1323.28	1292.13	1261.63	1231.17	1200.87	1170.71	1140.64	1100.00	1079.10	–	3
Serpentine	965.77	939.11	913.07	887.04	861.15	835.39	–	–	–	–	3

* 1 = this work; 2 = Mel'nik and Radchuk (1977b); 3 = Mel'nik (1972b); 4 = Robie and Waldbaum (1968).

TABLE XIII
Consistent thermodynamic constants at 298°K, 1 bar

Formula	Compound (name, state)	$\Delta G^0_{f\,298}$ (kcal/mol)	Source
Iron:			
FeS	Troilite (pyrrhotite)	−24.22	4
FeS	Mackinawite (synthetic)	−22.3	3
FeS	Freshly precipitated	−21.3	3
FeS_2	Pyrite	−38.30	4
FeS_2	Freshly precipitated	−34.5	3
Fe_3O_4	Greigite (synthetic)	−69.4	3
FeO	Wüstite, stoichiometric	−58.10	1, 2
$Fe_{0.947}O$	Wüstite, nonstoichiometric	−58.11	1, 2
Fe_3O_4	Magnetite	−240.61	1, 2
Fe_3O_4	Diagenetic	−237.5	1
Fe_3O_4	Freshly precipitated (hydromagnetite)	−234.7	1
Fe_2O_3	Hematite	−176.52	1, 2
$FeOOH$	Goethite	−116.76	1
$FeOOH$	Hydrogoethite	−115.9	1, 3
$Fe(OH)_3$	"Equilibrium" (inactive)	−171.4	1, 3
$Fe(OH)_3$	Amorphous (active)	−169.9	1, 3
$Fe(OH)_3$	Freshly precipitated	−168.9	1, 3
$Fe(OH)_2$	Amakinite	−117.3	1, 3
$Fe(OH)_2$	Freshly precipitated	−115.8	1, 3
$FeCO_3$	Siderite	−162.0	1, 2
$FeCO_3$	Spherulitic (diagenetic)	−161.0	1
$FeCO_3$	Freshly precipitated (finely divided)	−160.0	1
Fe_2SiO_4	Fayalite	−353.23	1, 2
$FeSiO_3$	Ferrosilite	−266.67	1, 2
$Fe_7Si_8O_{22}(OH)_2$	Grunerite	−2135.4	1, 2
$Fe_3Si_4O_{10}(OH)_2$	Minnesotaite	−1069.2	1
$Fe_3Si_2O_5(OH)_4$	Greenalite	−717.2	1
$Fe_3Si_2O_5(OH)_4$	Diagenetic	−715.0	1
$Fe_3Si_2O_5(OH)_4$	Freshly precipitated	−711.5	1
Fe^{2+}	Ion in aqueous solution	−21.54	3
$FeOH^+$	Ion in aqueous solution	−66.91	3
$Fe(OH)_2^0$	Ion in aqueous solution	−109.25	3
$Fe(OH)_3^-$	Ion in aqueous solution	−148.0	3
$Fe(OH)_4^{2-}$	Ion in aqueous solution	−185.0	3
Fe^{3+}	Ion in aqueous solution	−3.76	3
$FeOH^{2+}$	Ion in aqueous solution	−57.46	3
$Fe(OH)_2^+$	Ion in aqueous solution	−109.28	3
$Fe(OH)_3^0$	Ion in aqueous solution	−159.9	3
$Fe(OH)_4^-$	Ion in aqueous solution	−200.8	3
$Fe(OH)_2^{4+}$	Ion in aqueous solution	−117.0	3
$Fe_3(OH)_4^{5+}$	Ion in aqueous solution	−230.2	3
Magnesium:			
MgO	Periclase	−136.09	4
Mg(OH)2	Brucite	−200.05	3

TABLE XIII (continued)

Formula	Compound (name, state)	$\Delta G^0_{f\,298}$ (kcal/mol)	Source
$MgCO_3$	Magnesite	−246.11	4
$MgCO_3$	Freshly precipitated	−244.0	1, 3
Mg_2SiO_4	Forsterite	−491.94	4
$MgSiO_3$	Enstatite	−349.39	4
$Mg_7Si_8O_{22}(OH)_2$	Anthophyllite	−2716.97	3
$Mg_3Si_4O_{10}(OH)_2$	Talc	−1322.84	3
$Mg_3Si_4O_{10}(OH)_2$	Saponite	−1323.0	1, 3
$Mg_3Si_4O_{10}(OH)_2$	Freshly precipitated	−1313.0	1, 3
$Mg_3Si_2O_5(OH)_4$	Serpentine	−965.55	3
$Mg_3Si_2O_5(OH)_4$	Freshly precipitated	−959.0	1, 3
Mg^{2+}	Ion in aqueous solution	−108.8	3, 5
$MgOH^+$	Ion in aqueous solution	−152.5	3
$MgHCO_3^+$	Ion in aqueous solution	−251.1	3
Calcium:			
CaO	Crystalline	−144.35	4
$CaCO_3$	Calcite	−269.91	4
$CaCO_3$	Freshly precipitated	−268.0	1
$CaMg(CO_3)_2$	Dolomite	−518.73	4
$CaMg(CO_3)_2$	Freshly precipitated	−515.5	1
$CaSIO_3$	Wollastonite	−368.89	6
$Ca_6Si_6O_{17}(OH)_2$	Xonotlite	−2277.65	6
$Ca_6Si_6O_{17}(OH)_2$	Freshly precipitated	−2261.0	1
Ca^{2+}	Ion in aqueous solution	−132.10	5
$CaOH^+$	Ion in aqueous solution	−171.36	5
Silicon:			
SiO_2	α-quartz	−204.65	4
SiO_2	Chalcedony	−204.28	7
SiO_2	Amorphous, hydrogel	−202.89	7
$NaSi_7O_{13}(OH)_3 \cdot 3H_2O$	Magadiite	−1762.2	8
$NaSi_7O_{13}(OH)_3 \cdot 3H_2O$	Freshly precipitated	−1754.0	1
$H_4SiO_4^0$	Ion in aqueous solution	−312.57	3
$H_3SiO_4^-$	Ion in aqueous solution	−299.25	3
$H_2SiO_4^{2-}$	Ion in aqueous solution	−283.29	3
$HSiO_4^{3-}$	Ion in aqueous solution	−266.92	3
SiO_4^{4-}	Ion in aqueous solution	−250.55	3

1 = this work; 2 = Mel'nik and Radchuk (1977b); 3 = Mel'nik (1972b); 4 = Robie and Waldbaum (1968); 5 = Naumov et al. (1968); 6 = Mel'nik (1970); 7 = Walther and Helgeson (1977); 8 = Bricker (1969).

minerals (FeO and Fe_3O_4) and of reference substances (Fe, Si, SiO_2, O_2) were used. In turn, the values obtained for Fe_2SiO_4 were used to calculate the constants of $Fe_7Si_8O_{22}(OH)_2$, and so on. It should be kept in mind that if the constants of any compound, taken individually from the given system, are used in conjunction with the constants of other compounds from other sources, the consistency is disrupted. However, for the minerals occurring at the end of a chain it is easy to introduce corrections, if need be, for consistency with the whole system of constants.

Wüstite, FeO and $Fe_{1/x}O$

The structure of wüstite is defective due to an insufficiency of iron atoms and the formula is written $Fe_{1/x}O$, where the value $x = (O/Fe) > 1.0$ is different for the low-oxygen phase, in equilibrium with metallic iron, and the high-oxygen phase, in equilibrium with magnetite. The dependence of x on temperature and partial pressure of oxygen is known from experimental data (Darken and Gurry, 1945, 1946; Giddings and Gordon, 1973). At the triple point where eutectoid decomposition of wüstite into magnetite and iron occurs, $x = 1.5085$ at 562°C and 1 bar.

At pressures below 30 kbar FeO is not in equilibrium with respect to nonstoichiometric wüstite. However, in many experimental works and thermodynamic calculations the nonstoichiometry of wüstite is not taken into account, which leads to confusion and incompatibility of the results. As a result, $\Delta G^0_{f,T}$ values may differ by 2 to 3 kcal, as we showed earlier (Mel'nik, 1972b). Such large discrepancies preclude precise calculation of the parameters of wüstite stability.

Thus it is obvious that at atmospheric pressure the ΔG_T value of the reaction:

$$Fe_{1/x}O + (x-1)_{/x}Fe = FeO$$

should be positive. However, calculations from the tabulated constants (Robie and Waldbaum, 1968) give negative values of ΔG_T in the whole temperature range, as do those from the mutually consistent constants we recommended earlier (Mel'nik, 1972b). This inconsistency was eliminated in a new work (Mel'nik and Radchuk, 1977b), the main results of which are given below.

Stoichiometric wüstite, FeO

On the basis of a generalization of experimental data on the activity of oxygen in wüstite, Kurepin (1975) calculated for the reaction:

$$Fe + \tfrac{1}{2}O_2 = FeO$$

the value of the isobaric potential at high temperatures (T°K):

$$\Delta G_T = -62420 + 15.18 \cdot T, \text{ cal}$$

$\Delta G_{f,T}$ values for FeO in the 1000–1500°K range were calculated from this equation.

Calculation of $\Delta G_{f,T}$ from this equation for lower temperatures leads to an error of up to 0.2 kcal/mol, therefore from 1000 to 298°K a nonlinear extrapolation was made independently for the enthalpic ($\Delta H_{f,1000} = -62420$ cal) and entropic ($\Delta S_{f,1000} = +15.18$ cal·deg/mol) members using the tabulated values ($H_{1000} - H_{298}$) and ($S_{1000} - S_{298}$) of FeO, Fe, and O_2 according to Robie and Waldbaum (1968). As a result, $\Delta G_{f,T}$ values for FeO in the 298–1500°K range that were consistent with the experimental data were calculated.

Nonstoichiometric wüstite, $Fe_{1/x}O$

The density of wüstite has been determined fairly precisely, and for $Fe_{0.947}O$ is equal to 5.747 g/cm^3 (Robie et al., 1967), whence $V = 12.04$ cm^3/mol.

For the reaction:

$$Fe_{1/x}O + (x-1)_{/x}Fe = FeO$$

Kurepin (1975) calculated that the values of ΔG_T are positive, although very small (from +75 cal at 700°C to +105 cal at 1300°C), and at high temperatures can be expressed by the equation:

$$G_T + 19 + 0.056 \cdot T, \text{ cal}$$

Combining this equation with the equation $\Delta G_T = f(T)$ of the reaction of formation of FeO, we obtain

$$G_T = -62439 + 15.124 \cdot T, \text{ cal}$$

for the reaction of formation of nonstoichiometric wüstite:

$$(1/x)Fe + \tfrac{1}{2}O_2 = Fe_{1/x}O$$

For lower temperatures a nonlinear extrapolation from 1000°K (wüstite of the composition $Fe_{0.947}O$) to 298°K likewise is made.

Magnetite, Fe_3O_4

The density of pure Fe_3O_4 is 5.200 g/cm^3 and $V = 44.52$ cm^3/mol (Robie et al., 1967).

Analysis of experimental data on the stability of magnetite at high temperatures showed that the thermodynamic constants of Fe_3O_4 that were used contain a substantial error (up to 2–7 kcal/mol). Therefore we (Mel'nik, 1972b) suggested values of $\Delta G_{f,T}$ obtained by recalculating the measurements of the e.m.f. in a cell with solid electrolyte made by Charette and Flengas (1968) and Barbi (1964). Under standard conditions the value of $\Delta G_{f,298}$ of Fe_3O_4 came out to be -241.59 kcal/mol, considerably different from the value of -243.094 kcal/mol recommended by Robie and Waldbaum (1968).

Subsequent calculation of the high-temperature equilibrium of magnetite with wüstite, taking into account the values of the activity of the components of the solid solutions and the nonstoichiometry of the phases (Kurepin, 1975), made it possible to obtain the following equilibrium:

$$\Delta G_T = -258100 + 70.5 \cdot T, \text{ cal}$$

valid in the 1000–1500°K range for the reaction:

$$3\,Fe + 2\,O_2 = Fe_3O_4$$

To calculate $\Delta G_{f,T}$ of magnetite in the 298–1000°K range, the tabulated values of $(H_T - H_{298})$ and $(S_T - S_{298})$ for Fe_3O_4, Fe and O_2 (Robie and Waldbaum, 1968) and a new value of $\Delta G_{f,1000} = -187.60$ kcal/mol, calculated from the equation given, were used. The $\Delta G_{f,T}$ values obtained for Fe_3O_4 by that combined method (Table 11) differ by 2.5–2.9 kcal from those recommended by Robie and Waldbaum (1968) and by 1.0–2.7 kcal from those we adopted earlier (Mel'nik, 1972b), but their agreement with the high-temperature experimental data is considerably better.

Thus, one criterion of the reliability of the thermodynamic constants of solid phases in the system Fe–O can be the finding of the temperature of decomposition of wüstite (equilibrium of wüstite + magnetite + iron). Calculations based on generally accepted data for $T = 662$°C (Giddings and Gordon, 1973) give the following values: ~800°C (Wicks and Block, 1965), ~675°C (Robie and Waldbaum, 1968), ~585°C (Mel'nik, 1972b), and ~570°C (this work). As is seen, the agreement of the experimental and calculated data on the basis of the constants recommended by us is quite satisfactory, and with the calculations based on the widely used tabulated constants the discrepancy is inadmissibly great. In a later work, Haas and Robie (1973) treated 70 selected calorimetric, electrochemical, and equilibrium data on the system Fe–C–O–H and presented corrected values of

S_T, ΔH_f^0 and ΔG_f^0 for $Fe_{0.947}O$, Fe_3O_4 and Fe_2O_3. A calculation from these constants also gives a clearly too-high temperature of eutectoid decomposition (~700°C), which testifies to the advantage of critical analysis of experimental data as compared to statistical treatment.

Hydromagnetite, $Fe_3O_4 \cdot n\, H_2O$

When acid solutions containing a mixture of Fe^{2+} and Fe^{3+} ions are neutralized, a black finely dispersed deposit of $Fe_3O_4 \cdot n\, H_2O$, where n usually is about 2, is precipitated. This precipitate is different in color from partially oxidized $Fe(OH)_2$, of which various shades of green, brownish green, and brown are typical. An important feature of the hydrous ferrous-ferric iron oxide or hydromagnetite is the existence of magnetic properties—even fresh precipitates are strongly attracted by a magnet and affect a magnetic needle. Therefore $Fe_3O_4 \cdot n\, H_2O$ should be regarded as very finely divided crystalline magnetite containing a certain amount of water, apparently adsorptionally bound, rather than as an amorphous hydrate. Such compounds are also found in nature (Staschuk et al., 1964).

Lapteva (1958) experimentally determined the redox potential (E^0) in the neutralization of a mixture of Fe^{2+} and Fe^{3+} sulfates by an alkali—at a pH above 7 a black magnetite deposit was formed according to the reaction:

$$3\, Fe^{2+} + 4\, H_2O = Fe_3O_4 + 8\, H^+ + 2\, e$$

The value calculated taking into account the size of the ions, activity ratios, and complexing in sulfate environments (Mattoo, 1960) is $E^0 = +1.230$ V and $\Delta G_{298} = 56.7$ kcal/mol. Using the accepted values of $\Delta G_{f,298}$ of the Fe^{2+} ion and $(H_2O)_l$, we find (calculated to Fe_3O_4): $\Delta g_{f,298}$ $Fe_3O_4 \cdot H_2O$ (freshly precipitated) $= -234.7$ kcal/mol. An intermediate value for "diagenetic" magnetite is taken: -237.5 kcal/mol.

Hematite, Fe_2O_3

The density of pure α-Fe_2O_3 is 5.275 g/cm³ and $V = 30.27$ cm³/mol (Robie et al., 1967).

Correction of the thermodynamic constants of hematite raises the need for agreement with the new constants of magnetite. A combination of the linear equation $\Delta G = f(T)$ of the reaction of formation of Fe_3O_4 with the experimental measurements (Charette and Flengas, 1968) used earlier (Mel'nik,

1972b) yields:

$$\Delta G_T = -191750 + 58.29 \cdot T, \text{ cal}$$

for the reaction:

$$2\ Fe + \tfrac{3}{2}O_2 = Fe_2O_3$$

from which the $\Delta G_{f,T}$ values for Fe_2O_3 in the 1000–1500°K range are calculated. The $\Delta G_{f,T}$ values in the 298–1000°K range were obtained by nonlinear extrapolation and calculation by the usual procedure. The new values differ from the data of Robie and Waldbaum (1968) by 0.6–1.2 kcal, and from our data (Mel'nik, 1972b) by 0.2–1.8 kcal.

Goethite, α-FeOOH

In addition to α-FeOOH, (β, γ, δ)-FeOOH have been synthesized. The last varieties apparently are metastable and convert to hematite α-Fe_2O_3 when boiled, with α-FeOOH formed as an intermediate phase (Butler and Ison, 1965). All the other mineral species (limonite, turgite, hydrogoethite) are goethite with a variable amount of adsorbed water, or mixtures of goethite, amorphous iron hydroxides, and sometimes hematite.

The following values of ρ and V are accepted for pure crystalline goethite: $\rho = 4.269$ g/cm^3 and $V = 20.82$ cm^3/mol.

The conversion of goethite or iron hydroxides into hematite has been investigated experimentally in a great number of works, and the results obtained are rather diverse. Even at room temperature hematite often forms along with goethite, especially in acid and neutral environments. Apparently both these phases are characterized by similar thermodynamic properties and the ΔG_T value of the reaction of dehydration of α-FeOOH at moderate temperatures is not large.

From a critical review of the experimental data (Mel'nik, 1972b) it was concluded that the temperature of equilibrium conversion of goethite to hematite at the pressure of a vapor-saturated aqueous solution varies from 70 to 125°C. Experimental investigations at high pressures have been limited to P_{H_2O} up to 3 kbar (Smith and Kidd, 1949; Schmalz, 1959; Wefers, 1966), and their results are contradictory: the discrepancies in determinations of the goethite ⇄ hematite equilibrium temperature reached 50–60°C, and the slope of the *P-T* curve was not established.

We (Vorob'yeva and Mel'nik, 1977) investigated the phase transformations of freshly precipitated amorphous iron hydroxide at $P = 1$–9 kbar and $T = 100$–200°C in acid, neutral, and alkaline environments. It was established that under hydrothermal conditions only two stable crystalline phases are formed, goethite and hematite, and not only temperature and pressure affect the nature of the crystallization, but also the acid-alkali regime of the environment—an acid or, especially, an alkaline environment is more favorable for crystallization of goethite than a neutral one. Despite their considerable duration (up to 477 hours), in most of the experiments mixtures of hematite and goethite, sometimes with uncrystallized iron hydroxide, were obtained rather than pure phases, especially in a neutral environment. The diagram obtained (Fig. 99) obviously is similar to the diagram of the stability fields of these solid phases.

The experimental boundary of the field of synthesis of goethite and hematite was compared with the theoretical data. Constants of α-FeOOH obtained from thermochemical measurements: $S^0_{298} = 14.43 \pm 0.15$ cal/deg·mol; $C_p = 17.80 \pm 0.05$ cal/deg·mol (Kind and Weller, 1970); $\Delta H^0_{298} = -1.53$ kcal/mol, of the oxides (Ferrier, 1968), and the consistent constants of Fe_2O_3 and H_2O were used in the calculation.

Calculations taking into account a correction for the effect of pressure (ΔV_s) and more precise coefficients of water fugacity (Mel'nik and Radchuk, 1978) give the following equilibrium parameters: $T = 140°C$ at $P = 1$ kbar; $T = 153°C$ at $P = 2$ kbar; $T = 184°C$ at $P = 5$ kbar; and $T = 200°C$ at $P = 8$ kbar.

The agreement between the experimental and theoretical data is quite satisfactory, particularly if the experimental difficulties in attaining equilibrium and the small ΔG_T values of the reaction are taken into account. On this basis, in thermodynamic calculations of mineral equilibria in which crystalline goethite takes part it is advisable to use constants obtained by way of thermochemical measurements (King and Weller, 1970; Ferrier, 1965).

Iron hydroxide, $Fe(OH)_3$

Trivalent iron usually is precipitated from solution in the form of X-ray-amorphous brown precipitates, called amorphous, precipitated, or colloidal hydrate of Fe^{3+} oxide. The freshly precipitated hydroxide contains a large amount of water, part of which forms hydroxyl groups, but the rest of the water may retain a certain independence, being bound to the hydroxide by adsorption, capillary, or osmotic phenomena.

Freshly precipitated iron hydroxides are unstable; their properties (solubility, structure, morphology) vary with time. The physicochemical direction of the aging process consists of transformation of thermodynamically unstable active, i.e. easily reacting, varieties of X-ray-amorphous iron hydroxide into stable, inactive crystalline goethite.

Comparison of the solubility of iron hydroxides and goethite according to the data of Lengweiler et al. (1961) and Schindler et al. (1963) showed that there is a functional relationship between the variation in isobaric potential ($\Delta\Delta G$) during aging of the precipitate and aging time (Mel'nik, 1972b). This made it possible to calculate the isobaric potentials of formation of the compounds—from thermodynamically stable α-FeOOH to metastable freshly precipitated $Fe(OH)_3$-I ($\Delta G^0_{f,298}$, kcal/mol):

α-FeOOH, crystalline goethite	$= -116.7$
α-FeOOH, hydrogoethite	$= -115.9$
$Fe(OH)_3$-III, "equilibrium" (inactive) hydroxide	$= -171.4$
$Fe(OH)_3$-III, amorphous hydroxide (aged 200 hours)	$= -169.9$
$Fe(OH)_3$-I, freshly precipitated hydroxide	$= -169.0$

The transformation from freshly precipitated iron hydroxide into crystalline goethite is irreversible and is accompanied by liberation of energy ($\Delta G_{298} = 4.5$ kcal/mol) and a decrease in solubility by three orders of magnitude.

Hydrous ferrous oxide, $Fe(OH)_2$

Pure $Fe(OH)_2$ can be obtained only in the total absence of oxygen in highly reducing conditions. It is a bulky white precipitate which oxidizes rapidly in air to red-brown $Fe(OH)_3$; the intermediate products of oxidation are of a dirty green hue. The rare mineral amakinite occurs in nature, and formation of $Fe(OH)_2$ has been noted in bottom sediments (Stashchuk, 1968).

Comparison of the solubility of hydrous ferrous oxide made in the summary work by Feitknecht and Schindler (1963) makes it possible to calculate the values of the isobaric potentials of formation of precipitates ($\Delta G_{f,298}$, kcal/mol):

$Fe(OH)_2$, inactive (amakinite)	$= -117.3$
$Fe(OH)_2$, freshly precipitated	$= -115.8$

Thermodynamic analysis shows that amakinite is a stable phase with respect to iron oxides and liquid water. The low solubility of $Fe(OH)_2$ in nature is explained by its replacement by siderite, iron silicates and sulfides, and also by extreme ease of oxidation in the presence of traces of free oxygen.

Siderite, $FeCO_3$

The density of pure $FeCO_3$ is 3.944 g/cm^3 and $V = 29.38$ cm^3/mol (Robie et al., 1967).

Our calculations (Mel'nik, 1964, 1969) showed relatively low temperatures of thermal dissociation of $FeCO_3$ which disagreed appreciably with the experimental data (Weidner and Tuttle, 1965; French and Rosenberg, 1965). Recent experiments (Johannes, 1968; Sequin, 1968a) which yielded somewhat lower temperatures of equilibrium of $FeCO_3 + Fe_3O_4 + C$ ($T \approx 400°C$ at $P_f = 2$ kbar) were used to calculate the consistent thermodynamic constants of siderite (Mel'nik, 1972b). Here these constants are corrected in connection with refinement of the constants of magnetite and fugacity coefficients of CO_2 and CO (Mel'nik and Radchuk, 1977b, 1978). The value obtained for the isobaric potential of formation of crystalline siderite:

$$G^0_{f,298}\ FeCO_{3cr} = -162.0 \text{ kcal/mol}$$

is in good agreement with the $G^0_{f,298}$ $FeCO_3$ based on solubility investigations (Kelley and Anderson, 1935; Langmuir, 1969; Singer and Stumm, 1970).

The solubility of highly dispersed ("spherulitic") carbonate—the synthetic analog of diagenetic $FeCO_3$—is about 2 to 5 times greater than that of crystalline siderite (Kazakov et al., 1957); still more soluble is the fresh precipitate, which is a gel-like mass consisting of loose aggregates in a mixture with very tiny primary particles (Sokol et al., 1970). These aggregates gradually become compacted, acquire a spherical shape, and become crystalline.

Approximate values of $G^0_{f,298}$ of iron carbonate sediments are (in kcal/mol):

$FeCO_3$, finely divided	$= -160.0$
$FeCO_3$, diagenetic	$= -161.0$

Fayalite, Fe_2SiO_4

The density of pure Fe_2SiO_4 is 4.393 g/cm^3 and $V = 46.39$ cm^3/mol (Robie et al., 1967).

The constants of fayalite that we recommended earlier, based on experimental investigations of low-temperature equilibria (Teruo and Yukiyoshi, 1962), have been slightly revised in connection with refinement of the constants of Fe, O_2 and SiO_2. The values obtained for the isobaric potentials are close to the tabulated values (Robie and Waldbaum, 1968).

Ferrosilite, $FeSiO_3$

The density of synthetic α-$FeSiO_3$ (clinoferrosilite) is 4.005 g/cm^3 and $V = 32.94$ cm^3/mol, and of the high-temperature modification of α-$FeSiO_3$ (orthoferrosilite), 3.998 g/cm^3 and 33.00 cm^3/mol, respectively.

The thermodynamic constants of $FeSiO_3$ calculated by us previously (Mel'nik, 1972b) from experimental data on the stability of pure ferrosilite at high pressures (Muan et al., 1964; Muan, 1967; Lindsley et al., 1968; Akimoto and Syono, 1970) have been recalculated here in connection with the correction of the fayalite constants.

Grunerite, $Fe_7Si_8O_{22}(OH)_2$

The density of pure $Fe_7Si_8O_{22}(OH)_2$ is 3.603 g/cm^3 and $V = 277.96$ cm^3/mol (Robie et al., 1967).

Our experimental investigations on the synthesis and stability of this amphibole (Mel'nik and Radchuk, 1977a) made it possible to determine the *P-T* curve of the dehydration reaction (Fig. 99) for the first time:

$$2\ Fe_7Si_8O_{22}(OH)_2 = 7\ Fe_2SiO_4 + 9\ SiO_2 + 2\ H_2O$$

which runs through the points: 580°C—0.5 kbar; 608°C—1.0 kbar; 636°C—2.0 kbar; 650°C—3.0 kbar; and to obtain the equation:

$$\Delta G^0_{f,T} = -2286.5 + 0.507 \cdot T, \text{ kcal}$$

for the reaction of formation of grunerite:

$$7\ Fe + 8\ Si + 12\ O_2 + H_2 = Fe_7Si_8O_{22}(OH)_2$$

The values of $\Delta G^0_{f,T}$ at high temperatures calculated from this equation are close to the approximate values (Mel'nik, 1971) calculated from Schürmann's (1967) cummingtonite phase diagram, but at low temperatures the discrepancies reach 5.5 kcal.

Minnesotaite, $Fe_3Si_4O_{10}(OH)_2$

The density of pure $Fe_3Si_4O_{10}(OH)_2$ is 3.239 g/cm^3 and $V = 146.3$ cm^3/mol (Robie et al., 1967).

No thermochemical investigations have been made on minnesotaite, and the conditions of stability of iron-rich talcs likewise have been very inadequately studied. On Schürmann's (1967) experimental diagram, the upper temperature limit of the stability of minnesotaite at $P = 1000$ bar is shown as 450°C. However, in these experiments, with an iron content of 70–95% the association talc + olivine + quartz arose, and with an iron content of more than 95% olivine + quartz appeared instead of talc, which indicates that the phase transformations are not in equilibrium. On the basis of extrapolation of data on the stability of Fe-Mg talcs to the high-iron region, Forbes (1966) suggested that pure minnesotaite becomes unstable at $T = \sim 275$°C and $P_{H_2O} = 1000$ bar. In experiments with gels (Grubb, 1971) at 150–450°C, complex polymineralic mixtures arose, containing both stable and metastable phases. In particular, in mixtures of "riebeckitic" composition, minnesotaite was present in small amount in experiments conducted at $P = 1$ kbar and $T = 250$ and 350°C; at 450°C fayalite and magnetite were formed.

From the data given it can be presumed that at a pressure of 1–3 kbar and $T = 310$–350°C, in conditions of the beginning of the greenschist facies of metamorphism, equilibrium transformation of minnesotaite into grunerite takes place:

$$7\ Fe_3Si_4O_{10}(OH)_2 = 3\ Fe_7Si_8O_{22}(OH)_2 + 4\ SiO_2 + H_2O$$

Taking the average values of the equilibrium parameters ($P = 2$ kbar, $T = 600$°K $= 327$°C) and using the consistent thermodynamic constants of grunerite, quartz, and water, we find:

$$\Delta G^0_{f,600}\ Fe_3Si_4O_{10}(OH)_2 = -985.2\ \text{kcal/mol}$$

Using the principle of additivity, we calculate the approximate value of standard entropy:

$$S^0_{298}\ Fe_3Si_4O_{10}(OH)_2 = \Sigma S^0_{298}(FeO,\ SiO_2,\ H_2O') = 85.0\ \text{cal/deg} \cdot \text{mol}$$

where: S^0_{298} FeO = 14.15 (Mel'nik and Radchuk, 1977b); S^0_{298} $SiO_2 = 9.88$ (Robie and Waldbaum, 1968); and S^0_{298} $H_2O' = 3.0$—"partial" entropy of water in the talc structure, calculated from S^0_{298} $Mg_3Si_4O_{10}(OH)_2$.

In conjunction with the thermodynamic constants of the elements, these data are sufficient to obtain a linear equation:

$$\Delta G^0_{f,T} = -1152.0 + 0.298 \cdot T,\ \text{kcal}$$

for the reaction of formation of minnesotaite from simple substances:

$$3\ Fe + 4\ Si + 6\ O_2 + H_2 = Fe_3Si_4O_{10}(OH)_2$$

The standard value of the isobaric potential of formation ($\Delta G^0_{f,298} = -1069.2$ kcal/mol) is close to the value of 1069.9 calculated by us earlier (Mel'nik, 1972b) and used in many works.

Greenalite, $Fe_3Si_2O_5(OH)_4$

The density of the iron end member of the isomorphous series greenalite-serpentine is unknown; in mineralogic summaries only roughly approximate values—from 2.8 to 3.0 g/cm^3—are given for a mineral with the ratio $Fe^{2+} : Fe^{3+} = 9.2$.

If it is assumed that the volumetric effects (ΔV_s) of the reaction of dehydration of serpentine and greenalite are the same, then the calculated value of the density of pure $Fe_3Si_2O_5(OH)_4$ amounts to 3.1 g/cm^3 and $V = 120$ cm^3/mol.

From experimental data on the synthesis of hydrous iron silicates from gels (Grubb, 1971) and from petrographic observations (Dimroth and Chauvel, 1973; French, 1973; Klein and Bricker, 1977) it follows that in late diagenesis or early (low-rank) metamorphism, greenalite converts to minnesotaite. The equilibrium parameters of the conversion reaction:

$$Fe_3Si_2O_5(OH)_4 + 2\ SiO_2 = Fe_3Si_4O_{10}(OH)_2 + H_2O$$

cannot be established precisely as yet, for the reason that the degree of dispersion and imperfection of the coexisting solid phases has a considerable effect on the position of the *P-T* curve of dehydration. For crystalline greenalite and minnesotaite the equilibrium temperature apparently falls in the range of 190–230°C at $P = 1$–2 kbar. A substantial lowering of the low temperature limit of minnesotaite stability is hardly sound, as magnetite and hematite are usually present in slightly metamorphosed rocks of iron formations containing minnesotaite, but goethite, which is stable to 140–160°C, is not encountered.

A calculation similar to that made to determine the $\Delta G^0_{f,T}$ of minnesotaite gives the following linear equation:

$$\Delta G^0_{f,T} = -786.6 + 0.233 \cdot T, \text{ kcal}$$

for the reaction of formation of greenalite:

$$3\ Fe + 2\ Si + 4.5\ O_2 + 2\ H_2 = Fe_3Si_2O_5(OH)_4$$

In this case $S^0_{298}\ H_2O' = 8.0$ cal/deg·mol was calculated from the standard entropy of purely magnesian serpentine (Mel'nik, 1972b).

The standard isobaric potential of greenalite—$\Delta G^{0}_{f,298} = -717.2$ kcal/mol —is appreciably different from the value of 708.662 obtained by Eugster and I-Ming Chou (1973) by the method of comparison of standard enthalpies of layer silicates in conjunction with the principle of entropy additivity. ity.

The following approximate values of $\Delta G_{f,298}$ (in kcal/mol) were used for silicate iron-rich sediments:

$Fe_3Si_2O_5(OH)_4$, precipitated $= -711.5$
$Fe_3Si_2O_5(OH)_4$, diagenetic $= -715.0$

References

Akimoto, S.-I. and Syono, Y., 1970. High-pressure decomposition of the system $FeSiO_3$-$MgSiO_3$. Phys. Earth Planet. Inter., 3: 186–188.

Akshinskaya, N.V., Davydov, V.Ya., Zhuravlev, L.T., Kertoyz, D., Kiselev, A.V., Kuznetsov, B.V., Nikitin, Yu.S. and Rybina, V.V., 1964. The effect of hydrothermal treatment in an autoclave on the structure and adsorptional properties of silica gel. Koll. Zhurn. (Colloid J.), 26: 529–537 (in Russian).

Alexander, G.B., Heston, W.M. and Iler, R.K., 1954. The solubility of amorphous silica in water. J. Phys. Chem., 58: 453–455.

Alexandrov, E.A., 1955. Contribution to the studies of origin of Precambrian banded iron ores. Econ. Geol., 50: 459–468.

Alexandrov, E.A., 1973. The Precambrian banded iron-formations of the Soviet Union. Econ. Geol., 68: 1035–1062.

Allen, L.H. and Matijevic, E., 1969. Stability of colloidal silica. J. Colloid Interface Sci., 31: 287–396.

Anderson, A.T., 1967. The dimensions of oxygen isotopic equilibrium attainment during prograde metamorphism. J. Geol., 75: 323–331.

Antipov-Karatavev, I.N. and Tsyuryupa, I.G., 1961. On the forms and conditions of migration of substances in a soil profile. Pochvovedeniye (Soil Science), 8: 1–13 (in Russian).

Appel, P.U., 1974. On an unmetamorphosed iron-formation in the Early Precambrian of South-West Greenland. Miner. Deposita, 9: 75–82.

Astapova, S.D., Dobrovolskaya, I.A., Dromashko, S.G., Kovalev, V.K., Lukashev, K.I. and Shimanovich, S.L., 1969. Geochemical features of the secondary minerogenesis in the Quaternary depositions of Belorussia. In: Materialy seminara po geokhimii gipergeneza i kory vyvetrivaniya (Data of Seminar on the Geochemistry of Supergenesis and the Weathered Layer). Minsk, pp. 8–15 (in Russian).

Ayres, D.E., 1972. Genesis of iron-bearing minerals in banded iron formations mesobands in the Dales Gorge Member, Hamersley Group, Western Australia. Econ. Geol., 67: 1214–1233.

Backlund, H., 1949. The granitisation problem. In: Problemy obrazavaniya granitov (Problems of Granite Formation). Izd. I.L., Moscow (in Russian).

Banerji, A.K., 1977. On the Precambrian banded iron-formations and the manganese ores of the Singhbhum Region, Eastern India. Econ. Geol., 72: 90–98.

Barbi, G.B., 1964. Thermodynamic functions and phase stability limits by electromotive force measurements on solid electrolytic cells. J. Phys. Chem., 68: 1025–1029.

Barghoorn, E.S. and Tyler, S.A., 1965. Microorganisms from the Gunflint Chert. Science, 147: 563–577.

Barghoorn, E.S., Knoll, A.H., Dembicki, H.Jr. and Meinschein, W.G., 1977. Variation in stable carbon isotopes in organic matter from the Gunflint iron formation. Geochim. Cosmochim. Acta, 41: 425–430.

Barth, T., 1956. Theoretical Petrology. Izd. I.L., Moscow, 414 pp. (in Russian); J. Wiley & Sons, N.Y., London (1952, in English).

Bartholomé, P., 1962. Iron-magnesium ratio in associated pyroxenes and olivines. Geol. Soc. Am., Petrologic Studies: A volume in honour of A.F. Buddington, New York, pp. 1–20.

Basharina, L.A., 1963. Exhalations in the satellite craters of Klyuchevskiy volcano at different stages of lava cooling. In: Vulkanizm Kamchatki i nekotorykh drugikh rayonov SSSR (Volcanism of Kamchatka and Some Other Regions of the U.S.S.R.). Izd. Akad. Nauk S.S.S.R., Moscow, pp. 169–227 (in Russian).

Baykov, A.A., and Tumarev, A.S., 1937. Dissociation of natural carbonate salts on heating. Izv. Akad. Nauk S.S.S.R., Otd. Tekhn. Nauk (Proc. Acad. Sci. U.S.S.R., Sect. Techn. Sci.), 4: 565–592 (in Russian).

Bayley, R.W. and James, H.L., 1973. Precambrian iron-formations of the United States. Econ. Geol., 68: 934–959.

Becker, R.H., and Clayton, R.N., 1970. C^{13}/C^{12} ratios in a Precambrian banded iron-formation and their implications. Trans. Am. Geophys. Union (Abstr.), 51: 452.

Becker, R.H. and Clayton, R.N., 1976. Oxygen isotope study of a Precambrian banded iron-formation, Hamersley Range, Western Australia. Geochim. Cosmochim. Acta, 40: 1153–1165.

Belevtsev, R.Ya., 1975. Problems of metamorphic Zoning of the Precambrian. Izd. Naukova Dumka, Kiev, 230 pp. (in Russian).

Belevtsev, Ya.N., 1947. Deposition of the rocks of the Krivoy Rog group. Sov. Geologiya (Soviet Geology), 23: 44–53 (in Russian).

Belevtsev, Ya.N. (Editor-in-Chief), 1957. Geology and Iron Ores of the Krivoy Rog Basin. Gosgeoltekhizdat, Moscow, 280 pp. (in Russian).

Belevtsev, Ya.N. (Editor-in-Chief), 1959. Genesis of Iron Ores of the Krivoy Rog Basin. Izd. Akad. Nauk S.S.S.R., Kiev, 308 pp. (in Russian).

Belevtsev, Ya.N. (Editor-in-Chief), 1962. Geology of the Krivoy Rog Iron Ore Deposits. Izd. Akad. Nauk S.S.S.R., Kiev, Vol. I: 447 pp., Vol. II: 566 pp. (in Russian).

Belevtsev, Ya.N., 1964. Metallogeny of the Precambrian geosyncline of the Ukrainian Shield. Izv. Akad. Nauk S.S.S.R., Ser. Geol. (Proc. Acad. Sci. U.S.S.R., Geol. Ser.), 1: 8–20 (in Russian).

Belevtsev, Ya.N. and Goroshnikov, B.I., 1969. On the rationality of the terms "iron chert", "jaspilite", and "ferruginous quartzite". In: Problemy obrazovaniya zhelezistykh porod dokembriya (Problems of the Formation of the Precambrian Iron Formations). Izd. Naukova Dumka, Kiev, pp. 299–303 (in Russian).

Belevtsev, Ya.N., and Mel'nik, Yu.P., 1974. Genetic problems of Precambrian iron formations. Geologiya Rudnykh Mestorozhdeniy (Geology of Ore Deposits), 5: 66–76 (in Russian).

Belevtsev, Ya.N. and Mel'nik, Yu.P., 1976. Biogeochemical model of genesis of Precambrian ferruginous formations. Int. Geol. Congr., 25th, (Canberra), Abstr., 1: 152–153.

Belevtsev, Ya.N., Yepatko, Yu.M. and Vorob'yeva, K.A., 1966. Conditions of accumulation of iron and silica in the Precambrian from experimental data. In: Problemy teorii i eksperimenta v rudoobrazovanii. (Problems of Theory and Experiment in Ore Deposition). Izd. Naukova Dumka, Kiev, pp. 139–159 (in Russian).

Belevtsev, Ya.N., Lugovaya, I.P. and Mel'nik, Yu.P., 1969. Oxygen isotopic composition of the ore minerals of the Krivoy Rog iron-formations. In: Problemy obrazovaniya zhelezistykh porod dokembriya (Problems of the Formation of the Precambrian Iron Formations). Izd. Naukova Dumka, Kiev, pp. 271–279 (in Russian).

Berg, L.G. and Buzdov, K.A., 1961. Synthesis of ferrous carbonate and its thermal dissociation. Zhurn. Neorg. Khimii (J. Inorg. Chem.), 6: 2003–2008 (in Russian).

Berkner, L.V. and Marshall, L.C., 1965. History of major atmospheric components. Proc. Natl. Acad. Sci. USA, 53: 1215–1226.

Berner, R.A., 1964a. Distribution and diagenesis of sulfur in some sediments from the Gulf of California. Mar. Geol., 1: 117–140.

Berner, R.A., 1964b. Stability fields of iron minerals in anaerobic sediments. J. Geol., 72: 826–834.

Berner, R.A., 1967. Thermodynamic stability of sedimentary iron sulfides. Am. J. Sci., 265: 773–785.

Berner, R.A., 1969. Migration of iron and sulfur within anaerobic sediments during early diagenesis. Am. J. Sci., 267: 19–42.

Berner, R.A., 1970. Sedimentary pyrite formation. Am. J. Sci., 268: 1–23.

Besedin, V.V., 1957. Genesis of the quartz of the cherty iron-formations of the Krivoy Rog. Dokl. Akad. Nauk S.S.S.R. (Rep. Acad. Sci. U.S.S.R.), 112: 329–331 (in Russian).

Besedin, V.V., 1958. Problematical organic remains in rocks of the Krivoy Rog group. Trudy Krivorozskogo Gornorudnogo Instituta, 2: 17–22 (in Russian).

Beukes, N.J., 1973. Precambrian iron-formations of Southern Africa. Econ. Geol., 68: 960–1004.

Bezrukov, P.L., 1955. On the extent and rate of accumulation of siliceous sediments in the Sea of Okhotsk. Dokl. Akad. Nauk S.S.S.R. (Rep. Acad. Sci. U.S.S.R.), 103: 473–476 (in Russian).

Bien, G.S., Contois, D.E. and Thomas, W.H., 1958. The removal of soluble silica from fresh water entering the sea. Geochim. Cosmochim. Acta, 14: 35–54.

Bogdanov, Yu.V. and Golubchina, M.N., 1969. Sulfur isotopic composition as an indication of bacterial activity. Dokl. Akad. Nauk S.S.S.R. (Rep. Acad. Sci. U.S.S.R.), 189: 592–594 (in Russian).

Bogomolov, G.V., Plotnikova, G.N. and Titova, Ye.A., 1967. Silica in Thermal and Cold Waters. Izd. Nauka, Moscow, 112 pp. (in Russian).

Bogoyavlenskiy, A.N., 1966. Distribution and migration of dissolved silica in the oceans. In: Geokhimiya kremnezema (Geochemistry of Silica). Izd. Nauka, Moscow, pp. 11–36 (in Russian).

Bondarenko, L.P. and Dagelayskiy, V.B., 1968. Geology and Metamorphism of the Archean Rocks of the Central Part of the Kola Peninsula. Izd. Nauka, Leningrad, 168 pp. (in Russian).

Bonnichsen, B., 1969. Metamorphic pyroxenes and amphiboles in the Biwabik iron formation, Dunka River area, Minnesota. Miner. Soc. Am., Spec. Pap., 2: 217–239.

Bonnichsen, B., 1975. Geology of the Biwabik iron formation, Dunka River area, Minnesota. Econ. Geol., 70: 319–340.

Bordunov, I.N., 1964. Geology of the Kremenchug iron ore district. Izd. Naukova Dumka, Kiev, 218 pp. (in Russian).

Borisenkov, V.I., 1967. Some conditions of the migration of aluminium, iron and titanium in the postvolcanic hydrothermal process. Litologiya i poleznyye iskopayemyye (Lithology and Mineral Deposits), 4: 67–75 (in Russian).

Boychenko, Ye.A., 1967. On evolution of the oxidizing functions of the biosphere. Geokhimiya (Geochemistry), 8: 971–976 (in Russian).

Boychenko, Ye.A., Sayenko, G.N. and Udel'nova, T.M., 1968. The evolution of the concentration functions of the plants in the biosphere. Geokhimiya (Geochemistry), 10: 1260–1264 (in Russian).

Boychenko, Ye.A., Udel'nova, T.M. and Yuferova, S.G., 1969. The evolution of the reduction functions of the biosphere. Geokhimiya (Geochemistry), 11: 1392–1396 (in Russian).

Brandt, R.T., Van N. Dorr, II, J., Gross, G.A., Gruss, H. and Semenenko, N.P., 1972. Problems of nomenclature for banded ferruginous cherty sedimentary rocks and their metamorphic equivalents. Econ. Geol., 67: 682–684.

Bricker, O.P., 1969. Stability constants and Gibbs free energy of formation of magadiite and kenyaite. Am. Mineral., 54: 1026–1033.

Brock, T.D., 1967. Life at high tempertures. Science, 158: 1012–1019.

Brock, T.D., 1973. Lower pH limit for the existence of bluegreen algae, evolutionary and ecological consequences. Science, 179: 480–483.

Brooks, J., Muir, M.D. and Shaw, G., 1973. Chemistry and morphology of Precambrian microorganisms. Nature, 244: 215–217.

Brusilovskiy, S.A., 1963. On the forms of migration of the elements in natural waters. Gidrokhimicheskiye materialy (Hydrochemical Data). Izd. Akad. Nauk S.S.S.R., 35: 3–16 (in Russian).

Bruyevich, S.V., 1953. On the geochemistry of silica in the sea. Izv. Akad. Nauk S.S.S.R., ser. geol. (Proc. Acad. Sci. U.S.S.R., Geol. Ser.), 4: 67–69 (in Russian).

Bruyevich, S.V., Bogoyavlenskiy, A.N. and Mokiyevskaya, V.V., 1960. Hydrochemical characteristics of the Sea of Okhotsk. Trudy Inst. Okeanolog. Akad. Nauk S.S.S.R. (Trans. Inst. Oceanology Acad. Sci. U.S.S.R.), Moscow, 42: 125–198 (in Russian).

Butler, G. and Json, H.C.K., 1965. The thermal transformation of hydrated ferric oxides. Chem. Commun., 12: 264–265.

Button, A., 1976. Iron-formation as an end member in carbonate sedimentary cycles in the Transvaal supergroup, South Africa. Econ. Geol., 71: 193–201.

Butuzova, G.Yu., 1966. Iron ore sediments of the fumarole field of Santorin volcano, their composition and genesis. Dokl. Akad. Nauk S.S.S.R. (Rep. Acad. Sci. U.S.S.R.), 168: 1400–1404 (in Russian).

Butuzova, G.Yu., 1968. Recent sediments in the caldera of Santorin island and the effect of volcanism on their formation. In: Geokhimiya osadochnykh porod i rud (Geochemistry of Sedimentary Rocks and Ores). Izd. Nauka, Moscow, pp. 183–222 (in Russian).

Butuzova, G.Yu., 1969. Recent volcanogenic-sedimentary iron ore process in the caldera of Santorin island (Aegean Sea) and its effect on the geochemistry of the sediments. Izd. Nauka, Moscow, 112 pp. (in Russian).

Castano, J.R. and Garrels, R.M., 1950. Experiments on the deposition of iron with special reference to the Clinton iron ore deposits. Econ. Geol., 45: 755–770.

Chakraborty, K.L. and Taron, P.B., 1968. Thermal metamorphism of the banded iron formation of Orissa, India. Bull. Geol. Soc. Am., 79: 365–374.

Chalyy, V.P. and Lukachina, Ye.N., 1970. Phase transformations in the system of magnesium, manganese (II) and iron (III) hydroxides. Izv. Akad. Nauk S.S.S.R. Neorg. Materialy (Proc. Acad. Sci. U.S.S.R., Inorgan. Materials), 6: 1146–1150 (in Russian).

Charette, G.G. and Flengas, S.N., 1968. Thermodynamic properties of the oxides of Fe, Ni, Pb, Cu, and Mn by EMF measurement. J. Electrochem. Soc., 115: 796–804.

Chase, C.G. and Perry, E.C., 1972. The oceans; growth and oxygen isotope evolution. Science, 177: 992–994.

Chernov, V.M., Inina, K.A., Gor'kovets, V.Ya. and Rayevskaya, M.B., 1970. Volcanogenic cherty iron-formations of Karelia. Trudy Inst. Geol. Karel'skogo Filiala Akad. Nauk S.S.S.R. (Trans. Geol. Inst. Karelia Branch Acad. Sci. U.S.S.R.), Vol. 5, 285 pp. (in Russian).

Chertov, V.M., Dzhimbayeva, D.B., Plachinda, A.S. and Neymark, I.Ye., 1966. Hydroxyl groups on the surface of and within globules of silica gel obtained from hydrothermally treated hydrogels. Zhurn. Fiz. Khim. (J. Phys. Chem.), 40: 520–525 (in Russian).

Chukhrov, F.V., 1955. Colloids in the Earth's crust. Izv. Akad. Nauk S.S.S.R., Moscow, 671 pp. (in Russian).

Chukhrov, R.V., 1965. On the state of the problem of the role of colloids in ore formation. Sov. Geologiya (Soviet Geology), 2: 3–15 (in Russian).

Chukhrov, F.V., Yermilova, L.P. and Vinogradov, V.I., 1969. On the isotopic composition of sulphur as an indication of the possibility of some geochemical processes in the ancient Precambrian. Izv. Akad. Nauk S.S.S.R., Ser. Geol. (Proc. Acad. Sci. U.S.S.R., Geol. Ser.), 9: 50–60 (in Russian).

Chukhrov, F.V., Churikov, V.S., Yermilova, L.P. and Nosik, L.P., 1977. Isotopic composition of atmospheric sulphur and its possible evolution in the Earth's history. Izv. Akad. Nauk S.S.S.R., Ser. Geol. (Proc. Acad. Sci. U.S.S.R. Geol. Ser.), 7: 5–13 (in Russian).

Clayton, R.N. and Epstein, S., 1958. The relationship between O^{18}/O^{16} ratios in coexisting quartz, carbonate and iron oxides from various geological deposits. J. Geol., 66: 352–374.

Cloud, P.E. Jr., 1965a. Significance of the Gunflint (Precambrian) microflora. Science, 148: 27–35.

Cloud, P.E. Jr., 1965b. Symposium on the evolution of the Earth's atmosphere. Chairman's summary remarks. Proc. Natl. Acad. Sci. USA, 53: 1169–1172.

Cloud, P.E. Jr., 1968. Atmospheric and hydrospheric evolution on the primitive Earth. Science, 160: 729–736.

Cloud, P.E. Jr., 1969. The primitive Earth. New Scientist, 43: 325–327.

Cloud, P.E. Jr., 1973. Paleoecological significances of the banded iron-formation. Econ. Geol., 68: 1135–1143.

Cloud, P.E. Jr. and Licari, G.R., 1968. Microbiotas of the banded iron formations. Proc. Natl. Acad. Sci. USA, 61: 779–786.

Cloud, P.E. Jr., Gruner, J.W. and Hagen, H., 1965. Carbonaceous rocks of the Soudan iron formation. Science, 148: 1713–1716.

Commoner, B., 1965. Biochemical, biological and atmospheric evolution. Proc. Natl. Acad. Sci. USA, 53: 1183–1194.

Cooper, L.H.N., 1937. Some conditions governing the solubility of iron. Proc. R. Soc., London, 13 (124): 299–307.

Curtis, C.D. and Spears, D.A., 1968. The formation of sedimentary iron minerals. Econ. Geol., 63: 257–270.

Darken, L.S. and Gurry, R.W., 1945. The system iron-oxygen. I. The wüstite field and related equilibria. J. Am. Chem. Soc., 67: 1398–1412.

Darken, L.S. and Gurry, R.W., 1946. The system iron-oxygen. II. Equilibria and thermodynamics of liquid oxides and other phases. J. Am. Chem. Soc., 68: 798–816.

Davis, J.F., 1966. Petrology of Precambrian iron-formation and associated rocks, Palmer area, Marquette district, Michigan. Geol. Soc. Am., Spec. Pap., 87: 43.

Degens, E.T., 1967. Geochemistry of Sediments. Izd. Mir, Moscow, 299 pp. (in Russian). Prentice-Hall, Inc., New Jersey, 1965 (in English).

Diessel, C.F. and Offler, R., 1973. The relationship between graphite rank and metamorphic grade in the Mount Lofty Ranges, South Australia. Search, 4: 497–499.

Dimroth, E. and Chauvel, J.J., 1973. Petrography of the Sokoman Iron Formation in part of the central

Labrador Trough, Quebec, Canada. Bull. Geol. Soc. Am., 84: 111–134.

Dimroth, E. and Kimberley, M.M., 1976. Precambrian atmospheric oxygen: evidence in the sedimentary distributions of carbon, sulfur, uranium and iron. Can. J. Earth Sci., 13: 1161–1185.

Dinisenko, V.Yu. and Kanigin, L.I., 1970. On the sedimentary origin of the carbon of the graphitic gneiss of the Azov area. Dopovidi Akad. Nauk Ukr. R.S.R., 8: 679–681 (in Ukrainian).

Dobretsov, N.L. and Ponomareva, L.G., 1964. Paragenetic types of metamorphic pyroxenes and dependence of their composition on the composition and conditions of formation of their host rocks. Sov. Geologiya (Soviet Geology), 12: 39–58 (in Russian).

Dobrokhotov, M.N., 1964. Geology and Iron Ore Deposits of the Kremenchug District. Izd. Nedra, Moscow, 221 pp. (in Russian).

Dobrokhotov, M.N., 1969. On some problems of the Precambrian geology of the Krivoy Rog–Kremenchug structural-facies zone. Izv. Akad. Nauk S.S.S.R., Ser. Geol. (Proc. Acad. Sci. U.S.S.R., Geol. Ser.), 4: 16–34 (in Russian).

Dole, M. and Slobod, R.L., 1940. Isotopic composition of oxygen in carbonate rocks and iron oxide ores. J. Am. Chem. Soc., 62: 471–479.

Domarev, V.S., 1969. On the problem of the genesis of quartzites. In: Problemy obrazovaniya zhelezistykh porod dokembriya (Problems of the Formation of the Precambrian Iron Formation). Izd. Naukova Dumka, Kiev, pp. 110–114 (in Russian).

Dontsova, Ye.I., 1964. An attempt to use oxygen isotope ratios in geochemical investigations. In: Khimiya zemnoy kory (Chemistry of the Earth's Crust). Izd. Nauka, Moscow, pp. 631–637 (in Russian).

Dontsova, Ye.I., 1970. Oxygen isotopic exchange during the formation of rocks. Geokhimiya (Geochemistry), 8: 903–916 (in Russian).

Dorr, J. Van N., 1969. Physiographic, stratigraphic and structural development of the Quadrilatero Ferrifero, Minas Gerais, Brazil. U.S. Geol. Surv., Prof. Pap. 641-A, 109 pp.

Dorr, J. Van N., 1973. Iron-formation in South America. Econ. Geol., 68: 1005–1022.

Drever, J.I., 1974. Geochemical model for the origin of Precambrian banded iron formations. Bull. Geol. Soc. Am., 85: 1099–1106.

Dzotsenidze, G.S., 1965. Effect of volcanism on sedimentation. In: Vulkanogenno-osadochniye formatsii i polezniye iskopayemiye (Volcanogenic-Sedimentary Formations and Mineral Deposits), Izd. Nauka, Moscow. pp. 24–43 (in Russian).

Dzotsenidze, G.S., 1970. The problem of the effect of volcanism on the sedimentary process. In: Sostoyaniye i zadachi sovetskoy litologii (State and Problems of Soviet Lithology). Izd. Nauka, Moscow, 1: 103–117 (in Russian).

Edgell, H.S., 1964. Precambrian fossils from the Hamersley Range, Western Australia. J. Geol. Soc. Aust., 11: 235–262.

Eichmann, R. and Schidlowski, M., 1975. Isotopic fractionation between coexisting organic carbon-carbonate pairs in Precambrian sediments. Geochim. Cosmochim. Acta, 39: 585–595.

Eugster, H.P., 1961a. New experimental investigations of the stability and phase relationships of hydrous iron silicates. In: Fiziko-khimicheskiye problemy formirovaniya gornykh porod i rud (Physicochemical Problems of the Formation of Rocks and Ores), Izd. Akad. Nauk S.S.S.R., Moscow, 1: 622–640 (in Russian).

Eugster, H.P., 1961b. Reduction and oxidation in metamorphism. In: Geokhimicheskiye issledovaniya (Researches in Geochemistry). Izd. I.L., Moscow, pp. 495–527 (in Russian). Ph.H. Abelson (Editor), Wiley, New York, N.Y., 1959, Vol. 1, pp. 397–426.

Eugster, H.P. and Chou, I.-M., 1973. The depositional environments of Precambrian banded iron-formations. Econ. Geol., 68: 1144–1168.

Fed'kin, V.V., 1971. Mineral stability in the system $FeO–Al_2O_3–SiO_2–H_2O–O_2$. Izv. Akad. Nauk S.S.S.R., Ser. Geol. (Proc. Acad. Sci. U.S.S.R., Geol. Ser.), 6: 88–101 (in Russian).

Fedorchenko, V.S., 1969. Mineral composition and textural-structural features of Precambrian iron-formations of low-rank metamorphism (greenschist facies). In: Problemy obrazovaniya zhelezistykh porod dokembriya (Problems of Formation of the Precambrian Iron Formations). Izd. Naukova Dumka, Kiev, pp. 168–177 (in Russian).

Feitknecht, M. and Schindler, P., 1963. Solubility constants of metal oxides, metal hydroxides and metal hydroxide salts in aqueous solution. Pure Appl. Chem., 6: 130–199.

Feitknecht, W. and Michaelis, W., 1962. Über die Hydrolyse von Eisen (III)—perchlorat-Lösungen. Helv. Chim. Acta, 45: 212–224.
Ferrier, A., 1965. Influence de l'état de division de la goethite sur sa chaleur de déshydration. C.R. Acad. Sci., 261: 410–413.
Ferrier, A., 1968. Estimation quantitative de l'influence de l'enthalpie de surface, dans la thermochimie des poudres. Mém. Scient. Rév. Metall., 65: 477–480.
Fiebiger, W., 1975. Organische Substanzen in präkambrischen Itabiriten und deren Nebengesteinen. Geol. Rundsch., 64: 641–652.
Fisher, A.G., 1965. Fossils, early life and atmospheric history. Proc. Natl. Acad. Sci. USA, 53: 1205–1215.
Floran, R.J. and Papike, J.J., 1975. Petrology of the Gunflint Iron Formation, Ontario–Minnesota: Pt. I, The low grade rocks. Bull. Geol. Soc. Am., 86: 1169–1190.
Fonarev, V.I., Korol'kov, G.Ya. and Dokina, T.N., 1976. Stability of the association cummingtonite + quartz + magnetite. Geokhimiya (Geochemistry), 9: 1340–1352 (in Russian).
Forbes, W.C., 1966. Talc-minnesotaite system at 1 and 2 kb. Geol. Soc. Am. Spec. Pap., 87: 57–58.
Forbes, W.C., 1977. Stability relations of grunerite, $Fe_7Si_8O_{22}(OH)_2$. Am. J. Sci., 277: 735–749.
Formozova, L.N., 1963. Exhalative-sedimentary iron ore deposits of Lahn-Dill type and the geological environment of their formation. In: Vulkanogenno-osadochniye formatsii i terrigennyye formatsii (Volcanogenic-sedimentary and clastic formations), Moscow, pp. 161–208 (in Russian).
Formozova, L.N., 1965. Genesis and formational character of the Precambrian iron ores of Northern Sweden. Trudy Geol. Inst. Akad. Nauk S.S.S.R. (Trans. Geol. Inst. Acad. Sci. U.S.S.R.), Moscow, 141: 7–46 (in Russian).
Formozova, L.N., 1968. Regularities of formation of volcanogenic-sedimentary iron ores. Trudy Geol. Inst. Akad. Nauk S.S.S.R. (Trans. Geol. Inst. Acad. Sci. U.S.S.R.), Moscow, 196: 7–153 (in Russian).
French, B.M., 1966. Some geological implications of equilibrium between graphite and a C-H-O gas phase at high temperatures and pressures. J. Geophys. Res., 4: 223–253.
French, B.M., 1968. Progressive contact metamorphism of the Biwabik Iron-Formation, Mesabi Range, Minnesota. Minn. Geol. Surv. Bull., 45: 103.
French, B.M., 1973. Mineral assemblages in diagenetic and low-grade metamorphic iron-formation. Econ. Geol., 68: 1063–1074.
French, B.M. and Rosenberg, P.E., 1965. Siderite ($FeCO_3$): thermal decomposition in equilibrium with graphite. Science, 147: 1283–1284.
Fyfe, W.S., Turner, F.J. and Verhoogen, J., 1962. Metamorphic Reactions and Metamorphic Facies. Izd. I.L., Moscow: 414 pp. (in Russian). Geol. Soc. Am., Mem., 73, 259 pp. (in English).
Gair, J.E., 1973. Iron deposits of Michigan (United States of America). In: Genesis of Precambrian Iron and Manganese Deposits. Unesco, Paris, pp. 365–375.
Galimov, E.M.,1968. Geochemistry of the Stable Carbon Isotopes. Izd. Nedra, Moscow, 224 pp. (in Russian).
Galimov, E.M., Kuznetsova, N.G. and Prokhorov, V.S., 1968. The composition of the early atmosphere of the earth as indicated by isotopic analysis of Precambrian carbonates. Geokhimiya (Geochemistry), 11: 1376–1381 (in Russian). Geochem. Int., 5: 1126–1131 (in English).
Gameyeva, O.S., 1969. Physical and Colloidal Chemistry. 2nd Ed. Izd. Kiev Univ., 404 pp. (in Russian).
Garlick, G.D. and Epstein, S., 1967. Oxygen isotope ratios in coexisting minerals of regionally metamorphosed rocks. Geochim. Cosmochim. Acta, 31: 181–214.
Garrels, R.M., 1962. Mineral Equilibria at Low Temperature and Pressure. Izd. IL, Moscow, 306 pp. (in Russian); Harper & Brothers, N.Y. (in English).
Garrels, R.M. and Christ, Ch.L., 1968. Solutions, Minerals, and Equilibria. Izd. Mir, Moscow, 368 pp. (in Russian); Harper & Row, N.Y., 1955 (in English).
Garrels, R.M. and Mackenzie, F.T., 1974. Evolution of Sedimentary Rocks. Izd. Mir, Moscow, 272 pp. (in Russian); W.W. Norton and Company, New York (in English).
Garrels, R.M. and Naeser, C.R., 1958. Equilibrium distribution of dissolved sulfur species in water at 25°C and 1 atm total pressure. Geochim. Cosmochim. Acta, 15: 113–130.
Garrels, R.M., Perry, E.A. Jr. and Mackenzie, F.T., 1973. Genesis of Precambrian iron-formations and the development of atmospheric oxygen. Econ. Geol., 68.
Gershoyg, Yu.G., 1949. Phases of petrogenesis in the crystalline schists of the Krivoy Rog. Sov. Geologiya (Soviet Geology), 37: 26–40 (in Russian).

Gershoyg, Yu.G., 1957. Processes of formation of the iron ore formation and high-grade ores of the Krivoy Rog basin. Izv. Akad. Nauk S.S.S.R., Ser. Geol. (Proc. Acad. Sci. U.S.S.R., Geol. Ser.), 10: 25–38 (in Russian).

Gershoyg, Yu.G., 1965. Features of the banding of the Krivoy Rog iron formations. Izv. Akad. Nauk S.S.S.R., Ser. Geol. (Proc. Acad. Sci. U.S.S.R., Geol. Ser.), 8: 45–60 (in Russian).

Gershoyg, Yu.G., 1968. Genetic interpretation of the particulars of the textural-structural picture of the iron cherts and jaspilites of the Krivoy Rog. Geol. Rudnykh Mestorozhdeniy (Geology of Ore Deposits), 10: 90–101 (in Russian).

Ghose, S., 1962. The nature of Mg^{2+}–Fe^{2+} distribution in some ferromagnesian silicate minerals. Am. Mineral., 47: 388–395.

Giddings, R.A. and Gordon, R.S., 1973. Review of oxygen activities and phase boundaries in wüstite as determined by electromotive-force and gravimetric methods. J. Am. Ceram. Soc., 56 (3).

Ginzburg, I.I. and Kabanova, Ye.S., 1960. Silica content in natural waters and forms of its occurrence. In: Kora vyvetrivaniya (The Weathered Layer). Izd. Nauka, Moscow, 3: 313–342 (in Russian).

Glagolev, A.A., 1965. Metamorphism of the Precambrian Rocks of the KMA. Izd. Nauka, Moscow, 175 pp. (in Russian).

Glagoleva, M.A., 1959. Forms of migration of elements in river waters. In: K poznaniyu diageneza osadkov (On an Understanding of the Diagenesis of Sediments). Izd. Akad. Nauk S.S.S.R., Moscow, pp. 5–28 (in Russian).

Glazman, Yu.M., Strazhesko, D.N. and Bisikalova, N.A., 1958. Investigation of the coagulation of lyophobic sols by electrolytes by the method of radioactive tracers. Zhur. Neorg. Khimii (J. Inorg. Chem.), 3: 1–15.

Glevasskiy, Ye.B., Yesipchuk, K.Ye. and Tsukanov, V.A., 1971. Petrography and genesis of the ferruginous-siliceous schists of the Azov area. In: Rudoobrazovaniye i metallogeniya (Ore Formation and Metallogeny). Izd. Naukova Dumka, Kiev, pp. 160–173 (in Russian).

Goodwin, A.M., 1961. Genetic aspects of Michipicoten iron formation. Can. Min. Met. Bull., 54: 38–42.

Goodwin, A.M., 1962. Structure, stratigraphy and origin of iron formation, Michipicoten area, Algoma district, Ontario, Canada. Geol. Soc. Am. Bull., 73: 561–586.

Goodwin, A.M., 1973a. Archean iron-formations and tectonic basins of the Canadian Shield. Econ. Geol., 68: 915–933.

Goodwin, A.M., 1973b. Archaean volcanogenic iron-formation of the Canadian Shield. In: Genesis of Precambrian Iron and Manganese Deposits, Unesco, Paris, pp. 23–34.

Goodwin, A.M. and Ridler, R.H., 1970. The Abitibi orogenic belt. Geol. Surv. Can. Pap., 40: 1–24.

Goodwin, A.M., Monster, J. and Thode, H.G., 1976. Carbon and sulfur isotope abundances in Archean iron-formations and early Precambrian life. Econ. Geol., 71: 870–891.

Goryainov, P.M., 1969. Genetic relationships of the rocks of the cherty iron-formations of the Kola peninsula. In: Problemy obrazovaniya zhelezistykh porod dokembriya (Problems of the Formation of the Precambrian Iron Formations). Izd. Naukova Dumka, Kiev, pp. 129–139 (in Russian).

Govett, G.J.S., 1966. Origin of banded iron formations. Geol. Soc. Am. Bull., 77: 1191–1211.

Grant, J.A., 1963. Sur l'équilibre entre l'olivine et l'orthopyroxène. Ann. Soc. Géol. Belg., 86: 131–136.

Greenberg, S.A., 1957. Thermodynamic function for the solution of silica in water. J. Phys. Chem., 61: 196–197.

Grigor'yev, V.M., 1971. Regularities in the distribution and Conditions of Accumulation of Germanium in Iron Deposits (in Russian). Izd. Nauka, Moscow, 153 pp.

Grigor'yev, V.M. and Zelenov, K.K., 1965. On the source of germanium in iron ores. Geokhimiya (Geochemistry), 5: 582–586 (in Russian).

Grinberg, A.A., 1966. Introduction to the Chemistry of Complex Compounds. Izd. Khimiya, Moscow-Leningrad, 628 pp. (in Russian).

Gross, G.A., 1961. Metamorphism of iron-formations and its bearing on their beneficiation. Can. Min. Metall. Bull., 54: 30–37.

Gross, G.A., 1965. Geology of iron deposits in Canada. Volume I. General Geology and Evaluation of Iron Deposits. Econ. Geol. Rep. Surv. Can. 22: 181 pp.

Gross, G.A., 1973. The depositional environment of principal types of Precambrian iron-formations. In: Genesis of Precambrian Iron and Manganese Deposits, Unesco, Paris, pp. 15–21.

Grout, F.F. and Broderick, T.M., 1919. Organic structures in the Biwabik iron-bearing formation of the Huronian in Minnesota. Am. J. Sci., Ser. 4, 48: 199–205.

Grover, J.E. and Orville, P.M., 1969. The partitioning of cations between coexisting single- and multi-site phases with application to the assemblages: orthopyroxene-clinopyroxene and orthopyroxene-olivine. Geochim. Cosmochim. Acta, 33: 205–226.

Grubb, P.L.C., 1971. Silicates and their paragenesis in the Brockman Iron Formation. Econ. Geol., 66: 281–292.

Gruner, J.W., 1922. The origin of sedimentary iron formations: the Biwabik Formation of the Mesabi Range. Econ. Geol., 17: 407–460.

Gruner, J.W., 1923. Algae, believed to be Archean. J. Geol., 31: 146–148.

Gruner, J.W., 1925. Discovery of life in the Archean. J. Geol., 33: 151–152.

Gundersen, J.N. and Schwartz, G.M., 1962. The geology of the metamorphosed Biwabik iron-formation, Eastern Mesabi district, Minnesota. Univ. Minn. Minn. Geol. Surv. Bull., 43: 139 pp.

Gusel'nikov, V.N., 1972. Genetic Problems of the Iron Formations of the KMA [Kursk Magnetic Anomaly]. Izd. Nauka, Moscow, 228 pp. (in Russian).

Gusel'nikov, V.N., 1976. Ferruginous quartzites and volcanism. Geol. Rudnykh Mestorozhdeniy (Geology of Ore Deposits), 18/5: 110–116 (in Russian).

Gushchenko, I.I., 1965. The composition of young continental volcanic rocks and the conditions of removal of the iron group. In: Vulkanogenno-osadochnyye formatsii i poleznyye iskopayemyye (Volcanogenic-sedimentary Formations and Mineral Deposits). Izd. Nauka, Moscow, pp. 63–69 (in Russian).

Haas, J.L. and Robie, R.A., 1973. Thermodynamic data for wüstite, $Fe_{0.947}O$, magnetite, Fe_3O_4 and hematitite, Fe_2O_3 (abstr.). Trans. Am. Geophys. Union, 54: 483.

Harder, E.C., 1919. Iron-depositing bacteria and their geologic relations. U.S. Geol. Surv. Prof. Pap., 113: 89 pp.

Harder, H., 1965. Experimente zur "Ausfällung" der Kieselsäure. Geochim. Cosmochim. Acta, 29: 429–442.

Harder, H. and Flehmig, W., 1970. Quarzsynthese bei tiefen Temperaturen. Geochim. Cosmochim. Acta, 34: 295–305.

Harington, J.S. and Cilliers, J.J. le R., 1963. A possible origin of the primitive oils and amino acids isolated from amphibole asbestos and banded ironstone. Geochim. Cosmochim. Acta, 27: 411–418.

Harrison, C.G.D. and Peterson, M.N.A., 1965. A magnetic mineral from the Indian Ocean. Am. Mineral. 50: 704–712.

Hinrichsen, Th., 1967. Über den Stabilitätsbereich der Mg-Fe^{2+}-Al-Mischkristallreihe rhombischer Hornblenden. Teil I. Hydrothermale Untersuchungen der Anthophyllit – Ferroanthophyllit – Mischkristallreihe. Neues Jahrb. Mineral. Monatsh., N 9–10: 257–270.

Hoering, T.C., 1967. The Organic Geochemistry of Precambrian Rocks, Vol. 2. P.H. Abelson (Editor), John Wiley, New York, N.Y.

Holland, H.D., 1965a. Some applications of thermochemical data to problems of ore deposits. Econ. Geol., 60: 1101.

Holland, H.D., 1965b. The history of ocean water and its effect on the chemistry of the atmosphere. Proc. Natl. Acad. Sci. USA, 53: 1173–1182.

Holland, H.D., 1972. The geologic history of sea water—an attempt to solve the problem. Geochim. Cosmochim. Acta, 36: 637–651.

Holland, H.D., 1973. The Oceans: a possible source of iron-formations. Econ. Geol., 68: 1169–1172.

Hough, J.L., 1958. Fresh-water environment of deposition of Precambrian banded iron formations. J. Sediment. Petrol. 28: 414–430.

Hsu, L.C., 1968. Selected phase relationships in the system Al-Mn-Fe-Si-O-H: a model for garnet equilibria. J. Petrol., 9: 40–83.

Huber, N.K., 1960. The environmental control of sedimentary iron minerals. In: Termodynamika geokhimicheskikh protsessov (Thermodynamics of Geochemical Processes). Izd. I.L., Moscow, pp. 186–204 (in Russian); Econ. Geol., 53: 1958 (in English).

Iler, R.K., 1959. The Colloid Chemistry of Silica and Silicates. Gosstroyizdat, Moscow, 288 pp. (in Russian); Cornell University Press, Ithaca, N.Y., 324 pp. (in English).

Illarionov, A.A., 1965. Petrography and Mineralogy of the Ferruginous Quartzites of the Mikhaylovka Deposit of the Kursk Magnetic Anomaly. Izd. Nauka, Moscow, 164 pp. (in Russian).

Illarionov, A.A., 1969. Some considerations on the classification and nomenclature of the rocks of Precambrian cherty iron formations. In: Problemy obrazovaniya zhelezistykh porod dokembriya (Problems of the formation of the Precambrian iron formations). Izd. Naukova Dumka, Kiev, pp. 307–310 (in Russian).

Immega, I.P. and Klein, C. Jr., 1976. Mineralogy and petrology of some metamorphic Precambrian iron-formations in southwestern Montana. Am. Mineral., 61: 1117–1144 pp.

Ishchenko, A.M., 1957. On the geological age of the middle suite of the Krivoy Rog group. Dopovīdī Akad. Nauk Uk.R.S.R. (Rep. Acad. Sci. Ukrainian S.S.R.), 2: 176–178 (in Ukrainian).

Ivanov, V.V., 1956. Hydrotherms of the recent volcanic centers of Kamchatka and the Kurile Islands. Trudy Laboratorii Vulkanologii (Trans. laborat. Volcanology), Akad. Nauk S.S.S.R., is. 12: 197–217 (in Russian).

Ivantsīv, O.Ye. and Nenchuk, M.F., 1974. A graphite-like substance from the rocks of the Krivoy Rog series of the Ukrainian Shield. Dopodivi Akad. Nauk UkS.S.R. (Rep. Acad. Sci. Ukrainian S.S.R.), Ser. B, N 10: 884–885 (in Ukrainian).

James, H.L., 1954. Sedimentary facies of iron-formation. Econ. Geol., 49: 235–293.

James, H.L. and Clayton, R.N., 1962. Oxygen isotope fractionation in metamorphosed iron-formations of the Lake Superior region and in other iron-rich rocks. Geol. Soc. Am., Buddington Vol., pp. 217–239.

James, H.L. and Howland, A.L., 1955. Mineral facies in iron- and silica-rich rocks (Abstr.). Bull. Geol. Soc. Am., 66: 1580–1581.

James, H.L. and Sims, P.K., 1973. Introduction. Econ. Geol., 68: 913–914.

Johannes, W., 1968. Experimentelle Sideritbildung aus Calcit + $FeCl_2$. Contrib. Mineral. Petrol., 17: 155–164.

Johns, R.B., Belsky, T., McCarthy, E.D., Burlingame, A.L., Haug, Pat, Schnoes, H.K., Richter, W. and Calvin, M., 1966. The organic geochemistry of ancient sediments—part II. Geochim. Cosmochim. Acta, 30: 1191–1222.

Jones, O.A., 1963. The Bedded Chert, banded iron-formation problem. J. Geol. Soc. Aust., 10: 209–222.

Kablukov, I.A., Gapon, Ye.N. and Grindel', M.A., 1949. Physical and Colloidal Chemistry. Sel'khozizdat. Moscow, 464 pp. (in Russian).

Kalinin, A.S. and Kashik, S.A., 1968. On the problem of the forms of dissolved iron in hydrothermal solutions. Dokl Akad. Nauk S.S.S.R. (Rep. Acad. Sci. U.S.S.R.), 182: 190–192 (in Russian).

Kalugin, A.S., 1969. Formation of the volcanogenic-sedimentary banded iron ores in the Devonian deposits of the Altai. In: Problemy obrazovaniya zhelezistych porod dokembriya (Problems of Formation of the Precambrian Iron Formations). Izd. Naukova Dumka, Kiev, pp. 89–104 (in Russian).

Kalyayev, G.I., 1965. Precambrian Tectonics of the Ukrainian Iron Ore Province. Izd. Naukova Dumka, Kiev, 186 pp. (in Russian).

Kalyayev, G.I., Boychuk, M.D., Glevasskiy, Ye.B., Zaritskiy, A.I., Las'kov, V.A., Polinovskiy, R.M., Pyatnitsa, P.S. and Kharagezov, M.K., 1968. Stratigraphy of the host rocks and structure of the Mariupol iron ore deposit. Geol. Rudn. Mestorozhd. (Geology of Ore Deposits), 10: 3–16 (in Russian).

Kanibolotskiy, P.M., 1946. Petrogenesis of the Rocks and Ores of the Krivoy Rog Iron Ore Basin. Izd. Akad. Nauk Ukrain. S.S.R., Chernovtsy, 312 pp. (in Russian).

Kargin, V.A. and Kiseleva, V.L., 1938. On the structure of ferric hydroxide sols. Zhur. Fiz. Khim. Akad. Nauk S.S.S.R. (J. Phys. Chem.), 11: 461–466 (in Russian).

Karpov, I.K., Kashik, S.A. and Pampura, V.D., 1968. Constants for Thermodynamic Calculations in Geochemistry. Izd. Nauka, Moscow, 143 pp. (in Russian).

Karpov, I.K., Kiselev, A.I. and Letnikov, F.A., 1971. Chemical Thermodynamics in Petrology and Geochemistry. Izd. S.O. Akad. Nauk S.S.S.R., Irkutsk, 386 pp. (in Russian).

Kaukin, B.V., 1969. On the origin of ferruginous quartzites. In: Problemy obrazovaniya zelezistykh porod dokembriya (Problems of the Formation of the Precambrian Iron Formations). Izd. Naukova Dumka, Kiev, pp. 47–55 (in Russian).

Kaurichev, I.S., Kulàkov, Ye.V. and Nozdrunova, Ye.M., 1958. On the problem of the formation and migration of iron-organic compounds in soils. Pochvovedeniye (Soil Science), 12: 1–8 (in Russian).

Kazakov, A.V., Tikhomirova, M.M. and Plotnikova, V.I., 1957. The system $FeO-CO_2-H_2O$ and implications for the paragenesis of siderite and phosphorite. Trudy Inst. Geol. Nauk Akad. Nauk S.S.S.R. (Trans. Inst. Geol. Sci., Acad. Sci. U.S.S.R.), Ser. Geol., 64 (152) (in Russian).

Kazanskiy, Yu.P., Kazarinov, V.P. and Rezanova, N.M., 1968. Results of experiments with saturated solutions of silica and their geological significance. In: Fizicheskiye i khimicheskiye protsessy i fatsii (Physical and Chemical Processes and Facies). Izd. Nauka, Moscow, pp. 77–79 (in Russian).

Kelley, K.K. and Anderson, C.T., 1935. Contributions to the data on thermal metallurgy, Part IV. Bull. U.S. Bureau Mines, No. 384.

Khitarov, N.I., 1953. Experimental characteristics of quartz stability and silica migration under hydrothermal conditions. Trudy IV Soveshchaniya po eksperimentalnoy mineralogii i petrografii (Acta of the Fourth Conference on Experimental Mineralogy and Petrography). Izd. Akad. Nauk S.S.S.R., Moscow, 2: 180–186 (in Russian).

Khodyush, L.Ya., 1967a. Lithological interpretation of the banded structures of ferruginous quartzites. In: Problemy osadochnoy geologii dokembriya (Problems of Precambrian Sedimentary Geology). Izd. Nauka, Moscow, 2: 120–125 (in Russian).

Khodyush, L.Ya., 1967b. Problems of the lithogenesis of Precambrian ferruginous quartzites, illustrated by the Byelozerka iron-ore district. In: Prirodnyye i trudovyye resursy Levobereznoy Ukrainy i ikh ispolzovaniye (Natural and Labor Resources of Left-bank Ukraine and their Utilization), Kharkov, 2: 82–83 (in Russian).

Khodyush, L.Ya., 1969. The banded structure of ferruginous quartzites and its origin. In: Problemy obrazovaniya zhelezistykh porod dokembriya (Problems of Formation of the Precambrian Iron Formations). Izd. Naukova Dumka, Kiev, pp. 242–258 (in Russian).

Khodyush, L.Ya., 1971. Genetic classification of Precambrian iron formations. Litologiya i poleznyye iskopayemyye (Lithology and Mineral Deposits), 3: 47–56 (in Russian).

Khvorova, I.V., 1968. Silica accumulation in geosynclinal regions of the past. Trudy Geol. Inst. Akad. Nauk S.S.S.R. (Trans. Geol. Inst. Acad. Sci. U.S.S.R.), Moscow, 195: 9–136 (in Russian).

King, E.G. and Weller, W.W., 1970. Low-temperature heat capacities and entropies at 298, 15°K of goethite and pyrophyllite. U.S. Bur. Miner. Rept. Inv., N 7369.

Klein, C. Jr., 1966a. Mineralogy and petrology of the metamorphosed Wabush iron formation, Southwestern Labrador. J. Petrol., 7: 246–305.

Klein, C. Jr., 1966b. Petrology of a metamorphic iron formation in Southwestern Labrador, Canada. Geol. Soc. Am. Spec. Pap., 87: 88–89.

Klein, C. Jr., 1973. Changes in mineral assemblages with metamorphism of some banded Precambrian iron-formations. Econ. Geol., 68: 1075–1088.

Klein, C. Jr., 1974. Greenalite, stilpnomelane, minnesotaite, crocidolite and carbonates in a very low-grade metamorphic Precambrian iron-formation. Can. Mineral., 12: 475–498.

Klein, C. Jr. and Bricker, O.P., 1977. Some aspects of the sedimentary and diagenetic environment of Proterozoic banded iron-formation. Econ. Geol., 72: 1457–1470.

Klein, C. Jr. and Fink, R.P., 1976. Petrology of the Sokoman iron formation in the Howells River Area at the western edge of the Labrador Trough. Econ. Geol., 71: 453–487.

Knauth, L.P. and Epstein, S., 1976. Hydrogen and oxygen isotope ratios in nodular and bedded cherts. Geochim. Cosmochim. Acta, 40: 1095–1108.

Knyazev, G.I., Reshit'ko, V.A. and Fedorovskaya, L.I., 1969. On genesis of ferruginous quartzites. Ivz. Akad. Nauk S.S.S.R., Ser. Geol. (Proc. Acad. Sci. U.S.S.R., Geol. Ser.), 12: 33–46 (in Russian).

Kolesnikov, M.P. and Yegorov, I.A., 1977. Metal porphyrins in Precambrian deposits. Doklady Akad. Nauk S.S.S.R. (Rep. Acad. Sci. U.S.S.R.), 233: 483–486 (in Russian).

Kondakov, L.A. and Perunov, V.I., 1973. On the problem of the genesis of the ferruginous quartzites of Tanzania. Izv. Akad. Nauk S.S.S.R., Ser. Geol. (Proc. Acad. Sci. U.S.S.R., Geol. Ser.), 12: 136–138 (in Russian).

Konovalov, G.S., Ivanova, A.A. and Kolesnikov, T.Kh., 1968. Disseminated and rare elements dissolved in the water and contained in the suspended matter of the main rivers of U.S.S.R. In: Geokhimia osadochnykh porod i rud (Geochemistry of Sedimentary Rocks and Ores). Izd. Nauka, Moscow, pp. 72–87 (in Russian).

Kontorovich, A.E., 1968. Forms of migration of elements in the rivers of the humid zone. In: Geokhimiya osadochnykh porod i rud (Geochemistry of Sedimentary Rocks and Ores). Izd. Nauka, Moscow, pp. 88–101 (in Russian).

Kopeykin, V.A. and Mikhaylov, A.S., 1970. Solubility and forms of silica in dilute solutions under normal conditions. Dokl. Akad. Nauk S.S.S.R. (Rep. Acad. Sci. U.S.S.R.), 191: 917–920 (in Russian).

Korde, K.B., 1974. Development of the Earth's biosphere in the Precambrian and Early Paleozoic and cosmic factors. In: Kosmos i evolutsiya organizmov (The Cosmos and the evolution of Organisms). Moscow, pp. 231–248 (in Russian).

Kornilov, N.A., 1969. Thermodynamics of low-temperature metamorphism of Precambrian ferruginous quartzites. In: Problema metamorphogennogo rudoobrazovaniya (Problems of Metamorphic Ore Formation). Izd. Naukova Dumka, Kiev, pp. 211–213 (in Russian).

Korzhinskiy, D.S., 1940. Factors of mineral equilibria and mineralogical depth facies. Trudy Inst. Geol. Nauk Akad. Nauk S.S.S.R. (Trans. Inst. Geol. Sci. Acad. Sci. U.S.S.R.) 12: 100 pp. (in Russian).

Korzhinskiy, D.S., 1952. Granitization as magmatic replacement. Izv. Akad. Nauk S.S.S.R., Ser. Geol. (Proc. Acad. Sci. U.S.S.R., Geol. Ser.), 2: 56–69 (in Russian).

Korzhinskiy, D.S., 1957. Physicochemical Principles of Analysis of the Paragenesis of Minerals. Izd. Akad. Nauk S.S.S.R., Moscow, 184 pp. (in Russian).

Kovalev, V.A. and Generalova, V.A., 1969. Geochemical features of the migration of iron in the recent peat bogs of Belorussia. Geokhimiya (Geochemistry), 2: 210–220 (in Russian).

Kozlov, M.T., 1969. Metasomatic ferruginous quartzites of the Kola Peninsula. In: Problemy obrazovaniya zhelezistykh porod dokembriya (Problems of Formation of the Precambrian Iron Formations). Izd. Naukova Dumka, Kiev, pp. 105–110 (in Russian).

Kramer, J.R., 1965. History of sea water. Constant temperature-pressure equilibrium models compared to liquid inclusion analyses. Geochim. Cosmochim. Acta, 29: 921–945.

Kranck, S.H., 1961. A study of phase equilibria in a metamorphic iron formation. J. Petrol., 2: 137–184.

Krause, A. and Lezuchowska, J., 1962. Die Alterung des röntgenamorphen Eisen-III–hydroxid–Trockengels. Kolloid-Z. Polymere, 181: 69.

Krauskopf, K.B., 1956. Dissolution and precipitation of silica at low temperatures. Geochim. Cosmochim. Acta, 10: 1–26.

Krauskopf, K.B., 1959. The geochemistry of silica in sedimentary environments. In: Geokhimiya litogeneza (Geochemistry of Lithogenesis). Izd. I.L., Moscow, pp. 210–233 (in Russian). In Symp. "Silica in Sediments", Tulsa, 1959 (in English).

Kravchenko, G.L., 1971. Pyroxenes from the ferruginous quartzites of the Azov massif. Geol. Zhurn. Akad. Nauk Ukr. S.S.R. (Geol. J. Acad. Sci. Ukr. S.S.R.), 31: 21–30 (in Russian).

Kretz, P., 1961. Some applications of thermodynamics to coexisting minerals of variable compositions. J. Geol. 69: 361–387.

Krishnan, M.S., 1973. Occurrence and origin of the iron ores of India. In: Genesis of Precambrian Iron and Manganese Deposits. Unesco, Paris, pp. 69–76.

Kroyt, G.R., 1933. Colloids. Goskhimizdat, Moscow, 200 pp. (in Russian).

Kul'man, A.G., 1957. Physical and Colloid Chemistry. Pishchepromizdat, Moscow: 414 pp. (in Russian).

Kul'skiy, L.A., 1960. Chemistry and Technology of Water Treatment. Izd. Akad. Nauk Ukr. S.S.R., Kiev, 358 pp. (in Russian).

Kurepin, V.A., 1970. On the conditions of stability of pyroxenes of the $MgSiO_3$-$FeSiO_3$ series. Geokhimiya (Geochemistry), 8: 1000–1004 (in Russian).

Kurepin, V.A., 1975. Activity of the components, thermodynamic characteristics of the reactions and phase equilibria in the system Fe–O at high temperatures and pressures. Geokhimiya (Geochemistry), 10: 1475–1483 (in Russian).

Kuznetsov, V.V., 1964. Physical and Colloid Chemistry. Izd. Vysshaya Shkola, Moscow, 388 pp. (in Russian).

La Berge, G.L., 1964. Development of magnetite in iron-formations of the Lake Superior region. Econ. Geol., 59: 1313–1342.

La Berge, G.L., 1967. Microfossils and Precambrian iron-formations. Geol. Soc. Am. Bull., 78: 331–342.

La Berge, G.L., 1973. Possible biological origin of Precambrian iron-formations. Econ. Geol., 66: 1098–1109.

Landis, C.A., 1971. Graphitization of dispersed carbonaceous material in metamorphic rocks. Contrib. Mineral. Petrol., 30: 34–45.

Langmuir, D., 1969. The Gibbs free energies of substances in the system $Fe-O_2-H_2O-CO_2$ at 25°C. U.S. Geol. Surv. Prof. Pap., 650-B: 180–184.

Lapteva, O.N., 1958. On the dependence of the redox potential of a ferric-ferrous solution on the pH value. Zhurn. Prikl. khimii (J. Appl. Chem.), 31(8) (in Russian).

Larimer, J.W., 1968. Experimental studies on the system $Fe-MgO-SiO_2-O_2$ and their bearing on the petrology of chondritic meteorites. Geochim. Cosmochim. Acta, 32(11): 1187–1207.

Larsen, G. and Chilingar, G.V., 1967. Developments in Sedimentology: Diagenesis in Sediments. Elsevier, Amsterdam–New York, 551 pp.

Latimer, W.M., 1954. Oxidation Potentials. Izd. I.L., Moscow, 406 pp. (in Russian). Prentice-Hall, New York, 1952 (in English).

Lazarenko, E.K. (Editor-in-Chief), 1977. Mineralogy of the Krivoy Rog Basin. Izd. Naukova Dumka, Kiev, 544 pp. (in Russian).

Laz'ko, Ye.M., 1969. Particulars of sedimentation in the pre-geosynclinal stage of development of the Earth's crust. Geol. Zhur. Akad. Nauk Ukr. S.S.R. (Geol. J. Acad. Sci. Ukr. S.S.R.), 29: 13–22 (in Russian).

Leith, C.K., 1903. The Mesabi iron-bearing district of Minnesota. U.S. Geol. Surv. Monog., 43: 316 pp.

Lengweiler, H., Buser, W. und Feitknecht, W., 1961. Die Ermittlung der Löslichkeit von Eisen (III)-hydroxiden mit ^{59}Fe. Helv. Chim. Acta, 44: 796–805.

Lepp, H. and Goldich, S.S., 1959. Chemistry and origin of iron formation. Bull. Geol. Soc. Am., 70: 1637.

Lepp, H. and Goldich, S., 1964. Origin of Precambrian iron formations. Econ. Geol., 59: 1025–1060.

Levashkevich, G.A., 1968. Interaction of humic acids with ionic forms of iron and aluminum. In: Kora vyvetrivaniya (The Weathered Layer). Izd. Nauka, Moscow, 10: 249–259 (in Russian).

Levenshteyn, M.L. and Polunovskiy, R.M., 1971. On the conditions of accumulation and thermodynamic parameters of metamorphism of the Azov graphites. Sov. Geologiya (Soviet Geology), 1: 109–115 (in Russian).

Levina, S.A. and Yermolenko, N.F., 1955. Structure and adsorption activity of the aluminum, chrome and iron hydroxides depending on conditions of their formation. Koll. Zhurn. (Colloid J.), 17, N4: 287–294 (in Russian).

Lezuchowska, J., 1963. Systematyka wodorotlenków żlazowych. Wiadomośći Chemiczne, 17: 521–535 (in Polish).

Liddy, J.C., 1968. The jaspilite iron ores of Australia. Econ. Geol., 63: 815–821.

Lindsley, D.H., Speidel, D.H. and Nafziger, R.H., 1968. P-T-f_{O_2} relations for the system $Fe-O-SiO_2$. Am. J. Sci., 266: 342–360.

Lisitsyn, A.G., Belov, Yu.I., Bogdanov, Yu.A. and Bogoyavlenskiy, A.N., 1966. Regularities of the distribution and forms of silica suspended in the waters of the World Ocean. In: Geokhimiya kremnezema (Geochemistry of Silica). Izd. Nauka, Moscow, pp. 37–89 (in Russian).

Lopukhin, A.S., 1974. The primitive Earth and paleobiology. Sov. Geologiya (Soviet Geology), 8: 153–155 (in Russian).

Lovering, T.S., 1923. The leaching of iron protores: solution and precipitation of silica in cold water. Econ. Geol., 18: 523–540.

Mackenzie, F.T. and Garrels, R.M., 1966. Silica-bicarbonate balance in the ocean and early diagenesis. J. Sediment. Petrol., 36: 1075–1084.

Mackenzie, F.T. and Gees, R., 1971. Quartz: synthesis at earth-surface conditions. Science, 174: 533–535.

Macleod, W.N., 1966. The geology and iron deposits of the Hamersley Range area, Western Australia. West. Aust. Geol. Surv. Bull., 117.

Malakhovskiy, V.F., 1956. Geochemistry and main problems of the origin of the Kerch supergene iron ores. In: Voprosy mineralogii osadochnykh obrazovaniy (Problems of the Mineralogy of Sedimentary Formations). Izd. Lvov Univ., 3–4: 190–202 (in Russian).

Marakushev, A.A., 1968. Thermodynamics of Metamorphic Hydration of Minerals. Izd. Nauka, Moscow, 200 pp. (in Russian).

Marakushev, A.A. and Fed'nik, V.V., 1969. On the thermodynamics of carbonation processes. Geokhimiya (Geochemistry), 12: 1514–1520 (in Russian).

Marchenko, R.T., 1965. Physical and Colloidal Chemistry. Izd. Vysshaya Shkola, Moscow, 374 pp. (in Russian).
Markhinin, Ye.K., 1967a. Role of Volcanism in the Formation of the Earth's Crust. Izd. Nauka, Moscow: 255 pp. (in Russian).
Markhinin, Ye.K., 1967b. On the geochemistry of the Kurile hot springs. In: Regional'naya geotermiya i rasprostraneniye termal'nykh vod v S.S.S.R. (Regional Geothermy and Distribution of Thermal Waters in the U.S.S.R.). Izd. Nauka, Moscow, pp. 295–297 (in Russian).
Markov, M.S., 1959a. The jaspilite formation of the volcanogenic cherty series in the Karsakpay synclinorium. In: Zakomomernosti razmeshcheniya poleznykh iskopayemykh (Regularities in the Distribution of Mineral Deposits). Izd. Akad. Nauk S.S.S.R., 2: 221–243 (in Russian).
Markov, M.S., 1959b. On the stratigraphy and tectonic position of the jaspilites of the Karsakpay synclinorium. Izv. Akad. Nauk S.S.S.R., Ser. Geol. (Proc. Acad. Sci. U.S.S.R., Geol. Ser), 4: 30–46 (in Russian).
Marks, V.A., 1968. Conditions of equilibrium of iron-magnesium minerals in the metamorphic rocks of the greenstone zone of the Central Urals. In: Mineralog. Sb. Inst. Geol. i Geokimii Ural'skogo Fil. Akad. Nauk S.S.S.R. (Mineralogical Collected Articles, Inst. Geol. Geochemistry, Ural Branch Acad. Sci. U.S.S.R.), 8: 95–101 (in Russian).
Marshall, L.C. and George, N.F., 1969. The evolution of oxygen and nitrogen in the Earth's atmosphere: with notes on the evolution of life on Earth and the Martian atmosphere. School. Sci. Math., 69: 13–30.
Martynenko, L.I., 1950. On the problem of the formation of the high grade ores of Krivoy Rog. Uch. Zap. Chernovitsk. Univ., Ser. Geol., Geogr. (Sci. Mem. Chernovtsy Univ., Ser. Geol. Geogr.), 2: 65–87 (in Russian).
Mason, B., 1971. Principles of Geochemistry. Third edition. Izd. Nedra, Moscow, 311 pp. (in Russian). Wiley, New York, London, Sydney, 1966 (in English).
Mattoo, B.N., 1960. Characteristics of equilibria in systems of ferric-ferrous compounds on the basis of a new treatment of Lapteva's experimental data. Zhurn. Prikl. Khimii (J. App. Chem.), 33: 2015–2019 (in Russian).
McKirdy, D.M., 1974. Organic geochemistry in Precambrian research. Precambrian Res., 1: 75–137.
Medaris, L.G. Jr., 1969. Partitioning of Fe^{2+} and Mg^{2+} between coexisting synthetic olivine and orthopyroxene. Am. J. Sci., 267: 945–968.
Meleshko, P.S., 1959. On the higher ferric hydroxide. Dokl. Akad. Nauk S.S.S.R. (Rep. Acad. Sci. U.S.S.R.), 124: 1071 (in Russian).
Mel'nik, Yu.P., 1962a. Minerals of the iron-ore formations. In: Geologiya Krivorozhrhokikh zhelezorud-nykh mestorozhdeniy, (Geology of the Krivoy Rog Iron Ore Deposits). Izd. Akad. Nauk, Ukr. S.S.R., 1: 126–185 (in Russian).
Mel'nik, Yu.P., 1962b. Possible relation of the resistance of rocks to granitization and their equilibrium and energy anisotropy. Geol. Zhurn. Akad. Nauk Ukr. S.S.R. (Geol. J. Acad. Sci. Ukr. S.S.R.), 22: 10–21 (in Ukrainian).
Mel'nik, Yu.P., 1964. Thermodynamic analysis of the conditions of formation of the ore minerals in the metamorphism of the Precambrian iron ore formation. Geol. Rudn. Mestorozhdeniy (Geology of Ore Deposits), 5: 3–14 (in Russian).
Mel'nik, Yu.P., 1966a. Experimental investigations on the oxidation of siderite by water vapor. In: Issledovaniya prirodnogo i tekhicheskogo mineraloobrazovaniya (Investigations of Natural and Technological Minerogenesis). Izd. Nauka, Moscow, pp. 120–126 (in Russian).
Mel'nik, Yu.P., 1966b. Physicochemical analysis of the stability of rocks under the conditions of ultrametamorphism and granitization. In: Problemy teorii i eksperimenta v rudoobrazovanii (Problems of Theory and Experiment in Ore Deposition). Izd. Naukova Dumka, Kiev, pp. 114–138 (in Russian).
Mel'nik, Yu.P., 1966c. Theoretical and experimental study of the conditions of formation and stability of magnetite in the metamorphic iron formations and ores of the Krivoy Rog deposits. In: Problemy teorii i eksperimenta v rudoobrazovanii (Problems of Theory and Experiment in Ore Deposition). Izd. Naukova Dumka, Kiev, pp. 58–98 (in Russian).
Mel'nik, Yu.P., 1969. Particulars of the conversion of siderite to magnetite in the metamorphism of

Precambrian iron formations. In: Problemy obrazovaniya zhelezistykh porod dokembriya (Problems of Formation of the Precambrian Iron Formations). Izd. Naukova Dunka, Kiev, pp. 259–271 (in Russian).

Mel'nik, Yu.P., 1971. Thermodynamic properties of grunerite. In: Konstitutsiya i svoystva Mineralov (Constitution and Properties of Minerals). Izd. Naukova Dumka, Kiev, Vol. 5, pp. 84–89 (in Russian).

Mel'nik, Yu.P., 1972a. Thermodynamic model of the geochemical evolution of the atmosphere and ocean in the Precambrian. Geokhimiya (Geochemistry), 12: 1856–1864 (in Russian). Geochem. Int., 1975 (in English).

Mel'nik, Yu.P., 1972b. Thermodynamic Constants for Analysis of the Conditions of Formation of Iron Ores. Izd. Naukova Dumka, Kiev, 195 pp. (in Russian).

Mel'nik, Yu.P., 1972c. Thermodynamic parameters of compressed gases and some particulars of metamorphic reactions involving water and carbon dioxide. Geokhimiya (Geochemistry), 6: 654–662 (in Russian). Geochem. Int., 1973, 5–6: 419–426 (in English).

Mel'nik, Yu.P., 1973. On the origin of the Precambrian ferruginous quartzites. Geol. Zhurn. Akad. Nauk Ukr. S.S.R. (Geol. J. Acad. Sci. Ukr. S.S.R.), 33: 3–16 (in Russian).

Mel'nik, Yu.P., 1975. Sedimentary minerogenesis in the history of Krivoy Rog iron ore deposits. In: Osadochnoye mineraloobrazovaniye (Sedimentary Minerogenesis). Izd. Naukova Dumka, Kiev, Vol. 2, pp. 3–9 (in Russian).

Mel'nik, Yu.P., 1976. On the effect of pressure on metamorphic reactions involving water and carbon dioxide. In: Termodinamicheskiy rezhim metamorfisma (Thermodynamic Regime of Metamorphism). Izd. Nauka, Leningrad, pp. 86–97 (in Russian).

Mel'nik, Yu.P., 1978. Thermodynamic Properties of Gases under the Conditions of Deep Petrogenesis. Izd. Naukova Dumka, Kiev, 152 pp. (in Russian).

Mel'nik, Yu.P. and Kucher, V.M., 1967. On the effect of pressure on the rate of decarbonation reactions in metamorphism. Dopovidi Akad. Nauk Ukr. S.S.R. (Rep. Acad. Sci. Ukr. S.S.R.), 5: 416–419 (in Ukrainian).

Mel'nik, Yu.P. and Lugovaya, I.P., 1972. On the origin of the ore minerals of Precambrian ferruginous quartzites according to the data of a study of the oxygen isotopic composition. Geokhimiya (Geochemistry), 10: 1215–1225 (in Russian). Geochem. Int., 1973, 9(10): 808–817 (in English).

Mel'nik, Yu.P. and Radchuk, V.V., 1977a. The stability of grunerite—$Fe_7Si_8O_{22}(OH)_2$. Geokhimiya (Geochemistry), 5: 693–704 (in Russian).

Mel'nik, Yu.P. and Radchuk, V.V., 1977b. Thermodynamic model of metamorphism of sideritic ferruginous quartzites. Geol. rudnykh mestorozhdeniy (Geology of Ore Deposits), 5: 74–86 (in Russian).

Mel'nik, Yu.P. and Radchuk, V.V., 1978. Plotting of diagrams of mineral equilibria taking into account new data on the fugacities of H_2O and CO_2 at temperatures up to 1500°K and pressures up to 15 kbar. Geol. Zhurn. Akad. Nauk Ukr. S.S.R. (Geol. J. Acad. Sci. Ukr. S.S.R.), 38(2): 1–15 (in Russian).

Mel'nik, Yu.P. and Siroshtan, R.I., 1973. Physico-chemical conditions of the metamorphism of cherty-iron rocks. In: Genesis of Precambrian Iron and Manganese Deposits. Unesco, Paris, pp. 209–216.

Mel'nik, Yu.P. and Yaroshchuk, M.A., 1966. A thermodynamic analysis of the conditions of formation of the olivine-magnetite rocks and ores of the Volodarsk district of the Ukrainian shield. In: Problemy teorii i eksperimenta v rudoobrazovanii (Problems of Theory and Experiment in Ore Deposition). Izd. Naukova Dumka, Kiev, pp. 98–113 (in Russian). Geochem. Int. 1966, 3: 1218–1229 (in English).

Mel'nik, Yu.P. and Yaroshchuk, M.A., 1971. On some particulars of metamorphism of rocks under water-deficient conditions. Izd. Akad. Nauk S.S.S.R. (Proceed. Acad. Sci. U.S.S.R.), N 11: 75–84 (in Russian). Int. Geol. Rev., 1971, 13: 1439–1446 (in English).

Mel'nik, Yu.P., Drozdovskaya, A.A., Vorob'yova, K.A. and Kalinichenko, A.M., 1973. New experimental and theoretical data on the diagenesis of ferruginous sediments. Geol. Zhurn. Akad. Nauk Ukr. S.S.R. (Geol. J. Acad. Sci. Ukr. S.S.R.), 33(3): 32–40 (in Russian).

Mel'nik, Yu.P., Yaroshchuk, M.A., Rus'ko, Yu.A. and Gamarnik, M.Ya., 1975. Graphite in the cherty iron-formations of the Krivoy Rog and its evolution during metamorphism. Geol. Zhurn. Akad. Nauk Ukr. S.S.R. (Geol. J. Acad. Sci. Ukr. S.S.R.), 35(3): 24–31 (in Russian).

Mengel, J.T., 1973. Physical sedimentation in Precambrian cherty iron formations of the Lake Superior type. Int. Union Geol. Sci., A, 3: 179–193.

Miller, A.R., Densmore, C.D., Degens, E.T., Hathaway, J.C., Manheim, F.T., McFarlin, R.F., Pocklington, R. and Jokela, A., 1966. Hot brines and recent iron deposits in deeps of the Red Sea. Geochim. Cosmochim. Acta, 30: 341–360.

Mitsyuk, B.M., 1974. Interaction of Silica with Water in Hydrothermal Conditions. Izd. Naukova Dumka, Kiev, 88 pp. (in Russian).

Mitsyuk, B.M., Gorogodskaya, L.I. and Rastrenenko, A.I., 1976. The nature and properties of new varieties of silica. Geokhimiya (Geochemistry), 6: 803–814 (in Russian).

Miyake, Ya., 1969. Elements of Geochemistry. Izd. Nedra, Leningrad, 327 pp. (in Russian). Maruzed Comp., Tokyo, Japan, 1964.

Mokiyevskaya, V.V., 1959. On the problem of the geochemistry of iron in sea water. Trudy Inst. Okeanologii Akad. Nauk S.S.S.R. (Trans. Inst. Oceanology Acad. Sci. U.S.S.R.), 33: 114–125 (in Russian).

Mokiyevskaya, V.V., 1962. Methods of determining Fe in sea water and interstitial water. Trudy Inst. Okeanologii Akad. Nauk S.S.S.R. (Trans. Inst. Oceanology Acad. Sci. U.S.S.R.), 54: 115–124 (in Russian).

Moore, E., 1918. Iron-formation on Belcher Islands, Hudson Bay. J. Geol., 24(6): 412–438.

Moore, E. and Maynard, M., 1929. Solution, transportation and precipitation of iron and silica. Econ. Geol., No. 3–5.

Moorhouse, W.W. and Beales, F.W., 1962. Fossils from the Animikie, Port Arthur, Ontario. Trans. R. Soc. Can., 56: 97–110.

Morey, G.B., 1973. Mesabi, Gunflint and Cuyuna ranges, Minnesota. In: Genesis of Precambrian Iron and Manganese Deposits. Unesco, Paris, pp. 193–208.

Muan, A., 1967. Stabilities of oxide compounds and activity/composition relations in oxide solid solutions. Proc. Brit. Ceram. Soc., 8: 103–113.

Muan, A., Nafziger, R.H. and Roeder, R.L., 1964. A method for determination of the instability of ferrosilite. Nature, 202: 688–689.

Mueller, R.F., 1960. Compositional characteristics and equilibrium relations in mineral assemblages of a metamorphosed iron formation. Am. J. Sci., 258: 449–497.

Mueller, R.F., 1961a. Analysis of relations among Mg, Fe and Mn in certain metamorphic minerals. Geochim. Cosmochim. Acta, 25: 267–269.

Mueller, R.E., 1961b. Oxidation in high temperature petrogenesis. Am. J. Sci., 259: 460–480.

Naboko, S.I., 1963. Hydrothermal metamorphism of rocks in volcanic regions. Izd. Akad. Nauk S.S.S.R., Moscow, 172 pp. (in Russian).

Naboko, S.I., 1968. Chemistry of real minerogenic solutions. Geol. i Geofiz. (Geology and Geophysics) 3: 3–13 (in Russian).

Naboko, S.I. and Piyp, B.I., 1961. Recent metamorphism of the volcanic rocks in the region of Pauzhetka hot springs. Trudy Labor. Vulkanolog. Akad. Nauk S.S.S.R. (Trans. Labor Volcanology Acad. Sci. U.S.S.R.), 19: 99–114 (in Russian).

Naboko, S.I. and Sil'nichenko, V.G., 1960. The problem of the role of carbonic acid gas in the postvolcanic process. Trudy Labor. Vulkanology. Akad. Nauk S.S.S.R. (Trans. Labor. Volcanology Acad. Sci. U.S.S.R.), 18: 139–143 (in Russian).

Nagy, B., Nagy, L.A., Zumberge, J.E., Sklarew, D.S. and Anderson, Ph., 1975. Biological evolutionary trends and related aspects of carbon chemistry during the Precambrian between 3800 m.y. and 2300 m.y. Int. Geol. Congr., 25th, Canberra, Vol. 1: 33–34.

Naumov, G.B., Ryzhenko, B.N. and Khodakovskiy, I.L., 1971. Handbook of Thermodynamic Values. Atomizdat, Moscow, 239 pp. (in Russian).

Nikol'skiy, A.P., 1970. Geochemistry of volcanogenic-sedimentary formations. Geol. Zhurn. Akad. Nauk Ukr. S.S.R. (Geol. J. Acad. Sci. Ukr. S.S.R.), 30(1): 3–18 (in Russian).

Nikol'skiy, A.P. and Kaukin, B.V., 1969. Metamorphism of the Precambrian iron formations of the southwestern part of the Russian platform. In: Problemy obrazovaniya zhelezistykh porod dokembriya (Problems of the Formation of the Precambrian Iron Formations). Izd. Naukova Dumka, Kiev, pp. 145–155 (in Russian).

Novokhatsky, I.P., 1973. Precambrian cherty iron formations of Kazakhstan. In: Genesis of Precambrian Iron and Manganese Deposits. Unesco, Paris, pp. 153–158.

Nriagu, J.O., 1975. Thermochemical approximations for clay minerals. Am. Mineral., 60: 834–839.

Oehler, J.H., 1976. Hydrothermal crystallization of silica gel. Geol. Soc. Am. Bull., 87: 1143–1152.

Oftedahl, C., 1958. A theory of exhalative-sedimentary ores. Geol. Fören. Stockholm Förhandl., 80: 1–19.

O'Hara, M.J., 1963. Distribution of iron between coexisting olivines and calcium-poor pyroxenes in peridotites, gabbros, and other magnesian environments. Am. J. Sci., 261: 32–46.

Okamoto, G., Okura, T. and Goto, K., 1957. Properties of silica in water. Geochim. Cosmochim. Acta, 12: 123–132.

Okkerse, C. and de Boer, J.H., 1961. Reactivity of Solids. Amsterdam, 240 pp.

O'Neil, J.R. and Clayton, R.N., 1963. Oxygen isotopic geothermometry. In: Isotopic and Cosmic Chemistry, Amsterdam, pp. 157–168.

Oparin, A.I., 1957. On the Origin of Life on the Earth. Izd. Akad. Nauk S.S.S.R., Moscow, 459 pp. (in Russian).

Ostapenko, G.T., 1977. Thermodynamics of Nonhydrostatic Systems and its Application to the Theory of Metamorphism. Izd. Naukova Dumka, Kiev, 240 pp. (in Russian).

Ostroumov, E.A., 1955. Iron in the bottom sediments of the Sea of Okhotsk. Dokl. Akad. Nauk S.S.S.R. (Rep. Acad. Sci. U.S.S.R.), 102(1): 129–132 (in Russian).

Ostroumov, E.A. and Volkov, I.I., 1967. Geochemical behavior of sulphur in the bottom sediments of the Pacific Ocean. Trudy Inst. Okeanologii Akad. Nauk, S.S.S.R. (Trans. Inst. Oceanology Acad. Sci. U.S.S.R.), 83: 68–82 (in Russian).

Parak, T., 1975. Kiruna iron ores are not "Intrusive-magmatic ores of the Kiruna type". Econ. Geol., 70: 1242–1258.

Perchuk, L.L., 1968. Principle of the effect of temperature on the equilibria of natural iron-magnesian minerals. Izv. Akad. Nauk S.S.S.R., Ser. Geol. (Proc. Acad. Sci. U.S.S.R., Geol. Ser.), 12: 3–30 (in Russian).

Perchuk, L.L., 1970. Equilibria of Rock-forming Minerals. Izd. Nauka, Moscow, 391 pp. (in Russian).

Perel'man, A.I., 1968. Geochemistry of Epigenetic Processes (supergene zone), 3rd ed. Izd. Nedra, Moscow, 331 pp. (in Russian).

Perry, E.C. Jr., 1967. The oxygen isotope chemistry of ancient cherts. Earth Planet. Sci. Lett. 3: 62–66.

Perry, E.C. Jr. and Bonnichsen, B., 1966a. Oxygen isotope exchange between quartz and magnetite from the metamorphosed Biwabik iron formation, Minnesota. Trans. Am. Geophys. Union, 47: 212.

Perry, E.C. Jr. and Bonnichsen, B., 1966b. Quartz and magnetite: oxygen-18–oxygen-16 fractionation in metamorphosed Biwabik iron formation. Science, 153: 528–529.

Perry, E.C. and Tan, F.C., 1972. Significance of oxygen and carbon isotope variations in Early Precambrian cherts and carbonate rocks of Southern Africa. Geol. Soc. Am. Bull., 83: 647–664.

Perry, E.C. Jr. and Tan, F.C., 1973. Significance of carbon isotope variations in carbonates from the Biwabik Iron Formation, Minnesota. In: Genesis of Precambrian Iron and Manganese Deposits. Unesco, Paris, pp. 299–305.

Perry, E.C. Jr., Tan, F.C. and Morey, G.B., 1973. Geology and stable isotope geochemistry of the Biwabik Iron Formation, northern Minnesota. Econ. Geol., 68: 1110–1125.

Petrov, R.P., 1964. Some problems of the nomenclature of Precambrian rocks. In: Geologiya i poleznye iskopaemye Central'no-Chernozemnykh oblastey (Geology and Mineral Deposits of the Central Chernozem District). Izd. Voronezh Univ., pp. 78–79 (in Russian).

Petrova, M.A., 1958. On the source of free silica in volcanic regions. Trudy Mosk. Geol.-Razv. Inst. (Trans. Moscow Geol. Eng. Inst.), 32: 47–56 (in Russian).

Petrovskaya, N.V. (Editor-in-Chief), 1975. Supergene Iron Oxides in Geological Processes. Izd. Nauka, Moscow, 206 pp. (in Russian).

Pettijohn, F.J., 1957. Paleocurrents of Lake Superior Precambrian quartzites. Bull. Geol. Soc. Am., 68: 469–480.

Piatnitsky, P.P., 1924. The genetic relations of the Krivoy Rog ore deposits. Trudy Inst. Priklad. Mineralogii i Petrografii (Trans. Inst. Econ. Mineral. Petrography), 17: 42 pp (in Russian). Lithogaea, 1924, 9: 56 (in English).

Pirogov, B.I. and Pirogova, V.V., 1969. Mineral relationships and formation of the banding in the iron cherts of the Krivoy Rog. In: Problemy obrazovaniya zhelezistykh porod dokembriya (Problems of the Formation of the Precambrian Iron Formations). Izd. Naukova Dumka, Kiev, pp. 228–242 (in Russian).

Pirogov, B.I. and Shtoda, R.T., 1969. The composition and properties of pyroxenes from the ferruginous quartzites of the Petrov deposit. In: Konstitutsiya i svoystva mineralov (Constitution and Properties of Minerals). Izd. Naukova Dumka, Kiev, pp. 131–139 (in Russian).

Pisarenko, A.P., Pospelova, K.A. and Yakovlev, A.G., 1964. Course in Colloid Chemistry. 2nd Ed. Izd. Vysshaya Shkola, Moscow, 247 pp. (in Russian).

Plaksenko, N.A., 1966. Principal regularities of iron-ore sedimentation in the Precambrian. Izd. Voronezh Univ., 264 pp. (in Russian).

Plaksenko, N.A., 1969. Particulars of the paleogeographic setting of the formation of the cherty iron sediments of the Kursk group and problems of the theory of Precambrian iron-ore deposition. In: Problemy obrazovaniya zhelezistykh porod dokembriya (Problems of Formation of the Precambrian Iron Formations). Izd. Naukova Dumka, Kiev, pp. 11–27 (in Russian).

Polkanov, A.A., 1939. Pre-Quaternary geology of the Kola Peninsula and Karelia or of the easternmost parts of the Fennoscandian shield. Int. Geol. Congr., 17th, Moscow, Vol. 2 (in Russian).

Polovinkina, Yu.I., 1952. Mafic and Ultramafic Rocks of the Karsakpay in Connection with Problems of the Genesis of Ferruginous Quartzites. Gosgeolizdat, Moscow, 88 pp. (in Russian).

Polovko, N.I., 1956. Fayalitic quartzites of the Right-Bank group of magnetic anomalies. Dopovīdī Akad. Nauk. Ukr. R.S.R. (Rep. Acad. Sci. Ukr. S.S.R.), 4: 391–393 (in Ukrainian).

Presley, B.J. and Kaplan, I.R., 1968. Changes in dissolved sulfate, calcium and carbonate from interstitial water of near-shore sediments. Geochim. Cosmochim. Acta, 32: 1037–1048.

Rabinowitzh, A.I. and Kargin, W.A., 1928. Über die Elektrolytkoagulation der Kolloide. Z. Phys. Chem., 133: 203–232.

Ramberg, H. and De Vore, G., 1951. The distribution of Fe^{2+} and Mg^{2+} in coexisting olivines and pyroxenes. J. Geol., 59: 193–210.

Rankama, K., 1955. Geologic evidence of chemical composition of the Precambrian atmosphere. Geol. Soc. Am., Spec. Pap., 62: 651–664.

Rankama, K., 1957. Geological data on the chemical composition of the Precambrian atmosphere. In: Zemnaya kora (Crust of the Earth). Izd. IL, Moscow, pp. 672–686 (in Russian).

Read, H.H., 1949. Meditations on granite. In: Problemy obrazovaniya granitov (Problems of Granite Formation). Izd. IL, Moscow (in Russian). Proc. Geol. Assoc., 1, 54: 64–85 (1943); 55: 45–93 (1944) (in English).

Regnell, U., 1961. On pyrite in deep-sea sediments. Bull. Geol. Inst. Univ. Uppsala, 40: 305–314.

Robie, R.A. and Waldbaum, D.R., 1968. Thermodynamic properties of minerals and related substances at 298.15°K (25.0°C) and one atmosphere (1.013 bars) pressure and at higher temperatures. Bull. U.S. Geol. Surv., 1259, 252 pp.

Robie, R.A., Bethke, P.M. and Beardsley, K.M., 1967. Selected X-ray crystallographic data, molar volumes, and densities of minerals and related substances. Bull. U.S. Geol. Surv., 1248(IV), 87 pp.

Ronov, A.B., 1959. On the post-Precambrian geochemical history of the atmosphere and hydrosphere. Geokhimiya (Geochemistry), 5: 397–408 (in Russian).

Ronov, A.B., 1964. General tendency in the evolution of the composition of the Earth's crust, ocean and atmosphere. Geokhimiya (Geochemistry), 8: 715–743 (in Russian).

Ronov, A.B. and Migdisov, A.A., 1970. Evolution of the chemical composition of the rocks of shields and of the sedimentary cover of the Russian and North American platforms. Geokhimiya (Geochemistry), 4: 403–438 (in Russian).

Ronov, A.B. and Yarochevskiy, A.A., 1971. The chemical composition of the Earth's crust. In: Svyaz' poverkhnostnykh struktur zemnoy kory s glubinnymi (Relationship of Surface and Deep-Seated Structures of the Earth's Crust). Izd. Naukova Dumka, Kiev, pp. 192–207 (in Russian).

Rozanov, S.B., 1971. Structure of the Greenschist Ironore Beds of the Karsakpay Synclinorium. Diss., Moscow (in Russian).

Rozanov, S.B. and Filatova, L.I., 1967. On the Precambrian iron formations of the Ulutau (Central Kazakhstan). In: Problemy osadochnoy geologii dokembriya (Problems of Precambrian Sedimentary Geology). Izd. Nedra, Moscow, pp. 317–321 (in Russian).

Rubey, W.W., 1957. Development of the hydrosphere and atmosphere, with special reference to probable composition of the early atmosphere. In: Zemnaya Kora (Crust of the Earth). Izd. IL, Moscow, pp. 650–671 (in Russian). Geol. Soc. Am. Spec. Pap., 1955, 62: 631–650 (in English).

Ruckmick, J.C., 1963. The iron ores of Cerro Bolívar, Venezuela. Econ. Geol., 58: 218–236.

Rutten, M.G., 1962. Geological Aspects of the Origin of Life on Earth. Elsevier, New York–Amsterdam, 146 pp.

Sakamoto, T., 1950. The origin of the Precambrian banded iron ores. Am. J. Sci., 248: 449–474.

Sastry, T.H., 1967. Some structural features of Ongole banded magnetite quartzites. J. Indian Geosci. Assoc., 7: 67–84.

Saxena, S.K., 1969a. Silicate solid solutions and geothermometry. 1. Use of the regular solution model. Contrib. Mineral. Petrol., 21: 338–345.

Saxena, S.K., 1969b. Silicate solid solutions and geothermometry. 2. Distribution of Fe^{2+} and Mg^{2+} between coexisting olivine and pyroxene. Contrib. Mineral. Petrol., 22: 147–156.

Schidlowski, M., Eichmann, R. and Junge, C.E., 1975. Precambrian sedimentary carbonates: carbon and oxygen isotope geochemistry and implications for the terrestrial oxygen budget. Precambrian Res., 2: 1–69.

Schidlowski, M., Eichmann, R. and Fiebiger, W., 1976. Isotopic fractionation between organic carbon and carbonate carbon in Precambrian banded ironstone series from Brazil. Neues Jahrb. Mineral. Monatsh., 8: 344–353.

Schindler, P., Michaelis, W. und Feitknecht, W., 1963. Löslichkeitsprodukte von Metalloxiden und -hydroxiden. Helv. Chim. Acta, 46: 444–449.

Schmalz, R.F., 1959. A note on the system Fe_2O_3–H_2O. J. Geophys. Res., 64: 575–579.

Schürmann, K., 1967. Hydrothermale experimentelle Untersuchungen an metamorphen monoklinen Hornblenden. Teil I: Zur Stabilität der Cummingtonite. Neues Jahrb. Miner. Monatsh., 9–10: 270–278.

Seguin, M., 1968a. The Fe–C–O–S + (H_2O) system. Part I. Experimental results. Nat. Can., 95: 1195–1215.

Seguin, M., 1968b. The Fe–C–O–S + (H_2O) system. Part II. Applications to iron formations in the Michipicoten and Central Labrador areas. Nat. Can., 95: 1217–1230.

Seguin, M.K., 1971. Phase relations in the Fe–C–O–S–H_2O system and its geological application. Chem. Geol., 7: 5–18.

Semenenko, N.P., 1953. Metamorphism of the Ukrainian cristalline massif. Geol. Zhurn. Akad. Nauk Ukr. R.S.R. (Geol. J. Acad. Sci. Ukr. S.S.R.), 1: 20–28 (in Ukrainian).

Semenenko, N.P., 1966. Metamorphism of Mobile Zones. Izd. Naukova Dumka, Kiev, 298 pp. (in Russian).

Semenenko, N.P., 1973. The iron-chert formations of the Ukrainian shield. In: Genesis of Precambrian Iron and Manganese Deposits. Unesco, Paris, pp. 135–142.

Semenenko, N.P., 1975. The Continental Crust. Izd. Naukova Dumka, Kiev, 198 pp. (in Russian).

Semenenko, N.P., Polovko, N.I., Zhukov, G.V., Ladiyeva, V.D. and Makukhina, A.A., 1956. Petrography of the Cherty Iron-Formations of the Ukrainian S.S.R. Izd. Akad. Nauk Ukr. S.S.R., Kiev, 536 pp. (in Russian).

Semenenko, N.P., Polovko, N.I., Grizkov, Ya.M., Dobrokhotov, M.N., Makukhina, A.A., Ladiyeva, V.D., Zukov, G.V. and Nastenko, A.A., 1959. Geology of the Cherty Iron-Formations of the Ukraine. Izd. Akad. Nauk Ukr. S.S.R., 666 pp. (in Russian).

Semenenko, N.P., Boyko, V.L., Borgunov, I.N., Ladiyeva, V.D. and Makukhina, A.A., 1967. Geology of Sedimentary-volcanogenic Formations of the Ukrainian Shield. Izd. Naukova Dumka, Kiev, 378 pp. (in Russian).

Sen, S.K. and Rege, S.M., 1966. Distribution of magnesium and iron between metamorphic pyroxenes from Saltora, West Bengal, India. Mineral. Mag., 35: 759–762.

Senesi, N., Griffith, S.M., Schnitzer, M. and Townsend, M.G., 1977. Binding of Fe^{3+} by humic materials. Geochim. Cosmochim. Acta, 41: 969–976.

Sevast'yanov, V.F., 1968. Redistribution of chemical elements in redox processes in Mediterranean Sea sediments. Litologiya i Poleznyye Iskopayemyye (Lithology and Mineral Deposits), 1: 3–15 (in Russian).

Shapiro, J., 1973. Blue-green algae, why they become dominant. Science, 179: 382–384.

Sharma, T., Mueller, R.F. and Clayton, R.N., 1965. O^{18}/O^{16} ratios of minerals from the iron formations of Quebec. J. Geol., 73: 664–667.

Shatskiy, N.S., 1954. On manganesian formations and the metallogeny of manganese. Izv. Akad. Nauk S.S.S.R., Ser. Geol. (Proc. Acad. Sci. U.S.S.R., Geol. Ser.), 4: 3–37 (in Russian).

Shatskiy, N.S., 1957. On the problem of Paleozoic spores in the Krivoy Rog group of the Ukrainian Precambrian. Izv. Akad. Nauk S.S.S.R., Ser. Geol. (Proc. Acad. Sci. U.S.S.R. Geol. Ser.), 4: 119–120 (in Russian).

Shcherbina, V.V., 1956. Complex compounds and migration of chemical elements in the supergene zone. Geokhimiya (Geochemistry), 5: 54–60 (in Russian).

Sheynfayn, R.Yu., Linkind, B.A., Stal', O.P. and Neymark, I.Ye., 1964. On the problem of the mechanism of formation of porous structure of silica gel. Koll. Zhurn. (Colloid J.), 26(6): 734–738 (in Russian).

Shiikawa, M., 1970. Limonite deposits of volcanic origin in Japan. In: volcanism and Ore Genesis, Tokyo, pp. 295–300.

Shnyukov, Ye.F., 1965. Genesis of the Cimmerian Iron Ores of the Azov-Black Sea Ore Province. Izd. Naukova Dumka, Kiev, 195 pp. (in Russian).

Shnyukov, Ye.F. and Naumenko, P.I., 1964. Cimmerian Iron Ores of the Depressed Synclines of the Kerch Peninsula. Izd. Krym, Simferopol', 126 pp. (in Russian).

Sidorenko, A.V. and Borshchevskiy, Yu.A., 1977. General trends in the evolution of the isotopic composition of carbonates in the Precambrian and Phanerozoic. Dokl. Akad. Nauk S.S.S.R. (Rep. Acad. Sci. U.S.S.R.), 234(4): 892–895 (in Russian).

Sidorenko, A.V. and Sidorenko, S.A., 1968. On the abundance of supposedly biogenic carbon in the Precambrian. Dokl. Akad. Nauk S.S.S.R. (Rep. Acad. Sci. U.S.S.R.), 183: 181–184 (in Russian).

Sidorenko, A.V. and Sidorenko, S.A., 1970. On "hydrocarbon respiration" of Precambrian graphite-bearing beds. Dokl. Akad. Nauk S.S.S.R. (Rep. Acad. Sci. U.S.S.R.), 192(1): 184–187 (in Russian).

Sidorenko, A.V. and Sidorenko, S.A., 1971. Organic matter in Precambrian sedimentary-metamorphic rocks and some geological problems. Sov. Geologiya (Soviet Geology), 5: 3–20 (in Russian).

Sidorenko, A.V., Rozen, O.M., Gimmelfarb, G.B. and Teniyakov, V.A., 1969. On the problem of the abundance of carbonate deposits in the Precambrian. Dokl. Akad. Nauk S.S.S.R. (Rep. Acad. Sci. U.S.S.R.), 189(6): 1336–1339 (in Russian).

Sidorenko, A.V., Teniyakov, V.A., Rozen, O.M., Borshchevskiy, Yu.A. and Sidorenko, S.A., 1973. Main features of the geochemistry of supergenesis in the Early Precambrian. In: Mezhdunarodniy geokhimicheskiy kongress (Int. Geochem. Congr.) Moscow, 4th, 1: 9–27 (in Russian).

Sidorenko, S.A., 1971. Organic Matter in Precambrian Sedimentary-metamorphic Rocks. Diss., Moscow (in Russian).

Sidorenko, S.A. and Sidorenko, A.V., 1975. Organic Matter in Precambrian sedimentary-metamorphic rocks. Izd. Nauka, Moscow, 117 pp. (in Russian).

Sidorov, S.S., 1967. On the step-wise character of postvolcanic hydrothermal alteration of rocks, illustrated by alunitization. In: Trudy Inst. Vulkanologii Sibirsk. Otdeleniya Akad. Nauk S.S.S.R. (Trans. Inst. Volcanology Siberian Branch Acad. Sci. U.S.S.R.), 24: 51–65 (in Russian).

Siever, R., 1962. Silica solubility, 0–200°C, and diagenesis of siliceous sediments. J. Geol., 70: 127–150.

Sillen, L.G., 1961. The physical chemistry of sea water. In: M. Sears (Editor), Oceanography. Am. Assoc. Adv. Sci. Publ., 67: 549–581.

Silverman, S.R., 1954. The isotope geology of oxygen. In: Izotopy v geologii (Isotopes in Geology). Izd. IL, Moscow: 505–528 (in Russian). Geochim. Cosmochim. Acta, 1951, 2: 26–42 (in English).

Simons, L.H., Monaghan, P.H. and Taggart, M.S. Jr., 1953. Aluminum and iron in Atlantic and Gulf of Mexico surface waters. Anal. Chem., 25: 989–990.

Singer, P.S. and Stumm, W., 1970. The solubility of ferrous iron in carbonate-bearing waters. J. Am. Water Works Assoc., 62: 198–202.

Skopintsev, B.A., 1949. On the rate of decomposition of the organic matter of dead plankton. Trudy Vses. Gidrobiol. Ob-va (Trans. All-Union Hydrobiological Soc.), 1: 34–43 (in Russian).

Skopintsev, B.A. and Popova, T.P., 1960. Some results of determination of the content of iron, manganese and copper in Black Sea water. In: Trudy Morskogo Gidrofiz. Inst. Akad. Nauk S.S.S.R. (Trans. Marine Hydrophys. Inst. Acad. Sci. U.S.S.R.), 19: 21–30 (in Russian).

Skripchenko, N.S., 1969. On the stability of sulfides and oxides in low-temperature minerogenesis. Geokhimiya (Geochemistry), 3: 301–308 (in Russian).

Smetanin, D.A., 1959. Hydrochemistry of the region of the Kurile-Kamchatka deep-sea basin. In: Trudy Inst. Okeanologii Akad. Nauk S.S.S.R. (Trans. Inst. Oceanology Acad. Sci. U.S.S.R.), 33: 43–86 (in Russian).

Smith, F.G. and Kidd, D.J., 1949. Hematite-goethite relations in neutral and alkaline solutions under pressure. Am. Mineral., 34(5–6).

Snezko, A.M. and Berezovskiy, F.I., 1975. The nature of the graphite-bearing rocks of the Krivoy Rog group. Doklady Akad. Nauk S.S.S.R. (Rep. Acad. Sci. U.S.S.R.), 222(1): 197–200 (in Russian).

Sobolev, V.S., 1961. On pressure in metamorphic processes. In: Fiziko-khimicheskiye problemy formirovaniya gornykh porod i rud (Physicochemical Problems of the Formation of Rocks and Ores), Izd. Akad. Nauk S.S.S.R., Moscow, 1: 7–16 (in Russian).

Sobolev, V.S. (Editor-in-Chief), 1970. Facies of Metamorphism. Izd. Nedra, Moscow, 432 pp. (in Russian).

Sokol, V.A., Rokhlenko, D.A. and Bromberg, A.V., 1970. On the mechanism of formation of spherical particles of ferric oxide. Zhurn. Prikl. Khimii (J. Appl. Chem.), 43: 1897–1905 (in Russian).

Sokolov, B.S., 1975. Organic world of Earth on the way to Phanerozoic differentiation. Izd. Akad. Nauk S.S.S.R., Moscow, 20 pp. (in Russian).

Sretenskaya, N.T., 1970. On the state of silica in aqueous solutions. In: Eksperimental'nyye issledovaniya protsessov mineraloobrazovaniya (Experimental Investigations of the Processes of Minerogenesis). Izd. Nauka, Moscow, pp. 31–40 (in Russian).

Stashchuk, M.F., 1968. The redox potential in geology. Izd. Nedra, Moscow, 208 pp. (in Russian). Plenum, New York–London, 121 pp. (1972, in English).

Stashchuk, M.F., 1969. Paragenesis of the minerals: ferric hydroxide–siderite–magnetite–ferrous hydroxide in sedimentary rocks. Geol. Zhurn. Akad. Nauk Ukr. S.S.R. (Geol. J. Acad. Sci. Ukr. S.S.R.), 29(1): 89–96 (in Russian).

Stashchuk, M.F., Suprychev, V.L. and Khitraya, M.S., 1964. Mineralogy, Geology, and Conditions of Formation of the Bottom Sediments of the Siwash. Izd. Naukova Dumka, Kiev, 174 pp. (in Russian).

Strakhov, N.M., 1947. Iron Ore Facies and their Analogs in the History of the Earth. Izd. Akad. Nauk S.S.S.R., Moscow, 276 pp. (in Russian).

Strakhov, N.M., 1954. The general scheme of sedimentation in recent seas and lakes of low salinity. In: Obrazovaniye osadkov v sovremennykh vodoyemakh (Formation of Sediments in Recent Water Bodies). Izd. Akad. Nauk S.S.S.R., Moscow, pp. 275–377 (in Russian).

Strakhov, N.M., 1960. Principles of the Theory of Lithogenesis. Izd. Akad. Nauk S.S.S.R., Moscow, 212 pp. (in Russian).

Strakhov, N.M., 1962. Toward an understanding of lithogenesis of volcanogenic-sedimentary type. Izv. Akad. Nauk S.S.S.R., Ser. Geol. (Proc. Acad. Sci. U.S.S.R., Geol. Ser.), 5: 3–17 (in Russian).

Strakhov, N.M., 1963. Types of Lithogenesis and their Evolution in the History of the Earth. Gosgeoltekhizdat, Moscow, 535 pp. (in Russian).

Strakhov, N.M., 1965. Toward an understanding of submarine volcanogenic-sedimentary petrogenesis. In: Vulkanogenno-osadochnye formatsii (Volcanogenic-Sedimentary Formations). Izd. Nauka, Moscow, pp. 11–23 (in Russian).

Strakhov, N.M., 1966. On some problems of the geochemistry of silica. In: Geokhimiya kremnezema (Geochemistry of Silica). Izd. Nauka, Moscow, pp. 5–10 (in Russian).

Strygin, A.I., 1960. Types of metamorphism of the iron-formations of the Krivoy Rog basin. Trudy NIGRI (Trans. Geol. Prospecting Inst.), Krivoy Rog, 2: 115–131 (in Russian).

Strygin, A.I., 1969. Metamorphism of the iron-formations of the Ukrainian Shield. In: Problemy obrazovaniya zhelezistykh porod dokembriya (Problems of the Formation of the Precambrian Iron Formations). Izd. Naukova Dumka, Kiev, pp. 155–168 (in Russian).

Sudovikov, N.G., 1950. Metasomatic granites. Vestnik Leningr. Universiteta (Proc. Leningrad Univ.), 10: 78–110 (in Russian).

Sudovikov, N.G., 1955. Migmatites, their genesis and methods of study. In: Trudy Labor. Geologii Dokembriya Akad. Nauk S.S.S.R. (Trans. Labor. Geology Precambrian), Vol. 5 (in Russian).

Svital'skiy, N.I., 1924. The Krivoy Rog iron ore deposit and genesis of its ores. Izv. Geolkoma (Proc. Geol. Committee), 43(1): 18–39 (in Russian).

Svital'skiy, N.I., Polovinkina, Yu.I., Dubyaga, Yu.G., Lisovskiy, A.L., Muzylev, S.V., Dubrova, B.S. and Rabinovich, F.K., 1932. The Krivoy Rog Iron Ore Deposits. Trudy Vses. Geol.-Razv. Ob'edinen: NKTP (Trans. All-Union Geol.-Prosp. Soc.), 153: 283 pp. (in Russian).

Swain, F.M., Bratt, J.M. and Kirkwood, S., 1970. Carbohydrates from Precambrian and Cambrian rocks and fossils. Geol. Soc. Am. Bull., 81(2): 499–504.

Tanatar-Barash, Z.I., 1965. Minnesotaite—a contact mineral of the Belozerka iron ore district. Geol. Zhur. Akad. Nauk Ukr. R.S.R. (Geol. J. Acad. Sci. Ukr. S.S.R.), 4: 94–97 (in Ukrainian).

Tardy, Y. and Garrels, R.M., 1974. A method of estimating the Gibbs energies of formation of layer silicates. Geochim. Cosmochim. Acta, 38: 1101–1116.

Tardy, Y. and Garrels, R.M., 1976. Prediction of Gibbs energies of formation—I. Relationships among Gibbs energies of formation of hydroxides, oxides and aqueous ions. Geochim. Cosmochim. Acta, 40(9): 1051–1056.

Taylor, H.P. Jr., 1970. Oxygen isotope studies of hydrothermal mineral deposits. In: Geokhimiya gidrotermal'nykh rudnykh mestorozhdeniy (Geochemistry of Hydrothermal Ore Deposits). Izd. Mir, Moscow, pp. 101–128 (in Russian). H.L. Barnes (Editor), Pennsylvania Univ., Holt, Rinehart and Winston, Inc. (1967, pp. 109–142, in English).

Taylor, S., 1972. Geochemistry of andesites. In: Rasprostranennost' elementov v zemnoy kore (Origin and Distribution of the Elements). Izd. Mir, Moscow, pp. 16–39 (in Russian). Pergamon Press, London, N.Y. 1968, pp. 559–583 (in English).

Teruo, I. and Yukiyoshi, I., 1962. On formation of fayalite by Fe_3O_4–SiO_2 solid phase reaction during a sintering process. J. Iron Steel Inst. Japan, 7.

Timofeyev, B.V., 1966. Micropaleophytologic investigation of ancient suites. Izd. Nauka, Moscow–Leningrad, pp. 238 (in Russian).

Timofeyev, B.V., 1977. The Precambrian and its organic life. In: Problemy geologii rannego dokembriya (Problems of the Geology of the Early Precambrian). Izd. Nauka, Leningrad, pp. 137–139 (in Russian).

Tochilin, M.S., 1952. Relict features in the magnetite quartzites of the KMA and their genetic interpretation. In: Mineralog. Sb, LGO (Mineral. Coll. Articles Lvov Geol. Soc.), 6: 103–106 (in Russian).

Tochilin, M.S., 1963. Origin of Ferruginous Quartzites. Gosgeolizdat, Moscow, 168 pp. (in Russian).

Tochilin, M.S., 1969. Origin and factors in the evolution of jaspilites. In: Problemy obrazovaniya zhelezistykh porod dokembriya (Problems of the Formation of the Precambrian iron formations). Izd. Naukova Dumka, Kiev, pp. 40–46 (in Russian).

Tokhtuyev, G.V., 1967. Boudinage Textures and their Role in Ore Localization. Izd. Naukova Dumka, Kiev, 215 pp. (in Russian).

Tolbert, G.E., Tremaine, J.W., Melcher, G.C. and Gomes, C.B., 1973. Geology and iron ore deposits of Serra dos Carajás, Pará, Brazil. In: Genesis of Precambrian Iron and Manganese Deposits. Unesco, Paris, pp. 271–280.

Trendall, A.F., 1966. Altered pyroclastic rocks in iron formation in the Hamersley Range, Western Australia (Discussion). Econ. Geol., 61: 1451–1458.

Trendall, A.F., 1973a. Iron-formations of the Hamersley Group of Western Australia: type examples of varved Precambrian evaporites. In: Genesis of Precambrian Iron and Manganese Deposits. Unesco, Paris, pp. 257–270.

Trendall, A.F., 1973b Precambrian iron-formations of Australia. Econ. Geol., 68: 1023–1034.

Trendall, A.F., 1973c. Varve cycles in the Weeli Wolli Formation of the Precambrian Hamersley Group, Western Australia. Econ. Geol., 68: 1089–1097.

Troshchenko, V.N., 1970. Rhythmic bedding of the iron cherts of the New Krivoy Rog deposits. Geol. Rudn. Mestorozhdeniy (Geology of Ore Deposits), XII(6): 106–111 (in Russian).

Tugarinov, A.I., 1956. Epochs of Precambrian minerogenesis. Izv. Akad. Nauk S.S.S.R., Ser. Geol. (Proceed. Acad. Sci. U.S.S.R., Geological Series), 9: 3–26 (in Russian).

Tugarinov, A.I. and Grinenko, V.A., 1965. Conditions of deposition of Lower Proterozoic formations according to data on variations in sulfur isotopic composition in sulfides. In: Problemy geokhimii (Problems of Geochemistry). Izd. Nauka, Moscow, pp. 193–203 (in Russian).

Tugarinov, A.I. and Voytkevich, G.V., 1966. Precambrian Geochronology of the Continents. Izd. Nedra, Moscow, 434 pp. (in Russian).

Tugarinov, A.I., Bergman, I.A. and Gavrilova, L.K., 1973. The facies nature of the Krivoy Rog iron-formation. In: Genesis of Precambrian Iron and Manganese Deposits. Unesco, Paris, pp. 35–39.

Turobova, Z.V., 1969. On the origin of the banding in ferruginous quartzites. In: Voprosy geologii i mineralogii rudnykh mestorozdeniy Ukrainy, (Problems of the Geology and Mineralogy of the Ore Deposits of the Ukraine). Izd. Nedra, Moscow, 3: 157–171 (in Russian).

Twenhofel, W.H., 1926. Treatise on Sedimentation. Baltimore, 661 pp.

Tyapkin, K.F. and Fomenko, V.Yu., 1969. On the problem of the genesis of the Precambrian ferruginous quartzites. In: Problemy obrazovaniya zhelezistykh porod dokembriya (Problems of Formation of the Precambrian Iron Formations). Izd. Naukova Dumka, Kiev, pp. 114–118 (in Russian).

Tyler, S.A. and Barghoorn, E.S., 1954. Occurrence of structurally preserved plants in Precambrian rocks of the Canadian Shield. Science, 119: 606–608.

Urey, H.C., 1956. Regarding the early history of the Earth's atmosphere. Bull. Geol. Soc. Am., 67: 1125–1128.

Val'ter, A.A., 1969. On the distribution of magnesium and iron between coexisting olivine and orthopyroxene solid solutions with high iron content. Dokl. Akad. Nauk S.S.S.R. (Rep. Acad. Sci. U.S.S.R.), 187(2): 410–413 (in Russian).

Van-Hise C.R. and Leith, C.K., 1911. The geology of the Lake Superior region. Mon. USGS, LII.

Vernadskiy, V.I., 1940. Biogeochemical Essays, 1922–1923. Izd. Akad. Nauk S.S.S.R., Moscow, 250 pp. (in Russian).

Vernadskiy, V.I., 1967. The Biosphere. Izd. Mysl', Moscow, 376 pp. (in Russian).

Vinogradov, A.P., 1955. Regularities of Distribution of Elements in the Earth's Crust. Izd. Moscow Univ., 16 pp. (in Russian).

Vinogradov, A.P., 1959. Chemical Evolution of the Earth. Izd. Akad. Nauk S.S.S.R., Moscow, 44 pp. (in Russian).

Vinogradov, A.P., 1964. Gas regime of the Earth. In: Khimiya zemnoy kory (Chemistry of the Earth's Crust). Izd. Nauka, Moscow, pp. 5–21 (in Russian).

Vinogradov, A.P., 1967. Introduction to the Geochemistry of the Ocean. Izd. Nauka, Moscow, 215 pp. (in Russian).

Vinogradov, A.P. and Dontsova, Ye.I., 1952. Isotopic composition of the oxygen of skarn minerals. Dokl. Akad. Nauk S.S.S.R. (Rep. Acad. Sci. U.S.S.R.), 85: 1341–1343 (in Russian).

Vinogradov, V.I., Ivanov, I.B., Litsarev, M.A., Pertsev, N.N. and Shanin, L.L., 1969. On the age of the oxygenated atmosphere of the Earth. Dokl. Akad. Nauk S.S.S.R. (Rep. Acad. Sci. U.S.S.R.), 188: 1144–1147 (in Russian).

Volkov, I.I. and Sevast'yanov, V.F., 1968. Redistribution of the chemical elements in the diagenesis of the sediments of the Black Sea. In: Geokhimiya osadochnykh porod i rud (Geochemistry of Sedimentary Rocks and Ores). Izd. Nauka, Moscow, pp. 134–182 (in Russian).

Volkov, I.I. and Tikhomirova, A.A., 1966. The forms of iron in the sediments of the oxygenated zone of the Black Sea. Litholog. i polezn. iskop. (Lithology and Mineral Deposits), 4: 24–37 (in Russian).

Vologdin, A.G., 1969. On the paleontology of the Precambrian. Geol. Zhur. Akad. Nauk Ukr. S.S.R. (Geol. J. Acad. Sci. Ukr. S.S.R.), 29(4): 39–52 (in Russian).

Vologdin, A.G. and Strygin, A.I., 1969. Discovery of the remains of organisms in the Upper suite of the Krivoy Rog series of the Ukrainian Precambrian. Dokl. Akad. Nauk S.S.S.R. (Rep. Acad. Sci. U.S.S.R.), 188: 446–449 (in Russian).

Vorobyeva, K.A. and Mel'nik, Yu.P., 1977. Experimental investigation of the system $Fe_2O_3-H_2O$ at $T=100$–200°C and P up to 9 kbar. Geokhimiya (Geochemistry), 8: 1121–1128 (in Russian).

Voytkevich, G.V. and Belokrys, L.S., 1960. Traces of ancient life on the Earth. Sov. Geologiya (Sov. Geol.), 4: 3–22 (in Russian).

Voytkevich, G.V. and Zakrutkin, V.V., 1970. Principles of Geochemistry. Izd. Rostov. Univ., 241 pp. (in Russian).

Voytkevich, G.V., Zelenshchikova, K.Kh., Lebed'ko, G.I. and Poyarkova, I.F., 1975. The remains of the activity of organisms in the Precambrian of the Southeastern flank of the Voronezh crystalline massif. Doklady Akad. Nauk S.S.S.R. (Rep. Acad. Sci. U.S.S.R.), Vol. 224, 3: 689–690 (in Russian).

Voyutskiy, S.S., 1964. Course in Colloidal Chemistry. Izd. Khimiya, Moscow, 574 pp. (in Russian).

Wagman, D.D., Evans, W.H., Parker, V.B., Halow, J., Bailey, S.M. and Schumm, R.H., 1968. Selected values of chemical thermodynamic properties. U.S. Nat. Bureau Standards, pp. 270–274.

Walther, J.V. and Helgeson, H.C., 1977. Calculation of the thermodynamic properties of aqueous silica and the solubility of quartz and its polymorphs at high pressures and temperatures. Am. J. Sci., 277: 1315–1351.

Weber, J., 1965. On the problem of the $^{18}O/^{16}O$ ratio in the ancient oceans. Geokhimiya (Geochemistry), 6: 674–680 (in Russian).

Wefers, K., 1966a. Zum system $Fe_2O_3-H_2O$. 1. Teil. Ber. Dtsch. Keram. Ges., 43: 677–684.

Wefers, K., 1966b. Zum system $Fe_2O_3-H_2O$. 2. Teil. Ber. Dtsch. Keram. Ges., 43: 703–708.

Weidner, J.R. and Tuttle, O.F., 1965. Stability of siderite, $FeCO_3$. Geol. Soc. Am., Spec. Pap., 82: 220.

Weyl, W.A., 1951. A new approach to surface chemistry and to heterogeneous catalysis. Mineral Industries Exp. Sta., Bull., 57: 47–58.

White, D.A., 1954. The stratigraphy and structure of the Mesabi Range, Minnesota. Univ. Minn., Minn. Geol. Surv. Bull., 38.

White, D.E., Brannock, W.W. and Murata, K.J., 1956. Silica in hot-spring waters. Geochim. Cosmochim. Acta, 10: 27–59.

Wicks, C.E. and Block, F.E., 1965. Thermodynamic properties of 65 elements—their oxides, halides, carbides and nitrides. Izd. Metallurgiya, Moscow, 240 pp. (in Russian); U.S., Dep. Interior (in English).

Winkler, H.G.F., 1961. Genesen von Graniten und Migmatiten auf Grund neuer Experimente. Geol. Rundsch., 51: 347–364.

Winkler, H.G.F., 1969. Petrogenesis of Metamorphic Rocks. Izd. Mir, Moscow: 247 pp. (in Russian). 2 ed. Springer, Berlin, 218 pp. (1967, in English).

Yaroshchuk, M.A. and Onopriyenko, V.L., 1973. New data on the mineralogy and conditions of deposition of the cherty iron-formations of the sixth and seventh horizons of the Central Krivoy Rog. In: Geokhimiya i rudoobrazovaniye (Geochemistry and Ore Genesis). Izd. Naukova Dumka, Kiev, 2: 31–38 (in Russian).

Yaroshchuk, M.A., Burmistrova, V.V. and Onopriyenko, V.L., 1976. Development of talc in the evolution of the processes of metamorphic ore formation in the Krivoy Rog. Geol. Zhurn. Akad. Nauk Ukr. S.S.R. (Geol. J. Acad. Sci. Ukr. S.S.R.), 36(6): 79–84 (in Russian).

Yeliseyev, N.A., 1959. Metamorphism. Izd. Leningrad. Univ., 415 pp. (in Russian).

Yemel'yanov, Ye.M., 1968. Study of microquantities of iron, manganese and titanium in suspended matter in the Baltic Sea. Litolog. i Polezn. Iskop. (Lithology and Mineral Products), 6: 43–52 (in Russian).

Yepatko, Yu.M. and Mel'nik, Yu.P., 1965. Some theoretical and experimental data on the formation of goethite and hematite in the weathered layer of the Krivoy Rog iron ore basin. Geol. Zhur. Akad. Nauk Ukr. S.S.R. (Geol. J. Acad. Sci. Ukr. S.S.R.), XXV (2): 12–21 (in Ukrainian).

Yoi, S., 1966. Decomposition of siderite to magnetite at lower oxygen fugacities: a thermochemical interpretation and geological implications. Econ. Geol., 61: 768–776.

Young, G.M., 1967. Possible organic structures in Early Proterozoic (Huronian) rocks of Ontario. Can. J. Earth Sci., 4: 565–568.

Yurk, Yu.Yu. and Shnyukov, Ye.F., 1961. Iron Oxides of the Ukrainian Crystalline Shield. Izd. Akad. Nauk Ukr. S.S.R., Kiev, 108 pp. (in Russian).

Yurk, Yu.Yu., Shnyukov, Ye.F., Lebedev, Yu.S. and Kirichenko, O.N., 1960. Mineralogy of the Iron Ore Formation of the Kerch Basin. Krymizdat, Simferopol', 450 pp. (in Russian).

Zakgeym, A.Yu., 1966. On one limit of the applicability of the laws of thermodynamics. Zhurn. Fiz. Khim. (J. Phys. Chem.), 40: 1943–1944 (in Russian).

Zaytsev, Yu.S., 1966. The Mikhaylov series of the Belgorod iron ore district and some problems of the stratigraphy of Precambrian of the Voronezh anticlise. In: Trudy III soveshchaniya po problemam izucheniya Voronezhskoy anteklizy, (Trans. Third Conference on Problems of Study of the Voronezh Anteclise). Izd. Voronezh Univ., pp. 21–28 (in Russian).

Zelenov, K.K., 1958. On the carrying of dissolved iron to the Sea of Okhotsk by the hot springs of Ebeko volcano. Dokl. Akad. Nauk S.S.S.R. (Rep. Acad. Sci. U.S.S.R.), 120: 1089–1092 (in Russian).

Zelenov, K.K., 1960. Migration and accumulation of iron and aluminum in volcanic regions of the Pacific Ocean. Izv. Akad. Nauk S.S.S.R., Ser. Geol. (Proc. Acad. Sci. U.S.S.R., Geol. Ser.), 8: 58–74 (in Russian).

Zelenov, K.K., 1963. Submarine volcanism and its role in the formation of sedimentary rocks. Trudy Geol. Inst. Akad. Nauk S.S.S.R. (Trans. Geol. Inst. Acad. Sci. U.S.S.R.), 81: 30–86 (in Russian).

Zelenov, K.K., 1965. Some particulars of submarine volcanism as illustrated by Banua Wuhu volcano (Indonesia). In: Vulkanogenno-osadochnyye formatsii i poleznyye iskopayemyye (Volcanogenic-Sedimentary Formations and Mineral Deposits). Izd. Nauka, Moscow, pp. 70–79 (in Russian).

Zhdanov, V.V. and Malkova, T.T., 1974. Iron ore deposits of zones of regional basification (Petrology and genetic problems). Trudy Vsesoyuz. N.-I. Geol. Inst. (Trans. All Union Sci. Geol. Inst.), 208: 198 pp. (in Russian).

Zhizhchenko, B.P., 1959. Methods of Paleogeographic Investigation. Gostoptekhizdat, Moscow, 371 pp. (in Russian).

Zigmondi, R., 1933. Colloid Chemistry. Kharkov, 450 pp. (in Russian).

Subject index